**INTRODUCTION
TO OPERATIONAL
AMPLIFIER THEORY
AND APPLICATIONS**

**McGRAW-HILL
BOOK COMPANY**
New York
St. Louis
San Francisco
Auckland
Düsseldorf
Johannesburg
Kuala Lumpur
London
Mexico
Montreal
New Delhi
Panama
Paris
São Paulo
Singapore
Sydney
Tokyo
Toronto

JOHN V. WAIT
LAWRENCE P. HUELSMAN
GRANINO A. KORN
Department of Electrical Engineering
University of Arizona

Introduction to Operational Amplifier Theory and Applications

This book was set in Press Roman by Scripta Graphica.
The editors were Kenneth J. Bowman and Madelaine Eichberg;
the production supervisor was Leroy A. Young.
Kingsport Press, Inc., was printer and binder.

Library of Congress Cataloging in Publication Data

Wait, John V
    Introduction to operational amplifier theory and applications.

    Includes bibliographies.
    1. Operational amplifiers. 2. Electronic circuits.
I. Huelsman, Lawrence P., joint author. II. Korn,
Granino Arthur, date joint author. III. Title.
TK7871.58.06W3   621.3815'35   74-13989
ISBN 0-07-067765-4

**INTRODUCTION
TO OPERATIONAL
AMPLIFIER THEORY
AND APPLICATIONS**

3 4 5 6 7 8 9 0 K P K P 7 9 8 7

CONTENTS

PREFACE

During the 1960s transistorized solid-state amplifier technology evolved rapidly. Today such amplifiers in microminiature form have become basic building blocks that are used in a broad range of electronic circuit applications, e.g., amplification, filtering, nonlinear waveshaping, waveform generation, and switching. This text has been developed to be used with a course in operational amplifier circuit design for senior and graduate students in electrical engineering; it is intended to cover techniques of analysis and design of electronic circuits, using operational amplifiers, resistors, capacitors, diodes, and other special components as needed.

This book does not discuss the internal design of operational amplifiers themselves. Considerable attention is paid, however, to the ways in which nonideal properties that arise from internal design limitations affect circuit performance. The text assumes the typical background of an electrical engineering senior, including some prior knowledge of Laplace transforms, frequency response operators, and Bode plots, and a basic feeling for solid-state device characteristics.

This book should also be quite useful to the practicing electrical engineer as a self-teaching aid and as a source of applications ideas.

Chapter 1 provides a thorough presentation of *basic linear operational amplifier circuit design*, using the *ideal operational amplifier model*. This chapter also reviews important active circuit analysis techniques in general by using many examples. The ideal model is carefully defined, with an eye toward a treatment of nonideal effects in Chapter 2. The following sections present detailed discussions of the major circuit configurations, including the summer-inverter, the inverting summer-integrator (including mode control), the generalized inverting amplifier with two-port networks, the noninverting amplifier, and the differential amplifier. The eighth section is a general discussion of combined inverting and noninverting summer-amplifiers, which is not readily available elsewhere. The ninth section describes important first-order filter circuits; we postpone a detailed treatment of active *RC* filter circuits until Chap. 4, however. The last section presents a few of the well-known miscellaneous applications, such as instrumentation amplifiers, reference sources, and voltage-current converters. The reader will also find that the more advanced exercises will suggest many more useful applications.

Chapter 2 discusses means for analyzing the effects of *amplifier performance limitations* (nonidealness) on circuit performance. A general nonideal small-signal linear model is presented first, and we discuss techniques based on this model that may be used to estimate closed-loop circuit gain and circuit output and input impedance. Next we present a discussion of Bode plot techniques for estimating closed-loop circuit stability, and the problem of compensation is examined. Common-mode effects are described in the fourth section. Dc offsets and temperature drift effects are described in the fifth section, and standard compensation and balancing techniques are also described. A description of random noise effects, based upon a power spectral density model, is presented in the sixth section; the reader may want to skip this part on a first reading, since some appreciation of the principles of random process theory is required in order to apply the model. Large-signal limitations are discussed in the seventh section, e.g., slewing rate, full-power bandwidth, settling time, overload recovery time, and maximum common-mode voltage. The last three sections are devoted to some practical considerations involved in operational amplifier use. The eighth section discusses output power boosting techniques; the ninth section surveys power-supply requirements, grounding, and shielding; and finally, the last section describes the typical properties of passive components used in linear circuit design.

Chapter 3 is concerned with the design of *nonlinear circuits*. This chapter presents a number of nonlinear circuit design methods that are hopefully of general value. Piecewise-linear diode and transistor models are reviewed and applied to some simple examples. The second section discusses a variety of limiter and comparator circuits, and compares their merits. The use of hysteresis for noise-immune comparator design is included. General piecewise-linear-function generation methods are presented in the third section. The fourth section is devoted to logarithmic operators to help the reader evaluate these types of nonlinear operators in a specific application. Conventional methods for multiplying, dividing, and squaring are described and compared in the fifth section, and typical applications are presented. Finally, the sixth section presents a collection of well-known waveform generation circuits. Again, the examples and exercises suggest additional applications.

The remaining chapters cover advanced applications. Chapter 4 discusses one of the more important applications of operational amplifiers, namely, *active* RC *filters*. In the first section, some general design and analysis techniques are introduced. A discussion of sensitivity is included to permit the reader to evaluate different filter realizations. The following two sections present several of the most important filter configurations. These include the Sallen and Key structures and the infinite-gain multiple feedback configurations. The fourth section covers active RC filter realizations for the high-Q case. Both state-variable and Tarmy-Ghausi realizations are presented. The effect of finite amplifier bandwidth on the realizations is discussed. In the fifth section, the use of cascade techniques to realize filters of higher than second order is covered. The sixth and seventh sections discuss alternative methods of active RC filter realization, including the use of gyrators or negative-impedance converters, and the recently proposed technique of using frequency-dependent negative resistors as produced by generalized impedance converters. In all cases, design tables are provided which enable the reader to easily realize specific filter configurations for several of the commonest maximally flat magnitude (Butterworth), equal ripple (Chebychev), maximally flat delay (Thomson), and elliptic (Cauer) characteristics. Approximation tables for these characteristics are provided through the sixth order. In many cases alternative realizations are provided to make fabrication easier by specifying equal-valued R and C elements or to reduce sensitivity by using integer-valued gains. A discussion of tuning procedures for high-order realizations is also included. Many of the tables included in this chapter have never before appeared in the literature. Of special importance is the compilation of quadratic factors for a wide range of high-order filter characteristics.

Chapter 5 introduces up-to-date *electronic switching circuits,* including diode, transistor, JFET, MOSFET, and CMOS switches. Many examples show how to minimize circuit error due to leakage, ON resistance, offsets, and switching spikes. Electronic multiplexer circuits, sample-holds, and integrator-control switching are discussed, together with error sources and modern design practice.

Chapter 6 deals with *digital-to-analog and analog-to-digital converters,* both now low-cost, indispensable items in the system designer's bag of tricks. Separate sections describe digital codes, representation of negative voltages, and voltage, inverted, and current ladder circuits. The text then goes on to describe the design and use of current-switching monolithic digital-to-analog converters, error sources, and settling time. Descriptions of the principal types of analog-to-digital converters include the complete logic for a simple successive-approximation converter. A final section shows how to interpret function-module specifications.

There is clearly more material provided than one can thoroughly cover in a typical three-hour one-semester course. This has been done intentionally so that the book may be used for both an in-depth course on design fundamentals and a survey course on applications. We have been using class notes, from which this book originated, in a senior technical elective at the University of Arizona for four years, with considerable success. Our approach is to spend about 40 percent of the semester on the fundamental principles in Chaps. 1 and 2; we work many problems here. Then we use the material of Chap. 3 for about 20 percent of the course, with the primary goal of sharpening up the student's ability to design and analyze nonlinear circuits using primarily piecewise-linear diode and transistor models. The remaining 40 percent of the course is devoted to surveying the applications described in Chaps. 4 to 6.

Depending upon specific course goals, other instructors might choose a different emphasis. For example, Chaps. 1 to 3 form a nice introduction to design fundamentals; indeed, Chap. 1 could serve alone for a one-hour introductory course unit. Chapter 2 could be skipped if one wanted to put more emphasis on applications and less on detailed design. Chapters 1 and 4 together would support a full course in active RC filter design, with more time spent on the advanced problems provided at the end of Chap. 4. Chapters 1, 5 and 6 would similarly serve for a course in A/D, D/A, and data acquisition subsystems.

We have many friends who have contributed to the development of the book, and there is no room to thank them all. We do want to expressly thank Dr. Roy H. Mattson, Head of the Department of Electrical Engineering of the University of Arizona, for the assistance of the department in the course of this book's preparation. Thanks should also go to Dr. Gerald R. Peterson, who encouraged us to start the course from which this text has evolved. We are also particularly grateful to the Burr-Brown Research Corporation and Analog Devices, Inc., for providing technical specifications and applications information. Finally, we must very gratefully thank the many students in EE 226, who over the years have assisted in the proofreading and criticizing of the manuscript, including Mssrs. Yuchee Chih and Thomas Bruhns.

JOHN V. WAIT

LAWRENCE P. HUELSMAN

GRANINO A. KORN

**INTRODUCTION
TO OPERATIONAL
AMPLIFIER THEORY
AND APPLICATIONS**

BASIC OPERATIONAL AMPLIFIER CIRCUITS

During the 1960s the solid-state dc operational amplifier grew rapidly in importance to the electronic circuit designer. Today, such amplifiers, in microminiature form (Fig. 1.1), have become so *reliable, compact,* and *easy to use* that they are *basic electronic circuit building blocks* used in an extremely broad range of applications, including amplification, filtering, nonlinear waveshaping, wave generation, and switching. Today, instead of assembling an amplifier from dozens of components, the wise circuit designer frequently develops his circuit around a few standard, commercially available *prepackaged operational amplifiers,* which *comprise the active elements;* the rest of the circuit will be made up of resistors, capacitors, diodes, and other special components as needed.

The use of such standard active elements eliminates the need for detailed design of individual transistor stages. When they are properly used, the overall transfer characteristics of a circuit (gain, frequency response, etc.) can be precisely controlled by stable passive elements (e.g., resistors, capacitors, diodes). *Feedback techniques* are used to suppress any nonideal properties of the operational amplifier so that individual variations

FIGURE 1.1
Operational amplifier packaging. (*a*) Epoxy-encapsulated amplifier made up of discrete components; (*b*) interior construction of discrete-component units; (*c*) typical packaging of integrated-circuit units; (*d*) hybrid construction.

in particular amplifiers have a negligible effect on final circuit performance. Hence circuit designs based on the use of operational amplifiers usually have *highly predictable performance.*

When most of the active elements of a circuit are concentrated into the small space occupied by a modern solid-state operational amplifier, *miniaturization* of electronic circuits *is facilitated.* Moreover, today's operational amplifiers are *highly reliable,* and the circuit designer can expect that production copies of his final system will closely emulate the performance of the prototype, and will require a minimum of initial debugging and little long-term maintenance (if he has chosen operational amplifiers supplied by a competent manufacturer).

1.1   GENERAL CHARACTERISTICS OF OPERATIONAL AMPLIFIERS

Fig. 1.1 shows some typical operational amplifier modules. Two basic methods of fabrication are employed: discrete and integrated-circuit. In Fig. 1.1*a*, we see an epoxy-encapsulated amplifier made up of discrete elements, viz., transistors and resistors. In such a form, the elements are usually mounted on small printed-circuit boards (Fig. 1.1*b*). Connecting pins are attached to one of the boards, and the final unit is surrounded by the encapsulating material. Both bipolar and field-effect transistors (FETs) may be employed. Integrated-circuit operational amplifiers are currently made using conventional monolithic techniques, and packaged either in a round TO-99 type of transistor case or in an epoxy-clad dual-in-line package (Fig. 1.1*c*). Hybrid thin-film, thick-film, and/or monolithic IC techniques may also be used (Fig. 1.1*d*).

In this book we will not discuss the techniques for designing the operational amplifier itself; the reader is referred to Refs. 1 to 3 for information about the internal design of modern solid-state operational amplifiers. We are primarily interested here in *applications* of operational amplifiers to signal-processing tasks (see also Refs. 1, 3 to 6, and 12 as further sources of application ideas).

Figure 1.2 illustrates the typical environment in which an operational amplifier is used. Often several operational amplifiers in a given system are supplied from a single pair of matched positive and negative sources of regulated dc power (for example, ±15 V) via low-impedance dc buses. An internal ground lead is sometimes provided on the amplifier package, although typically *all signals* are merely *referenced* to the *power* supply common-ground potential (see Sec. 2.9 for details of power and grounding systems). Operational amplifiers usually have two input terminals (inverting and noninverting) and an output terminal. In addition to these main signal terminals, connector pins are sometimes provided for connection of frequency response compensation and dc offset balancing networks.

Figure 1.3 shows the conventional operational amplifier symbol that will be used throughout this text. In this symbol, only the principal signal terminals are illustrated; the other necessary connections to the amplifier (for power, etc.) are assumed to be made as specified by the amplifier manufacturer.

Table 1.1 lists the features normally required in an operational amplifier.

1.2   THE IDEAL AMPLIFIER MODEL

In this chapter we will present the so-called ideal operational amplifier model (Fig. 1.4*b* or *c*); in the next chapter, the important nonideal properties of actual amplifiers will be described, and methods for estimating their effect on circuit performance will be treated. Here, however, we will neglect almost all the nonideal properties and describe an operational amplifier in terms of how we would like it to behave. First of all, we will assume there are no internal dc offsets or nonlinearities associated with the amplifier. With this assumption, we can represent an operational amplifier fairly completely by the idealized linear model of Fig. 1.4*a*. Here we see that the operational amplifier is an active

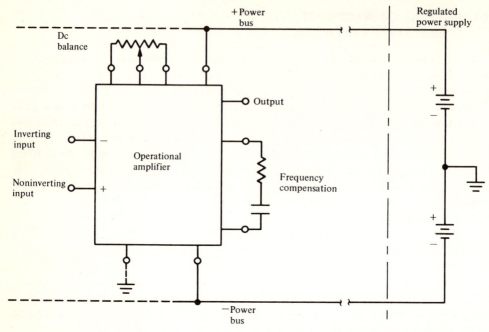

FIGURE 1.2

Typical operational amplifier environment. This diagram illustrates a typical manner in which power is supplied and external balancing and compensation networks are connected (not always required). Several amplifiers may be powered from a single pair of regulated power supplies.

device. The Thévenin equivalent circuit associated with the output terminal is modeled by an ideal controlled voltage source $-Ae_g$ in series with an equivalent output impedance Z_0. The input terminals are associated with a passive equivalent circuit representing the *common-mode impedances* Z_{cm1} and Z_{cm2} between each input and the common-signal ground, and the *differential input impedance* Z_i between the input terminals.

In this chapter, *a yet simpler model suffices,* as depicted in Fig. 1.4b and c. In Fig. 1.4b, we have assumed that the internal impedances associated with the input terminals are negligibly large, and that the internal output impedance is negligibly small. All that

FIGURE 1.3

Conventional operational amplifier symbol; only active signal lines are shown, and all signals are referenced to ground.

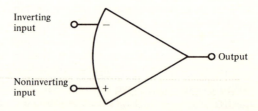

remains is the active voltage source $-Ae_g$ at the output terminal. Throughout the rest of this chapter, we will adopt the more convenient notation of Fig. 1.4c, where the amplifier terminal voltages e_1, e_2, and e_0 are assumed to be referenced to a common ground.

In summary, we can describe the ideal operational amplifier model as follows:

1 The ideal operational amplifier is a *linear voltage-controlled voltage source* (VCVS), with

$$e_0 = A(e_2 - e_1) = -Ae_g \qquad (1.1a)$$

where

$$e_g = e_1 - e_2 \qquad (1.1b)$$

e_g is the differential voltage between the amplifier input terminals.

2 The amplifier *open-loop voltage gain* A is assumed to be a *very large constant*, that is,

$$A \ggg 1 \qquad \text{essentially infinite} \qquad (1.2)$$

3 Implicit in (1) is the assumption that the amplifier input terminals are essentially open-circuit control nodes, i.e., the impedance between terminals e_1 and e_2 is infinite, and the impedance between each input terminal and the ground is infinite. Thus the input terminal currents are zero.

4 Also implicit in (1) is the assumption that the amplifier output voltage is unaffected by external loads.

NOTE: We have made the choice of polarity indicated in Fig. 1.4 so that the amplifier *open-loop voltage gain* A can be treated as a large *positive* constant at low frequencies. The polarity relationship between the respective input terminals is as specified in (1.1); therefore with the sign marking on the Thévenin voltage

---

Table 1.1 FEATURES OF OPERATIONAL AMPLIFIERS

---

1 *High,* relatively linear, *voltage gain,* down to and including direct current; open-loop gain at direct current may be 10^7 or greater.

2 There must be a *polarity inversion* between input and output. In some cases the noninverting input is not externally available, but it is preferable for a *differential input* to be provided, with both an *inverting* and *noninverting* terminal.

3 *Direct current offsets should be minimized;* i.e., the input voltage (e_g, Fig. 1.4c) should be near zero when the output voltage is zero (good direct current *balance*). *Temperature compensation* techniques (in some cases chopper stabilization systems) should be employed to provide long-term stability to this balance. Drifts due to power-supply variations should also be minimized.

4 There should be *careful control of high-frequency response,* so that the amplifier will accommodate a large amount of negative feedback.

5 The *differential impedance* between the input terminals and the *common-mode impedance* between the terminals and ground should be *high.*

6 The amplifier *output impedance* should be *low.*

7 The amplifier *output stage should have the ability to deliver specified maximum current to or absorb it from* an output load over some nominal *bipolar* (±) *voltage range,* e.g., ±10 V.

8 When a differential input is provided, there should be *good common-mode rejection*; i.e., the output depends only on the *difference* between the input voltages, and is not dependent on the magnitude of either input voltage.

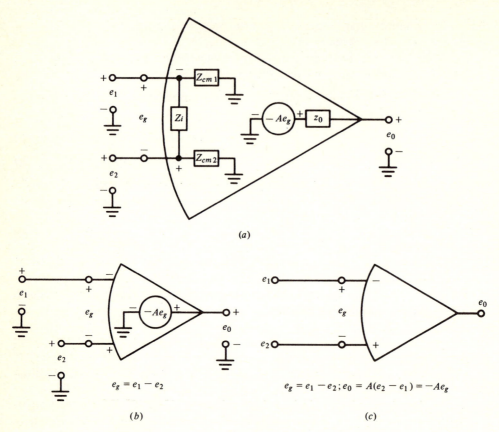

(a)

(b)

$$e_g = e_1 - e_2$$

(c)

$$e_g = e_1 - e_2 ; e_0 = A(e_2 - e_1) = -Ae_g$$

FIGURE 1.4
Operational amplifier circuit models. (a) Detailed linear model; (b) simplified model, neglecting internal impedances; (c) abbreviated notation for simplified model; voltages e_1, e_2, and e_0 are assumed to be referenced to ground.

generator in Fig. 1.4a or b, the value of the generator is $-Ae_g$. The reader is cautioned to study this thoroughly; it is easy to make mistakes in writing circuit equations. It must be remembered that e_1 is associated with the inverting terminal, and e_2 with the noninverting terminal.

As mentioned in Sec. 1.1, the operational amplifier is usually embedded in a circuit that provides *large amounts* of *negative feedback*, especially at low frequencies. The effect of the feedback is to make the amplifier behavior more ideal, viz., dc offsets are suppressed, the amplifier characteristics are linearized, the amplifier is made less sensitive to external output loads, and the differential input voltage e_g is forced to be close to zero, so that input currents are indeed very low.

If the circuit designer will accept the model in good faith, and use a few basic design practices to ensure that a reasonable amount of negative feedback is employed in

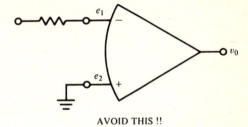

FIGURE 1.5
This circuit condition is normally to be avoided; operational amplifiers are not designed to run open-loop, without feedback.

AVOID THIS !!

the overall circuit design, he will normally find that the ideal model is quite valid for his initial design efforts.

NOTE: Even though the model described by (1.1), (1.2), and Fig. 1.4*b* is often adequate for analysis, it is frequently necessary to know the allowable deviations of the terminal voltages from ground, especially the maximum output voltage swing, which is typically ±10 V. In special applications, the maximum allowable input swing for voltages e_1 and e_2 is also important. Also the available amplifier output current often must be considered.

Of course, the ideal model is not truly attainable; Table 1.1 hints at the myriad considerations one may have to include if the ideal model is not sufficiently complete to fully predict circuit performance. The reader may well ask how big the amplifier voltage gain A actually is. Indeed, this is a very important question. Typically, at low frequencies the gain may be very high, perhaps as large as 10^6 or 10^8. Of course, at higher frequencies, the voltage gain will be less, and will include phase shift. In Chap. 2, we will explore the ramifications of the variation of A with frequency in some detail, but in many simple applications, the assumptions of (1.1) and (1.2) suffice, at least to make a first try at a circuit design.

1.3 THE SUMMER-INVERTER

Operational amplifiers are normally *not operated open-loop*; some form of *feedback* is used to control the overall circuit transfer characteristics. If the amplifier were energized in an open-loop configuration (Fig. 1.5), the output voltage would saturate (amplitude-overload) with even a very small input signal because of the high open-loop gain; in fact, small internal dc offsets which are always present would normally be sufficient to cause overload without feedback.

Figure 1.6*a* shows a typical inverting operational amplifier configuration; the noninverting terminal is grounded for single-ended operation, either directly or through a drift-compensating resistor (see Sec. 2.5). The simplified diagram of Fig. 1.6*b*, where the noninverting terminal is assumed to be effectively at ground potential and all voltages are assumed to be referred to this same ground potential, is often used. Here also we use v's for the voltages on the external nodes of the complete circuit.

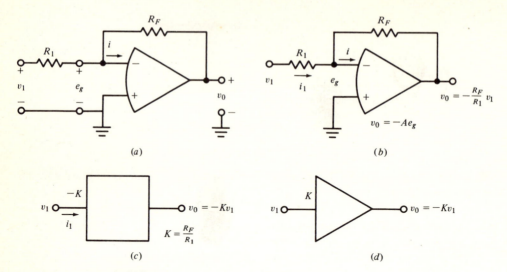

FIGURE 1.6
Simple resistive inverter-amplifier. (*a*) Complete circuit; (*b*) same circuit with all voltages referenced to ground; (*c*) shorthand or block-diagram symbol for complete circuit; (*d*) analog-computer symbol.

We make three basic assumptions in deriving the transfer function equations:

1 The amplifier output voltage v_0 is related to the *operational amplifier* input terminal voltage e_g (the summing junction) by

$$v_0 = -A e_g$$

where A is termed the *operational amplifier open-loop gain* (assumed here to be a large + constant). Typically $A > 10^4$. *Thus, so long as the amplifier output is within its rated range (for example, ±10 V), we will be able to assume* $e_g = 0$ *without serious error.*

2 The output voltage v_0 is not materially affected by loads applied to the amplifier output, including any feedback elements.

3 The amplifier input impedance is sufficiently high to permit us to assume that the amplifier input current i is approximately zero (since from the first assumption e_g is very small, the amplifier input impedance does not really have to be exceptionally high to assure validity of the final results).

In the simplified circuit of Fig. 1.6*b*, we can write a current-balance equation (nodal equation)† at node e_g:

$$\frac{v_0 - e_g}{R_F} + \frac{v_1 - e_g}{R_1} = i \qquad (1.3)$$

†For the reader whose background in undergraduate circuit theory is weak, Ref. 7 gives a modern presentation of circuit analysis techniques.

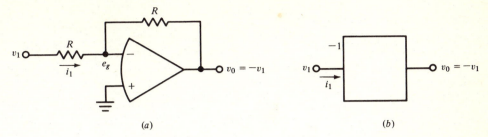

FIGURE 1.7
A simple unity gain inverter. (*a*) Detailed circuit; (*b*) block-diagram symbol.

Now, if the circuit is working properly (no amplifier overload), we assume $e_g \cong 0, i \cong 0$, so that the above equation becomes

$$\frac{v_0}{R_F} + \frac{v_1}{R_1} \cong 0 \qquad (1.4)$$

A general rule to remember in writing equations for amplifier performance is that *the sum of the currents entering the inverting terminal is approximately zero.* This, of course, is not strictly true, and we will discuss more exact expressions later.

Thus, the transfer function for the circuit of Fig. 1.6 is simply

$$\frac{v_0}{v_1} = -\frac{R_F}{R_1} \qquad (1.5)$$

We note that the gain is essentially controlled by the passive elements R_1 and R_F; *don't forget the minus sign!* Equation (1.5) is accurate to the extent that the above assumptions are valid. Resistance R_F is usually called the *feedback resistance*; R_1 is often called the *input resistance.*

Figure 1.6*c* shows a block-diagram symbol that is used in this book for the simple inverting amplifier; here we denote the voltage gain of the circuit by $-K$, that is,

$$\frac{v_0}{v_1} = -K \qquad K = \frac{R_F}{R_1} \qquad (1.6)$$

When describing an inverting amplifier, we shall use the minus sign explicitly; later on we shall be describing noninverting amplifiers as well.

Figure 1.6*d* shows a conventional analog-computer symbol for an inverting amplifier; in analog-computer diagrams, the minus sign is assumed and not specified explicitly. To avoid confusion, we shall use the symbol of Fig. 1.6*c* here.

Note also that if $R_F = R_1 = R$, we have a *simple unity gain inverter* (Fig. 1.7), with

$$v_0 = -v_1 \qquad R_F = R_1 = R \qquad (1.7)$$

Typical values for the resistances lie in the range from 100 to 10^6 Ω, and values of the ratio K are typically between 0.1 and 100. Note also that the input impedance of the circuit of Fig. 1.6 is

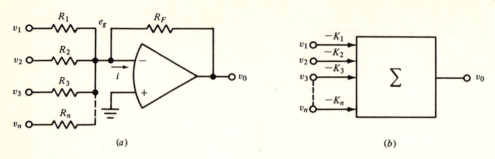

FIGURE 1.8
The summer-inverter circuit. (*a*) Complete circuit; (*b*) block-diagram symbol.

$$Z_{in} = \frac{v_1}{i_1} = R_1 \qquad (1.8)$$

since e_g is assumed to be zero. The output impedance is also predicted to be zero, since the ideal operational amplifier itself has zero output impedance. Later on, in Chap. 2, we will see that even for the nonideal case, these results are quite accurate for most purposes.

EXAMPLE 1.1   In the circuit of Fig. 1.6, let

$$R_1 = 2000 \ \Omega$$

and

$$R_F = 50,000 \ \Omega$$

Then we have

$$K = \frac{50 \times 10^3}{2 \times 10^3} = 25$$

and

$$v_0 = -25v_1$$

We might describe this circuit as either an *inverting amplifier* with a (voltage) gain of *magnitude* 25, or alternatively, an amplifier with a voltage gain of -25. Both descriptions are commonly used.   ////

Because we have used the ideal model of Fig. 1.4*c* to represent the operational amplifier, and have assumed that the input and feedback resistances are pure resistive elements, with no parasitic capacitance or inductance, (1.6) predicts that the circuit of Fig. 1.6 has a gain that is constant and independent of frequency. This of course is not strictly possible, and we will see later on that there are methods for predicting the frequency-dependent behavior of the circuit. However, if the resistance values are kept fairly low (e.g., less than 100,000 Ω) and a relatively fast (broadband) amplifier is used, then (1.6) usually will be a useful description of the inverting amplifier for frequencies in the audio range or below (e.g., direct current to 10 kHz). Also, we should note that (1.6) indicates that the only important thing is the *ratio* of R_F to R_1. This, too, will be essentially true so long as the individual resistance values fall within a reasonable range (for example, 100 to 1,000,000 Ω). Of course, if R_1 is made too low, the circuit of Fig. 1.6 may excessively load whatever system is providing its input signal. The reader again is reminded that (1.5) will hold only so long as the magnitude of v_0 is less than the specified maximum for the amplifier (typically ±10 V), and the magnitude of the total amplifier output current, including the current through R_F, does not exceed the rated value.

Figure 1.8a shows the circuit of a typical *summer-inverter.* If we write Kirchhoff's current law at node e_g, we have

$$\frac{v_1 - e_g}{R_1} + \frac{v_2 - e_g}{R_2} + \cdots + \frac{v_n - e_g}{R_n} + \frac{v_0 - e_g}{R_F} = i \qquad (1.9)$$

but since we assume $e_g \cong 0$ and $i \cong 0$, we can write

$$\frac{v_1}{R_1} + \frac{v_2}{R_2} + \cdots + \frac{v_n}{R_n} \cong \frac{-v_0}{R_F} \qquad (1.10)$$

or

$$v_0 = -R_F\left(\frac{v_1}{R_1} + \frac{v_2}{R_2} + \cdots + \frac{v_n}{R_n}\right) \qquad (1.11a)$$

$$= -(K_1 v_1 + K_2 v_2 + \cdots + K_n v_n) \qquad K_i = \frac{R_F}{R_i} \qquad i = 1, 2, \ldots, n \qquad (1.11b)$$

Node e_g is often called the *summing junction,* and engineers often describe the results of the above analysis by saying that the summing junction current (labeled i in Fig. 1.8a) is essentially zero. In view of the result of (1.11), we can see why the circuit of Fig. 1.8 is called a *summer-inverter.* Each of the input signals is weighted by a gain constant K_i, a weighted sum is formed internally, and the output is equal to this weighted sum but inverted in polarity; Fig. 1.8b shows a simplified block-diagram symbol.

The summer-inverter circuit is generally very useful. In analog computers, it is used to sum computer variables precisely; in this application, fixed, precise resistor ratios are used to provide a standard set of gains.

EXAMPLE 1.2 Figure 1.9 shows a two-input summer-inverter circuit. Assuming an ideal operational amplifier, we can predict the circuit performance from the resistor values by

$$v_0 = -\frac{R_F}{R_1}v_1 - \frac{R_F}{R_2}v_2$$

If $R_F = 20$ kΩ, $R_1 = 5$ kΩ, and $R_2 = 10$ kΩ, then

$$v_0 = -4v_1 - 2v_2$$

Suppose we also know that the output voltage of the amplifier should never exceed 10 V in magnitude; that is,

$$|v_0| \leqslant 10 \text{ V}$$

Thus, in using the circuit, we must assure that

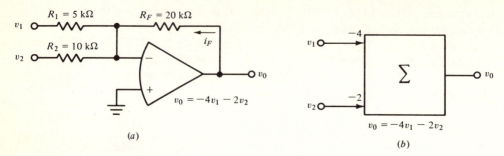

FIGURE 1.9
A two-input summer-inverter circuit (Example 1.2). (*a*) Complete circuit; (*b*) block-diagram symbol.

$$|-4v_1 - 2v_2| \leqslant 10 \qquad (1.12)$$

If we have no prior information about the polarity of the individual input voltages, then to assure that (1.12) is satisfied in the worst-case situation, we must assure that

$$4|v_1| + 2|v_2| \leqslant 10$$

(Note that it is not sufficient to limit v_1 to 2.5 V and v_2 to 5 V magnitude separately; why?)

Suppose also that we know our amplifier has a maximum rated output current of 5 mA; then R_F may require

$$|i_F| \leqslant \frac{10 \text{ V}}{20,000 \text{ } \Omega} = 0.5 \text{ mA}$$

and thus only 4.5 mA is available to drive an external load.   ////

In Chap. 2, we will discuss the nonideal operational amplifier and how the ideal equations such as (1.11) are affected by such things as dc offset, finite bandwidth and gain, etc.

1.4 THE INVERTING SUMMER-INTEGRATOR

Figure 1.10 shows the essential features of the *inverting summer-integrator* circuit, commonly called an *integrator*. The principal difference between the integrator circuit of Fig. 1.10 and the summer-inverter circuit of Fig. 1.8 is that the *feedback element* is now a *capacitor C.* Again, as in Sec. 1.3, we assume an ideal amplifier with very high open-loop gain, and thus we can assume $e_g \cong 0$, $i \cong 0$. Kirchhoff's current equation at node e_g is

$$\frac{v_1 - e_g}{R_1} + \frac{v_2 - e_g}{R_2} + \cdots + \frac{v_n - e_g}{R_n} + i_c = i \qquad (1.13)$$

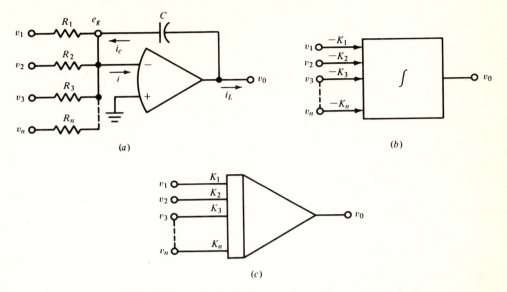

FIGURE 1.10
The summer-integrator circuit. (*a*) Complete circuit; (*b*) block-diagram symbol;
(*c*) analog-computer symbol.

but, of course, the terminal properties of a capacitance require that (again e_g is assumed to be very small)

$$i_c = C\left(\frac{dv_0}{dt} - \frac{de_g}{dt}\right) \cong C\frac{dv_0}{dt} \qquad (1.14)$$

Substituting (1.14) in (1.13), and setting $e_g \cong 0$ and $i \cong 0$, we have

$$\frac{v_1}{R_1} + \frac{v_2}{R_2} + \cdots + \frac{v_n}{R_n} + C\frac{dv_0}{dt} = 0 \qquad (1.15a)$$

or

$$C\frac{dv_0}{dt} = -\left(\frac{v_1}{R_1} + \frac{v_2}{R_2} + \cdots + \frac{v_n}{R_n}\right) \qquad (1.15b)$$

and thus

$$\frac{dv_0}{dt} = -\frac{1}{C}\left(\frac{v_1}{R_1} + \frac{v_2}{R_2} + \cdots + \frac{v_n}{R_n}\right) \qquad (1.16)$$

integrating both sides of (1.16), we have

$$v_0(t) = -\frac{1}{C}\int_0^t \frac{v_1}{R_1} + \frac{v_2}{R_2} + \cdots + \frac{v_n}{R_n}\; dt + v_0(0) \qquad (1.17)$$

or

$$v_0(t) = -\int_0^t (K_1 v_1 + K_2 v_2 + \cdots + K_n v_n)\, dt + v_0(0) \qquad (1.18)$$

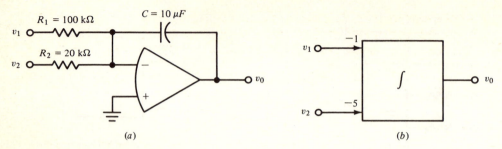

FIGURE 1.11
A two-input summer-integrator circuit (Example 1.3). (a) Complete circuit; (b) block-diagram symbol.

where
$$K_i = \frac{1}{CR_i} \quad \mathrm{s}^{-1} \quad i = 1, 2, \ldots, n \qquad (1.19)$$

and $v_0(0)$ is the initial voltage on the capacitance at $t = 0$ (more on initial-condition setting later).

Figure 1.10*b* shows the block-diagram symbol we intend to use in this book; again the polarity inversions are depicted explicitly. Figure 1.10*c* shows the symbol conventionally used in analog-computer block diagrams, where the polarity inversion is assumed.

EXAMPLE 1.3 Consider the circuit of Fig. 1.11. We can use (1.18) and (1.19) to obtain
$$v_0 = -\int_0^t (K_1 v_1 + K_2 v_2) \, dt + v_0(0)$$

with
$$K_1 = \frac{1}{R_1 C} = 1 \ \mathrm{s}^{-1}$$

and
$$K_2 = \frac{1}{R_2 C} = 5 \ \mathrm{s}^{-1}$$

or
$$v_0 = -\int_0^t (v_1 + 5v_2) \, dt + v_0(0) \qquad (1.20) \qquad ////$$

The factors K_i calculated by (1.19) are called the *integrator gain constants*. More precisely, we may say that each input to the integrator circuit of Fig. 1.10 has either a gain of $-K_i$ or an inverting gain of magnitude K_i. The gain constant K_i is obviously equal to the reciprocal of the time constant $R_i C$; thus K_i has units of s^{-1}. The concept of integrator gain is to some extent a shorthand way of specifying the (reciprocal) time constants associated with an integrator circuit.

We see also that one can achieve a particular integrator gain in many ways; viz., it is the time constant that is important, not the individual values of R and C. In Example 1.3, we could also obtain the same circuit of performance, as predicted by (1.20), with

$$R_1 = 1,000,000 \ \Omega$$
$$R_2 = 200,000 \ \Omega$$

and
$$C = 1 \ \mu F$$

In actuality, there would be some difference in circuit performance, which we will be in a better position to predict after the presentation of Chap. 2. Of course, one does not have complete freedom in the choice of component values. Again, the input impedance at each input terminal of Fig. 1.10 is

$$Z_i = R_i \quad i = 1, 2, \ldots, n \quad (1.21)$$

so that the circuit will excessively load its driving sources if the resistors are made too small. With regard to practical values of C, we should note that very small values cannot be easily obtained accurately (because of the stray capacity of wiring, etc.). On the other hand, very large values of capacity cannot be obtained inexpensively. Electrolytic capacitors do not permit bipolar potentials to be applied; they also have a large amount of dissipation, i.e., they have an appreciable leakage conductance. We will have more to say about component limitations and selection later (Sec. 2.11). For most applications, one should select low-loss capacitors (Mylar or polystyrene dielectric). Typical practical capacity values range from 0.001 to 10 μF. Resistor values should again range from perhaps 1000 to 1,000,000 Ω. Thus we see that a relatively wide range of integrator gains is possible.

A handy, relatively error-free way of calculating integrator gains is to work in megohms and microfarads, thereby obtaining a time constant directly in seconds, and thus the reciprocal of the time constant in seconds is the gain in s^{-1}.

EXAMPLE 1.4   Given an integrator with

$$C = 1000 \ \text{pF}$$

and
$$R = 10,000 \ \Omega$$

we can find the gain as follows:

$$C = 10^{-3} \ \mu F$$
$$R = 10^{-2} \ M\Omega$$

Thus the time constant is

$$T = 10^{-2} \times 10^{-3} = 10^{-5}$$

and the integrator gain constant is

$$K = \frac{1}{T} = 100,000 \ s^{-1} \qquad ////$$

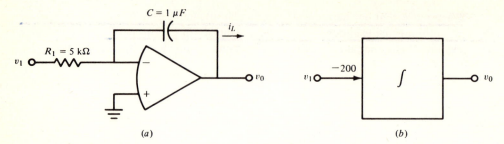

FIGURE 1.12
An inverting gain-of-200 integrator (Example 1.5).

Some additional practical considerations that should be mentioned are again output limitations. Of course the amplifier output voltage should never be allowed to exceed its nominal rated magnitude; this requires some knowledge of the nature of the signals to be integrated. For example, one cannot integrate a signal with a nonzero average value indefinitely.

A more subtle point is that the capacitor charging current must be considered in estimating the total output current required from the amplifier. If the amplifier must supply a current of i_L to some external load, then we must assure that

$$\left| C\frac{dv_0}{dt} + i_L \right| \leqslant i_{max} \qquad (1.22)$$

where i_{max} is the rated amplifier output current.

From (1.15*b*) we see also that the capacitor current equals minus the sum of the input resistor currents, e.g., in Fig. 1.10

$$\left| i_L - \left(\frac{v_1}{R_1} + \frac{v_2}{R_2} + \cdots + \frac{v_n}{R_n} \right) \right| \leqslant i_{max} \qquad (1.23)$$

Thus there are rate limitations on the integrator output voltage indicated by (1.22), which in turn dictate limitations on the instantaneous magnitude of the input signals (1.23).

EXAMPLE 1.5   Consider the integrator circuit of Fig. 1.12, with a gain of -200. Suppose the amplifier is rated so that

$$|v_0| \leqslant 10 \text{ V} \qquad (1.24a)$$
$$|i_{max}| \leqslant 10 \text{ mA} \qquad (1.24b)$$

Then from (1.22)

$$\left| C\frac{dv_0}{dt} + i_L \right| \leqslant 10 \text{ mA} \qquad (1.25)$$

and from (1.23) this means that

$$\left| -\frac{v_1}{R_1} + i_L \right| \leqslant 10 \text{ mA} \qquad (1.26)$$

If we want to reserve enough output current so that i_L can be as large as 5 mA, then

$$\left| \frac{v_1}{R_1} \right| \leqslant (10 - 5) \text{ mA} = 5 \times 10^{-3} \text{ A}$$

Thus with $R_1 = 5 \text{ k}\Omega$,

$$|2 \times 10^{-4} \, v_1| \leqslant 5 \times 10^{-3} \text{ A}$$

or

$$|v_1| \leqslant 25 \text{ V} \qquad (1.27)$$

which is not a very strict requirement.

Suppose, however, that

$$v_1 = A \cos 2\pi f t = A \cos \omega t$$

that is, v_1 is a sinusoid of *peak* (not rms) value A, frequency f Hz, and $\omega = 2\pi f$ rad/s. Assuming $v_0(0) = 0$, then

$$v_0(t) = -200 \int_0^t A \cos \omega t \, dt = \frac{-200A}{\omega} \sin \omega t$$

Then to prevent a *voltage overload* at the output,

$$\left| \frac{200A}{\omega} \sin \omega t \right| < 10 \text{ V}$$

but $\cos \omega t \leqslant 1$, so that

$$\frac{200A}{\omega} \leqslant 10$$

or

$$\frac{A}{f} \leqslant \frac{2\pi}{20} = 0.314 \text{ V.s} \qquad (1.28)$$

Suppose we combine the results of (1.27) and (1.28). We find that if

$$v_1 = A \cos 2\pi f t$$

then for the circuit of Fig. 1.12, if 5 mA of output current is required, A is limited by

$$A \leqslant 0.314 f \quad \text{V} \quad \text{in the range } 0 < f \leqslant 79.6 \text{ Hz}$$

due to output voltage limitation (1.24a), and

$$A \leqslant 25 \text{ V} \quad \text{in the range } f \geqslant 79.6 \text{ Hz}$$

due to amplifier output current limitation (1.24b).   ////

We can gain more insight into the dynamic behavior of the integrator by frequency domain analysis. Taking the Laplace transform of (1.18), we obtain [with $v_0(0) = 0$]

$$v_0(s) = \frac{-1}{s} [K_1 V_1(s) + K_2 V_2(s) + \cdots + K_n V_n(s)] \qquad (1.29)$$

Thus we see that the integrator transfer function (relating any input to the output) has a pole at $s = 0$; indeed, the operational amplifier integrator circuit provides a convenient way of implementing nearly ideal integration, with a transfer function having a pole *very close* to the origin of the s plane. Indeed, it is the ability of the operational amplifier circuit of Fig. 1.10*a* to implement the integration operation accurately that has made the electronic analog computer possible (see Ref. 3 for a good presentation of the role of operational amplifiers in analog computation).

It is interesting to examine the frequency response of the integrator. Let us represent sinusoidally varying quantities by a conventional phasor notation (Ref. 7, sec. 7.3); that is, if a voltage has the form

$$v(t) = A \cos (\omega t + \theta)$$

we represent it by the notation

$$\mho = A \angle \theta$$

where A is the magnitude and θ the angle (in radians) of a phasor $\mho$.

We can then easily find the phasor operator relating input and output (sinusoidal) voltages by setting $s = j\omega$ in the Laplace transform transfer function. Thus for sinusoidal inputs, (1.29) becomes

$$\mho_0 = -\frac{1}{j\omega} (K_1 \mho_1 + K_2 \mho_2 + \cdots + K_n \mho_n)$$

$$= \frac{1}{\omega} \angle \frac{\pi}{2} (K_1 \mho_1 + K_2 \mho_2 + \cdots + K_n \mho_n) \qquad (1.30)$$

Here we see that (in the steady state) the ideal integrator frequency response varies inversely with frequency, and is infinite at direct current! Also, inputs are shifted $+\pi/2$ radians, or $+90°$.

EXAMPLE 1.6 Consider again the circuit of Fig. 1.12 (an inverting integrator with a gain of 200). We can describe the integrator frequency response by

$$\frac{\mho_0}{\mho_1} = H(j\omega) = \frac{-200}{j\omega} = \frac{200}{\omega} \angle + \frac{\pi}{2}$$

Thus we arrive again at the conclusion of Example 1.5: if the circuit output is limited to 10 V peak, and the input is a sinusoid of peak magnitude A, then

$$\frac{200A}{\omega} \leqslant 10$$

and with $\omega = 2\pi f$, we have

$$\frac{A}{f} \leqslant \frac{2\pi}{20} = 0.314\,\text{V·s}$$

which is the same result as (1.28).   ////

We can summarize the results of the last two examples as follows:

1 At low frequencies, the amplitude of the input signal to an integrator is in
general limited by output voltage limitations. This is not too surprising, since an
integrator with gain constant $-K$ has a sinusoidal frequency response.

$$\frac{\mathcal{V}_0}{\mathcal{V}_1} = H(j\omega) = \frac{-K}{\omega}\bigg/\!-\frac{\pi}{2} = \frac{K}{\omega}\bigg/\frac{-3\pi}{2}$$

2 At higher frequencies, there may be an additional limitation on input signal
magnitude resulting from the amplifier output current required to charge the
integrating capacitor.

The above examples have been worked under the assumption that the input signals
were sinusoids. For more complex waveforms, careful analysis must be made, considering
the effects of output voltage and output current limitations separately, to determine
whether a given signal can be accommodated.

Initial-condition Setting and Mode Control

It is often necessary to establish an initial condition (voltage) on an integrator.
Figure 1.13*a* shows one commonly used circuit which implements what is often called
three-mode control. It requires two single-pole double-throw switches, S_A and S_B, and
the initial-condition-setting network, R_A and R_B. The *negative* of the desired
initial-condition voltage is established at node v_X. Potentiometer P is used to permit
establishment of an adjustable initial-condition value. The potentiometer is supplied with
either $+10$ or $-10\,\text{V}$, depending upon the desired initial condition. The operation of the
circuit is best understood by analyzing it separately in three different modes (switch
settings):

Position 1: RESET mode. In this state (Fig. 1.13*b*), we *acquire an initial condition*.
Noting that node Y is forced by the amplifier feedback to be near ground potential, we
can describe the circuit by

$$\frac{v_X}{R} + \frac{v_0}{R} + C\frac{dv_0}{dt} = i = 0 \qquad (1.31)$$

which may be rewritten

$$\frac{dv_0}{dt} = -\frac{1}{RC}v_0 - \frac{1}{RC}v_X \qquad (1.32)$$

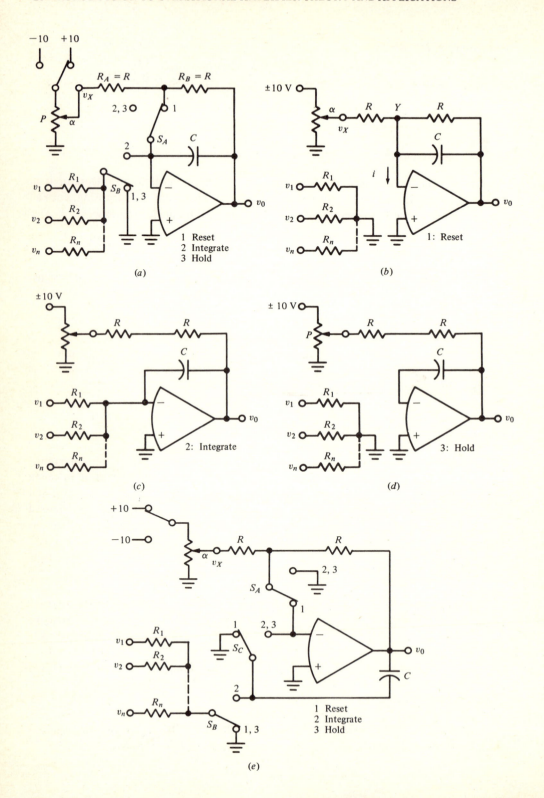

(a)

(b)

(c)

(d)

(e)

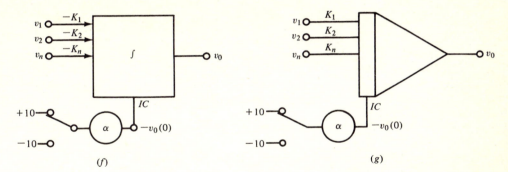

FIGURE 1.13
Three-mode integrator control circuit. (a) Complete circuit; (b) position 1, RESET or initial-condition mode; (c) position 2, INTEGRATE mode; (d) position 3, HOLD mode; (e) three-switch alternative to a which provides faster reset; (f) integrator block-diagram symbol; (g) corresponding analog-computer symbol.

We see that (1.32) is a simple time-invariant linear first-order ordinary differential equation. If we define α to be the loaded potentiometer ratio, viz.,

$$v_x = \pm 10\alpha = \text{constant}$$

then v_0 will approach a steady-state value, which we can find from (1.32); setting

$$\frac{dv_0}{dt} = 0$$

in (1.32), we obtain

$$v_{0SS} = -v_x \qquad \text{steady-state} \qquad (1.33)$$

Thus we see that the steady-state value of v_0 is the *negative* of the initial-condition voltage established by the potentiometer (or any other source of voltage for v_x). This polarity inversion, while often annoying, can often be accepted. We note also that (1.32) indicates that the basic time constant of the initial-condition-setting circuit is

$$\tau = RC \, \text{s}$$

Thus we can estimate the time required for initial-condition acquisition; e.g., five time constants are usually adequate.

Position 2: INTEGRATE mode. In this state (Fig. 1.13c) we have a conventional integrator circuit with inputs v_1 through v_n. The initial-condition-setting network is merely a low-current resistive load on the amplifier output, and the integrator equations derived previously apply for this mode. Now, however, if we go quickly from position 1 (RESET) to position 2 (INTEGRATE), the value of v_0 established in the RESET mode serves as the initial condition for the integration time interval.

Position 3: HOLD mode. The presence of the two switches S_A and S_B permits a third useful mode of operation (Fig. 1.13d), wherein the amplifier equation becomes

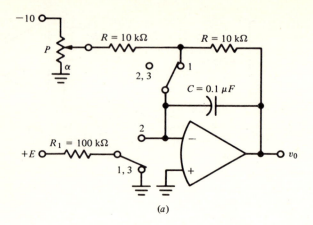

(a)

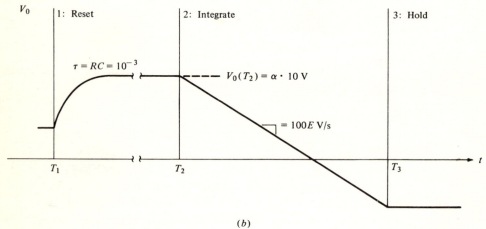

(b)

FIGURE 1.14
An inverting gain-of-100 integrator with mode control (Example 1.7). (a) Circuit;
(b) typical output waveform.

$$C\frac{dv_0}{dt} = 0 \qquad (1.34)$$

or

$$v_0 = \text{constant} \qquad (1.35)$$

In this mode, the circuit has *memory* of the value of v_0 present when the HOLD mode
was established.

EXAMPLE 1.7 Figure 1.14 shows a typical sequence of integrator modes; in this case, the
initial-condition-setting circuit is supplied with -10 V dc. Therefore, the initial condition
can be set anywhere between 0 and $+10$ V, depending on the setting of the

potentiometer. Beginning at time T_2, a positive constant voltage E is integrated with a gain of -100, so that in the INTEGRATE mode, the output decreases linearly at the rate of $100E$ V/s. If the circuit is put into the HOLD mode at time T_3, the value of v_0 will be held, or stored.   ////

We should mention that the ability of the integrator circuit of Fig. 1.13 to hold an output value in the HOLD mode is dependent on internal offsets in the amplifier (and possible switch leakage current, if the switches are solid-state devices); we have assumed here an ideal operational amplifier and switches. In practice, it may only be possible to hold the integrator output for a short time; more will be said about this in Chap. 2.

Figure 1.13*e* shows an alternative three-mode integrator control circuit which permits faster setting of the initial conditions. This circuit requires three single-pole double-throw switches. The operation is essentially the same as that of the circuit of Fig. 1.13*a* in HOLD and INTEGRATE, but differs in RESET, where the capacitor is now made the output load of the operational amplifier (see Prob. 1.9). Some operational amplifiers will not tolerate a large capacitive output load, and cannot be used with this last circuit; it is preferred, however, where fast initial-condition setting is desirable. (See also Chap. 5 for a discussion of high-speed analog signal switching and the design of sample-hold circuits based on this circuit.)

1.5   THE GENERAL INVERTING AMPLIFIER

Figure 1.15 shows a general (linear) inverting (and summing) operational amplifier circuit; here we allow the input summing networks and the feedback network to be generalized linear two-port (three-terminal) networks. In each network $Z_i(s)$ is defined to be the short-circuit transfer impedance. If the particular network is a two-terminal (one-port) network, then $Z_i(s)$ is merely the conventional terminal driving-point impedance. If the network is the more general three-terminal (two-port) network shown in Fig. 1.16, then we define the *short-circuit transfer impedance* as

$$\left| Z_{sc}(s) = \frac{V_1(s)}{I_2(s)} \right|_{V_2(s)=0} \tag{1.36}$$

NOTE: We have not used the conventional network-theory definition, since our port currents are directed outward. However, our choice of direction eliminates a lot of minus signs in subsequent analyses; e.g., it would be unfortunate for a simple resistor to have a negative impedance.

Note that a tee network has the same value of Z_{sc} regardless of which terminal is used for input and which for output. In Fig. 1.17 the value of Z_{sc} can be shown to be

$$Z_{sc} = \frac{Z_1 Z_2 + Z_2 Z_3 + Z_3 Z_1}{Z_2} = \left. \frac{V_1}{I_2} \right|_{V_2=0} = \left. \frac{V_2}{I_1} \right|_{V_1=0} \tag{1.37}$$

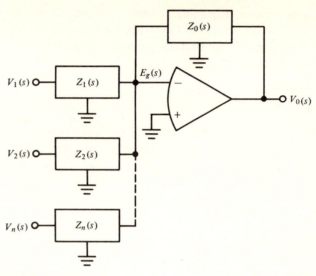

FIGURE 1.15
General inverting amplifier circuit with three-terminal (two-port) networks.

We again assume that the action of the amplifier, via the feedback network, is such that the summing-point voltage E_g is held at a virtual ground potential, that is, $E_g \cong 0$ and $I \cong 0$. With this assumption (using Laplace transform notation)

$$\frac{V_0(s)}{Z_0(s)} + \frac{V_1(s)}{Z_1(s)} + \frac{V_2(s)}{Z_2(s)} + \cdots + \frac{V_n}{Z_n(s)} = I(s) \cong 0 \qquad (1.38)$$

or

$$V_0(s) = -Z_0(s)\left[\frac{V_1(s)}{Z_1(s)} + \frac{V_2(s)}{Z_2(s)} + \cdots + \frac{V_n(s)}{Z_n(s)}\right] \qquad (1.39)$$

EXAMPLE 1.8 Consider the single-input inverting amplifier circuit of Fig. 1.18. Suppose we want an inverting gain of 100. We also want to use fairly low resistance values

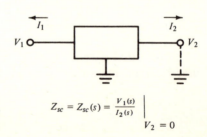

FIGURE 1.16
Definition of short-circuit transfer impedance for a three-terminal network.

$$Z_{sc} = Z_{sc}(s) = \left.\frac{V_1(s)}{I_2(s)}\right|_{V_2 = 0}$$

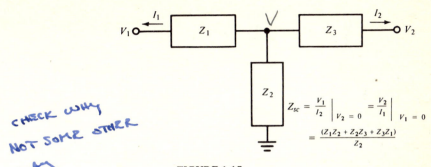

$$Z_{sc} = \frac{V_1}{I_2}\bigg|_{V_2 = 0} = \frac{V_2}{I_1}\bigg|_{V_1 = 0}$$

$$= \frac{(Z_1 Z_2 + Z_2 Z_3 + Z_3 Z_1)}{Z_2}$$

FIGURE 1.17
The short-circuit transfer impedance of a tee network.

throughout, so that the effects of stray wiring capacitances will be suppressed. We also want the circuit input impedance to be 10 kΩ. This latter requirement can be met by setting the input resistor $R_A = 10$ kΩ. From (1.39) we now see that the short-circuit feedback resistance should be

$$Z_0 = R_0 = 100 \times 10^4 = 10^6 \ \Omega \qquad (1.40)$$

The tee configuration consisting of R_1, R_2, and R_3, can be used to obtain a short-circuit transfer resistance that is considerably larger than any of the individual resistances. With $R_1 = R_3 = 10$ kΩ we can use (1.37) to find that

$$R_2 = 102.0408 \ \Omega$$

will make the short-circuit feedback resistance 10^6. In practice, one might purchase three nominal 10 kΩ resistors and use an adjustable resistor for R_2, so that the voltage gain could be set to exactly −100.   ////

We now have the capability to implement rather general transfer function operators with the circuit of Fig. 1.15. In Chap. 4 we will see even more general configurations

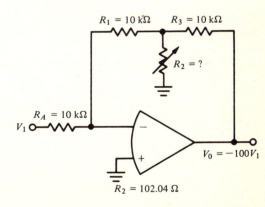

FIGURE 1.18
High-gain inverting amplifier using relatively low resistance values.

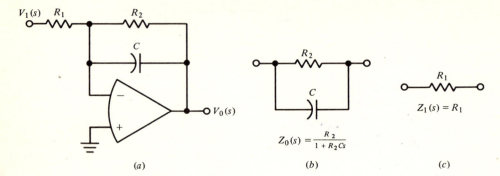

FIGURE 1.19
Simple inverting low-pass filter circuit. (*a*) Complete circuit; (*b*) feedback network; (*c*) input network.

specifically designed to implement transfer functions in an optimum fashion, but (1.39) is often sufficient if the designer can find the proper networks to use in the input and feedback paths to obtain a desired transfer function. Note again that it is the short-circuit transfer impedance that is important; Table A-1 of Ref. 3 provides a large collection of *RC* networks and describes their short-circuit transfer impedances. We will only discuss a few important examples in this chapter, and will approach the problem of general transfer function synthesis more thoroughly in Chap. 4.

EXAMPLE 1.9 Consider the circuit of Fig. 1.19*a*. We can analyze its performance using (1.39) by treating the feedback and input networks separately. Clearly

$$Z_1(s) = R_1$$

and $Z_0(s)$ is the parallel combination of R_2 and C, that is,

$$Z_0(s) = \frac{R_2}{1 + R_2 Cs}$$

Thus the transfer function of the circuit of Fig. 1.19*a* is

$$\frac{V_0(s)}{V_1(s)} = -\frac{R_2}{R_1}\frac{1}{1 + Ts} \qquad (1.41)$$

where

$$T = R_2 C$$

Hence the circuit implements a *first-order low-pass filter* transfer function with

$$\text{Dc steady-state gain} = -\frac{R_2}{R_1} \qquad (1.42)$$

and the −3-dB cutoff frequency

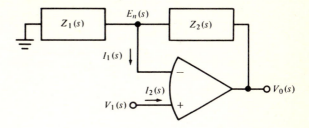

FIGURE 1.20
General noninverting amplifier circuit (two-terminal networks assumed).

$$\omega_{-3 \text{ dB}} = \frac{1}{T} = \frac{1}{R_2 C} \quad \text{rad/s} \quad (1.43)$$

This circuit is discussed further in Sec. 1.9. ////

1.6 THE NONINVERTING AMPLIFIER

So far we have always connected the noninverting input of the operational amplifier to ground. Most modern operational amplifiers, however, have this terminal available, as well as the inverting input, i.e., one usually deals with a differential input type of operational amplifier. If the noninverting input is available, then a broader variety of circuit applications is possible. In particular, one can easily implement the noninverting amplifier circuit of Fig. 1.20; here the + terminal is utilized along with the - terminal. If the amplifier has the ideal properties described in Sec. 1.3, we can assume that consequently $I_1(s)$ and $I_2(s)$ are zero, and that

$$V_1(s) \cong E_n(s)$$

Thus

$$\frac{E_n(s)}{Z_1(s)} + \frac{E_n(s) - V_0(s)}{Z_2(s)} = -I_1(s) = 0 \quad (1.44)$$

or

$$\frac{V_0(s)}{V_1(s)} = \frac{Z_1(s) + Z_2(s)}{Z_1(s)} = 1 + \frac{Z_2(s)}{Z_1(s)} \quad (1.45)$$

Note that

$$\left| \frac{V_0(s)}{V_1(s)} \right| \geqslant 1 \quad (1.46)$$

We also see that there is no polarity inversion in this circuit, and the magnitude of the gain is always greater than unity. Above, (1.45) was derived for general two-terminal (one-port) impedances; some special cases merit attention. In Fig. 1.21a, we see the circuit of a simple noninverting amplifier circuit using only resistances; in this case

$$\frac{V_0(s)}{V_1(s)} = \frac{R_1 + R_2}{R_1} = 1 + \frac{R_2}{R_1} \geqslant 1 \quad (1.47)$$

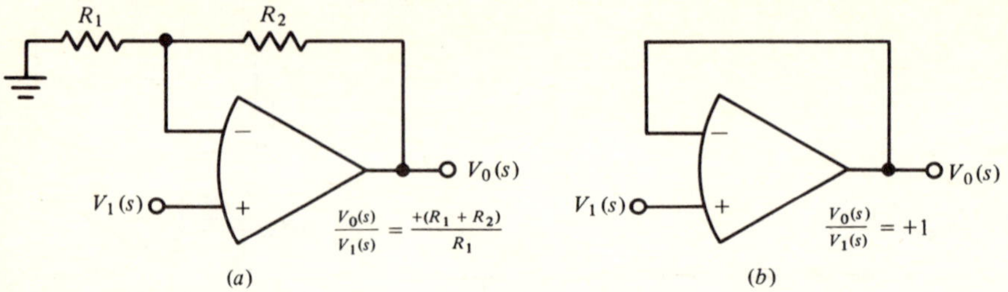

FIGURE 1.21
Noninverting amplifier circuit with resistive elements. (a) General circuit; (b) simple unity gain follower.

The disarmingly simple circuit of Fig. 1.21b is that of a *unity gain follower*. Its performance is predicted by setting $R_2 = 0$ and/or letting R_1 approach infinity. The voltage transfer function for the circuit of Fig. 1.21b is

$$\frac{V_0(s)}{V_1(s)} = 1 \qquad (1.48)$$

Note that the noninverting circuit of Fig. 1.21 has a very high input impedance (for example 10^{12} Ω with an FET-input operational amplifier); on the other hand, the output impedance is very low (for example, less than 0.1 Ω). Thus the noninverting circuit usually does not load the signal source driving it, and is therefore a suitable choice in many situations where a high Thévenin-equivalent source impedance is associated with some signal to be amplified. As mentioned earlier, the *inverting* amplifier circuit has an input impedance governed by the value of the input summing impedances associated with each input terminal (Fig. 1.6 or 1.15). Hence the simple unity gain follower of Fig. 1.21b serves a useful purpose as a buffer or unloading circuit to permit observing a high-impedance signal with some instrument that requires considerable driving current. In Chap. 2, when we discuss nonideal amplifiers, we will see how to estimate more precisely just how high the input impedance is, and how low the output impedance is.

More complex noninverting arrangements are possible. For example, Fig. 1.22 shows a useful noninverting integrator circuit with the transfer function

$$V_0(s) = \frac{2}{sRC} V_1(s) \qquad (1.49)$$

The reader should attempt to derive (1.49) for himself (Prob. 1.16); in particular, it is important to note that resistor imbalance can yield a transfer function with undesirable properties (e.g., the circuit may be unstable!).

EXAMPLE 1.10 Consider the instrumentation problem posed in Fig. 1.23. Here we have a signal source with a resistive internal impedance of 100,000 Ω. We want to amplify e_s by 100 (polarity not important) without distorting the signal level by loading the signal

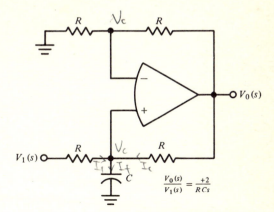

FIGURE 1.22
A noninverting integrator circuit (see also Prob. 1.16).

$$\frac{V_0(s)}{V_1(s)} = \frac{+2}{RCs}$$

source by more than 1 percent. We want to keep all circuit resistances less than 10,000 Ω in order to reduce internal amplifier dc offsets (see Sec. 2.5 for a discussion of this consideration). We can achieve the required gain of 100 (no polarity inversion) using the noninverting circuit of Fig. 1.23a merely by selecting

$$R_2 = 99R_1$$

for example, $R_2 = 9900$ Ω and $R_1 = 100$ Ω; assuming an ideal operational amplifier, the circuit of Fig. 1.23a would not load the signal source (we will see in Chap. 2 that some loading will occur, but probably not enough to be important in this situation). Suppose the inverting circuit of Fig. 1.23b is selected; then in order to keep the circuit input impedance sufficiently high, we are forced to make

$$R_0 \geqslant 100 \times 100,000 = 10 \ M\Omega$$

and already we have violated our requirement that all resistors be less than 10,000 Ω in value.   ////

Simple Kirchhoff summing may be employed with the noninverting amplifier circuit as well. In Fig. 1.24 we see a two-input noninverting summing amplifier circuit. Clearly, from (1.47),

$$V_0(s) = \left(1 + \frac{R_F}{R_S}\right) V_S(s) \qquad (1.50)$$

Since the amplifier does not load node V_s,

$$V_S(s) = \frac{R_P}{R_1} V_1(s) + \frac{R_P}{R_2} V_2(s) \qquad (1.51)$$

where R_P is the parallel combination of R_1 and R_2 (this last equation can be easily shown to be the Kirchhoff summing equation at node V_S, assuming $I(s) = 0$). Thus, combining (1.50) and (1.51), we have

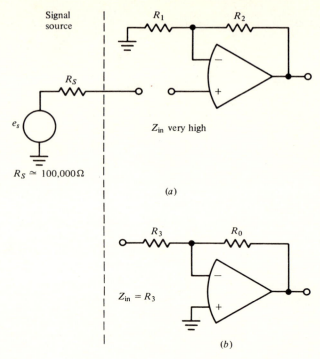

FIGURE 1.23
Circuit of Example 1.10. (*a*) Noninverting amplifier; (*b*) inverting amplifier.

$$V_0(s) = \left(1 + \frac{R_F}{R_S}\right)[K_1 V_1(s) + K_2 V_2(s)] \qquad (1.52a)$$

with

$$K_1 = \frac{R_P}{R_1} \qquad K_2 = \frac{R_P}{R_2} \qquad (1.52b)$$

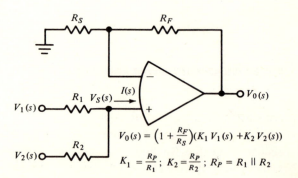

FIGURE 1.24
A simple two-input noninverting sum-ming amplifier circuit.

$$V_0(s) = \left(1 + \frac{R_F}{R_S}\right)(K_1 V_1(s) + K_2 V_2(s))$$

$$K_1 = \frac{R_P}{R_1}; \quad K_2 = \frac{R_P}{R_2}; \quad R_P = R_1 \parallel R_2$$

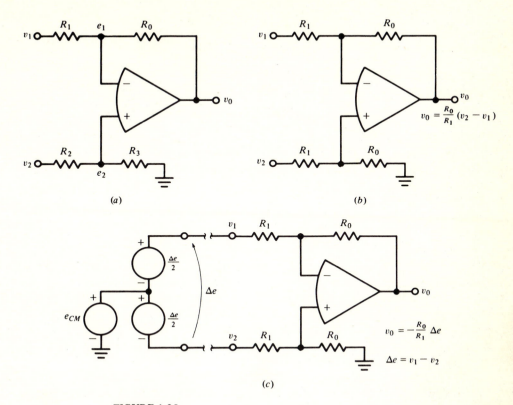

FIGURE 1.25
Single-output differential-input amplifier circuit. (*a*) General circuit; (*b*) circuit with balanced resistance ratios; (*c*) common-mode rejection.

and

$$R_P = R_1 \| R_2 \qquad (1.52c)$$

Note also that $K_1 \leqslant 1$ and $K_2 \leqslant 1$, and that (1.52) is derived assuming zero source impedance for each input signal. Signal source impedances will affect the gain factors K_1 and K_2.

1.7 THE DIFFERENTIAL AMPLIFIER

Figure 1.25 shows a special operational amplifier circuit commonly termed a *differential amplifier*, or more precisely, a differential-input single-ended-output amplifier. This circuit features a balanced (differential or push-pull) input and an unbalanced (single-ended) output. Assuming an ideal operational amplifier and ideal resistances, we can write time domain circuit equations,

$$e_1 \left(\frac{1}{R_1} + \frac{1}{R_0} \right) - \frac{v_0}{R_0} - \frac{v_1}{R_1} = 0 \qquad (1.53a)$$

and
$$e_2\left(\frac{1}{R_2} + \frac{1}{R_3}\right) - \frac{v_2}{R_2} = 0 \qquad (1.53b)$$

Combining (1.53a) and (1.53b), assuming $e_1 \cong e_2$, we have

$$v_0 = \frac{(R_1 + R_0)R_3}{(R_2 + R_3)R_1}v_2 - \frac{R_0}{R_1}v_1 \qquad (1.54)$$

(The reader should verify the above.)

Note that (1.54) has been derived under the assumption that we have an ideal operational amplifier. In Chap. 2, we will take nonideal behavior into account in more detail. In most situations, the special case of Fig. 1.25b is employed, that is, $R_3 = R_0$ and $R_2 = R_1$; in this case, (1.54) becomes

$$v_0 = \frac{R_0}{R_1}(v_2 - v_1) \qquad (1.55)$$

Differential amplifiers such as Fig. 1.25b thus respond only to the *difference* between the input signals v_2 and v_1; they are said to have *good common-mode rejection* of noise induced in equal amounts in both signal sources. Figure 1.25c shows a typical situation where differential amplifiers are commonly employed. Here we are trying to observe, and usually also amplify, a small signal Δe, which may be generated in the presence of a relatively large common-mode signal e_{cm}. By using a differential amplifier, the common-mode signal is ideally rejected.

This type of situation is common in many instrumentation systems, where Δe originates from some low-level transducer; e_{cm} may be a dc offset in some push-pull signal source or some form of noise (for example, 60-Hz interference or some other crosstalk from a nearby electrical system)—in other words, an unwanted signal component which will hopefully be ignored by the amplifier.

In Fig. 1.25c, we assume

$$v_1 = e_{cm} + \frac{\Delta e}{2} \qquad (1.56a)$$

$$v_2 = e_{cm} - \frac{\Delta e}{2} \qquad (1.56b)$$

that is,
$$v_1 - v_2 = \Delta e \qquad (1.57)$$

and thus (1.55) yields, for the case of Fig. 1.25c,

$$v_0 = -\frac{R_0}{R_1}\Delta e \qquad (1.58)$$

independent of e_{cm}.

Note that the differential input resistance of the circuit of Fig. 1.25b is

$$R_{in} = 2R_1 \qquad (1.59)$$

which creates a loading on the differential signal source that may attenuate it.

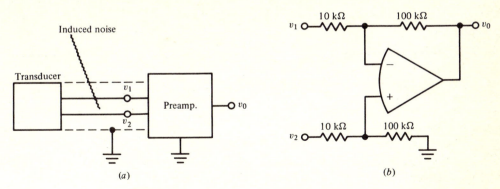

FIGURE 1.26
Circuit of Example 1.11. (*a*) Signal source with noise; (*b*) preamplifier circuit.

EXAMPLE 1.11 Consider Fig. 1.26*a*; this depicts a low-level transducer whose signal is carried over a shielded cable with two floating conductors for some distance to a preamplifier. Induced noise (usually 60 Hz from the ac power lines in the vicinity of the transducer) is to be rejected, while the transducer signal itself is to be amplified by 10. The desired signal may be considered a differential signal in the presence of an unwanted common-mode noise signal. We have the further restriction that the transducer should be terminated in a total load resistance of 20 kΩ between terminals. The differential amplifier circuit of Fig. 1.26*b* meets all the requirements stated above. It has a differential voltage gain of 10, supposedly a high common-mode rejection, if the amplifier and resistors are well balanced, and a total input impedance of 20 kΩ. The output signal is single-ended, and may be subsequently processed by single-ended amplifiers for equalization and further amplification. ////

Note that if the common-mode voltage (e_{cm} in Fig. 1.25*c*) is fairly large, then the operational amplifier input terminals will have a similar voltage impressed upon them. Most operational amplifiers have a *maximum allowable common-mode input voltage* that is slightly less than the magnitude of the power-supply voltages.

Figure 1.27 shows a *differential-input differential-output* amplifier circuit. This maintains a balanced (or push-pull) mode at both input and output. This circuit requires an operational amplifier with a differential output stage. Such amplifiers are less common than the unbalanced-output variety we have been discussing, but they are available from some suppliers. This configuration is mentioned here primarily for completeness; of course such amplifiers are useful for amplifying signals for delivery to balanced transmission lines, etc. The balanced circuit of Fig. 1.27 has a voltage gain described by

$$v_0 = v_4 - v_3 = -\frac{R_0}{R_1}(v_1 - v_2) = -\frac{R_0}{R_1}v_i \qquad (1.60)$$

where v_0 and v_i are the differential output and input voltages, respectively.

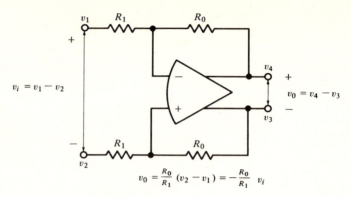

FIGURE 1.27
Differential-output differential-input amplifier circuit.

1.8 THE COMBINED INVERTING AND NONINVERTING AMPLIFIER

Figure 1.28 shows a general amplifier circuit that permits summation with inversion and without. Such a general circuit provides considerable flexibility. Our method of analysis will be instructive when we come to similar circuits in Chap. 2. We assume that all the impedances are simple two-terminal driving-point impedances. We assume that there are m inverting inputs $V_{I1}(s), \ldots, V_{Im}(s)$ and n noninverting inputs $V_{N1}(s), \ldots, V_{Nn}(s)$, all obtained from ideal voltage sources. In addition, we have added two shunt impedances $Z_{IS}(s)$ and $Z_{NS}(s)$ for the sake of generality. These elements can be adjusted to affect the transfer function gains. The typical situation is for all the impedances to be resistances, but general fixed linear networks can also be used.

The Kirchhoff nodal equations at nodes $E_A(s)$ and $E_B(s)$ are

$$E_A(s)\left[\frac{1}{Z_{IS}(s)} + \frac{1}{Z_{I1}(s)} + \cdots + \frac{1}{Z_{Im}(s)} + \frac{1}{Z_0(s)}\right] = \frac{V_0(s)}{Z_0(s)} + \sum_{i=1}^{m} \frac{V_{Ii}(s)}{Z_{Ii}(s)} \qquad (1.61a)$$

and

$$E_B(s)\left[\frac{1}{Z_{NS}(s)} + \frac{1}{Z_{N1}(s)} + \cdots + \frac{1}{Z_{Nn}(s)}\right] = \sum_{j=1}^{n} \frac{V_{Nj}(s)}{Z_{Nj}(s)} \qquad (1.61b)$$

Let us define

$$Z_A(s) = Z_0(s) \| Z_{IS}(s) \| Z_{I1}(s) \| \cdots \| Z_{Im}(s)$$

$$= \left[\frac{1}{Z_0(s)} + \frac{1}{Z_{IS}(s)} + \sum_{i=1}^{m} \frac{1}{Z_{Ii}(s)}\right]^{-1} \qquad (1.62a)$$

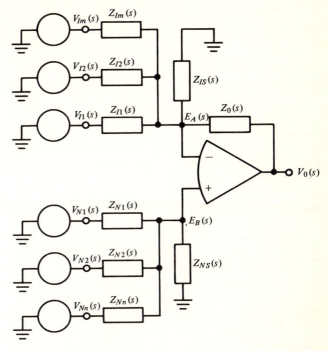

FIGURE 1.28
General inverting and noninverting summing amplifier.

and

$$Z_B = Z_{N1}(s) \, \| Z_{N2}(s) \, \| \cdots \, \| Z_{Nn}(s) \, \| Z_{NS}(s) = \left[\sum_{j=1}^{n} \frac{1}{Z_{Nj}(s)} + \frac{1}{Z_{NS}(s)} \right]^{-1} \qquad (1.62b)$$

For an ideal operational amplifier,

$$E_A(s) \cong E_B(s) \qquad (1.63)$$

Using (1.62) in (1.61), we obtain

$$\frac{E_A(s)}{Z_A(s)} = \frac{V_0(s)}{Z_0(s)} + \sum_{i=1}^{m} \frac{V_{Ii}(s)}{Z_{Ii}(s)} \qquad (1.64a)$$

and

$$\frac{E_B(s)}{Z_B(s)} = \sum_{j=1}^{n} \frac{V_{Nj}(s)}{Z_{Nj}(s)} \qquad (1.64b)$$

These last equations are generally of great importance to us in this text; they describe what we shall call *Kirchhoff voltage summing* at a node. Consider Fig. 1.29, which illustrates a general Kirchhoff summing network. We assume that there is no current

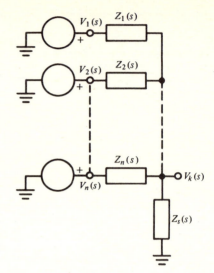

FIGURE 1.29
A general Kirchhoff voltage summing
network.

flowing out of node V_k except via the branches indicated. (This is the situation for nodes
E_A and E_B in Fig. 1.28, since we assume no current flows into either amplifier terminal.)
A nodal analysis of Fig. 1.29 yields

$$V_k(s)\left[\frac{1}{Z_1(s)} + \frac{1}{Z_2(s)} + \cdots + \frac{1}{Z_n(s)} + \frac{1}{Z_s(s)}\right] = \frac{V_1(s)}{Z_1(s)} + \cdots + \frac{V_n(s)}{Z_n(s)} \qquad (1.65)$$

If we define $Z_p =$ parallel combination of all impedances at node V_k.

$$Z_p = \left[\frac{1}{Z_1(s)} + \cdots + \frac{1}{Z_n(s)} + \frac{1}{Z_s(s)}\right]^{-1} \qquad (1.66)$$

then (1.65) becomes

$$\frac{V_k(s)}{Z_p(s)} = \frac{V_1(s)}{Z_1(s)} + \cdots + \frac{V_n(s)}{Z_n(s)} \qquad (1.67a)$$

or

$$V_k(s) = Z_p(s) \sum_{i=1}^{n} \frac{V_i(s)}{Z_i(s)} \qquad (1.67b)$$

Thus in the circuit of Fig. 1.29 the voltages V_i, $i = 1, \ldots, n$, are summed at V_k, but with
a weighting factor Z_p/Z_i. This is the general nature of such summing networks, and
equations of the form of (1.67) will arise frequently in our work.
   Employing (1.63) in (1.64a), we have

$$\frac{V_0(s)}{Z_0(s)} = -\sum_{i=1}^{m} \frac{V_{Ii}(s)}{Z_{Ii}(s)} + \frac{E_B(s)}{Z_A(s)} \qquad (1.68)$$

Then combining (1.64b) and (1.68), we obtain

$$V_0(s) = Z_0(s)\left[\frac{Z_B(s)}{Z_A(s)} \sum_{j=1}^{n} \frac{V_{Nj}(s)}{Z_{Nj}(s)} - \sum_{i=1}^{m} \frac{V_{Ii}(s)}{Z_{Ii}(s)}\right] \qquad (1.69)$$

This last equation permits us to design combined inverting and noninverting summing amplifiers. Relative values of the inverting gains can be selected by the proper choice of the $Z_{Ii}(s)$ and relative values of the noninverting gains by the $Z_{Nj}(s)$. Overall inverting and noninverting gains can then be adjusted by the proper choice of $Z_0(s)$, $Z_{IS}(s)$, and $Z_{NS}(s)$. Of course the differential amplifier circuit of Fig. 1.25 is included in this general circuit, along with the more common inverting and noninverting circuits of Figs. 1.8 and 1.20.

Let us define $K_{Ii}(s)$ as the transfer function between $V_{Ii}(s)$ and $V_0(s)$ with all other inputs set to zero. *Warning!* To do this properly, it is conceptually necessary to set all other inputs to zero by replacing them with short circuits to ground, since all the development in this section has been predicated on the inputs $V_{Ii}(s)$ and $V_{Nj}(s)$ being ideal voltage sources. Thus in calculating the total shunt impedances at summing nodes, such as E_A and E_B in Fig. 1.28, the signal sources must be considered to have zero internal impedance. Let us also define $K_{Nj}(s)$ as the transfer function between $V_{Nj}(s)$ and $V_0(s)$. We note from (1.69) that

$$K_{Ii}(s) = -\frac{Z_0(s)}{Z_{Ii}(s)} \qquad i = 1, \ldots, m \qquad (1.70a)$$

and

$$K_{Nj}(s) = +\frac{Z_B(s)}{Z_A(s)} \frac{Z_0(s)}{Z_{Nj}(s)} \qquad j = 1, \ldots, n \qquad (1.70b)$$

These last equations facilitate the design process. Often we can pick $Z_{NS}(s)$ and/or $Z_{IS}(s)$ such that $Z_A(s) = Z_B(s)$, so that

$$K_{Nj}(s) = \frac{+Z_0(s)}{Z_{Nj}(s)} \qquad Z_A(s) = Z_B(s) \qquad (1.71)$$

and the input summing impedances can all be designed individually. (Reference 9 provides some useful design guides and design examples for resistive summing amplifiers.)

Example 1.12 Consider the circuit of Fig. 1.30: Here we have a general summing amplifier with two inverting inputs and two noninverting inputs. Thus the general form of the amplifier gain equation is

$$V_0 = K_{I1}V_{I1} + K_{I2}V_{I2} + K_{N1}V_{N1} + K_{N2}V_{N2} \qquad (1.72)$$

From (1.70),

$$K_{I1} = \frac{-R_0}{R_{I1}} \qquad (1.73a)$$

$$K_{I2} = \frac{-R_0}{R_{I2}} \qquad (1.73b)$$

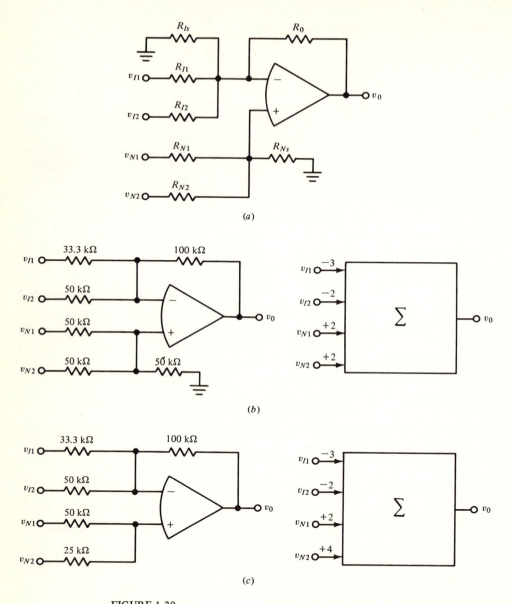

FIGURE 1.30

Circuit of Example 1.12. (*a*) General circuit with two inverting and two noninverting inputs; (*b*) circuit for gains of -3, -2, $+2$, and $+2$; (*c*) circuit for gains of -3, -2, $+2$, and $+4$.

$$K_{N1} = \frac{R_B}{R_A}\frac{R_0}{R_{N1}} \qquad (1.73c)$$

$$K_{N2} = \frac{R_B}{R_A}\frac{R_0}{R_{N2}} \qquad (1.73d)$$

where

$$R_A = R_0\,\|R_{I1}\|R_{I2}\|R_{Is} \qquad (1.74a)$$

and

$$R_B = R_{N1}\|R_{N2}\|R_{Ns} \qquad (1.74b)$$

We can make $R_B = R_A$ by the proper choice of R_{Is} and R_{Ns} (usually one is made infinite, i.e., not used).

Suppose we want (Fig. 1.30b)

$$v_0 = -(3v_{I1} + 2v_{I2}) + 2v_{N1} + 2v_{N2}$$

$$\text{Pick } R_0 = 100 \text{ k}\Omega \qquad (1.75)$$

Then from (1.73),

$$R_{I1} = \frac{100}{3} \text{ k}\Omega \qquad (1.76a)$$

$$R_{I2} = 50 \text{ k}\Omega \qquad (1.76b)$$

$$R_{N1} = 50 \text{ k}\Omega \quad \text{if } R_A = R_B \qquad (1.76c)$$

and

$$R_{N2} = 50 \text{ k}\Omega \quad \text{if } R_A = R_B \qquad (1.76d)$$

From (1.74),

$$R_A = 16.67 \text{ k}\Omega \, \| R_{Is}$$

$$R_B = 25 \text{ k}\Omega \, \| R_{Ns}$$

If we let $R_{Is} = \infty$,

$$R_A = 16.67 \text{ k}\Omega \qquad (1.77)$$

We can then make $R_B = R_A$ with

$$R_{Ns} = 50 \text{ k}\Omega \qquad (1.78)$$

Figure 1.30b shows the final circuit.

Suppose instead we want

$$v_0 = -(3v_{I1} + 2v_{I2}) + 2v_{N1} + 4v_{N2}$$

First of all, this changes R_{N2} to 25 kΩ. We now have

$$R_A = 16.67 \text{ k}\Omega \, \| R_{Is} \qquad (1.79a)$$

$$R_B = 16.67 \text{ k}\Omega \, \| R_{Ns} \qquad (1.79b)$$

Now we need neither R_{Is} nor R_{Ns} to make $R_A = R_B$. Figure 1.30c shows the final circuit. Note that changing the gain for v_{N2} did not affect the values of the feedback resistor or the other summing resistors. Other examples will lead to a requirement for finite R_{IS} but not R_{NS}, etc. (see Prob. 1.22).   ////

1.9 SIMPLE FIRST-ORDER TRANSFER FUNCTIONS

In order to illustrate the use of the basic linear amplifier design equations we have thus far developed, this section will present some basic configurations for implementing first-order (one-pole) linear transfer functions. Methods for designing higher-order transfer functions will be presented in Chap. 4 (Active RC Filters). We have shown how to implement the basic linear operations of inversion (Fig. 1.7), multiplying by a constant (Figs. 1.6 and 1.21), summing (Figs. 1.8, 1.24, and 1.30), and integration (Figs. 1.10 and 1.22). Note that with a collection of these types of circuits, one can implement any linear transfer function by analog-computer programming techniques (see Ref. 3, pp. 10 to 16, and Ref. 8, Chap. 10). This approach yields designs with a relatively large number of components. We will attempt to present here and in Chap. 4 methods of designing transfer function operators that require relatively few components and still provide accurate implementations.

First-Order Low-Pass

Consider the first-order low-pass transfer function

$$\frac{V_0(s)}{V_1(s)} = \frac{-H_0\omega_0}{s + \omega_0} = \frac{-H_0}{1 + s/\omega_0} \qquad (1.80)$$

As already suggested in Example 1.9, this transfer function can be implemented by the circuit of Fig. 1.19a, redrawn in Fig. 1.31a. We have already shown in Example 1.9 that we can use the approach of Sec. 1.5, and by means of (1.39), it follows that

$$\frac{V_0(s)}{V_1(s)} = \frac{-Z_0(s)}{Z_1(s)} \qquad (1.81)$$

where

$$Z_0(s) = \frac{R_2}{1 + R_2 Cs} \qquad (1.82a)$$

and

$$Z_1(s) = R_1 \qquad (1.82b)$$

or

$$\frac{V_0(s)}{V_1(s)} = \frac{-R_2/R_1}{1 + R_2 Cs} = \frac{-1/R_1 C}{s + 1/R_2 C} \qquad (1.83)$$

Thus we have matched the transfer function of (1.80) with

$$H_0 = \frac{R_2}{R_1} = \text{dc steady-state gain} \qquad (1.84a)$$

$$\omega_0 = \frac{1}{R_2 C} = \text{3-dB cutoff frequency} \qquad (1.84b)$$

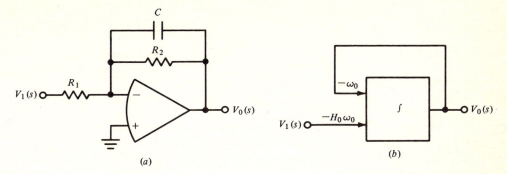

FIGURE 1.31
First-order low-pass filter circuit. (*a*) Circuit components; (*b*) circuit block diagram.

As an alternative approach, let us write the time domain equation that corresponds to the frequency domain transfer function (1.80), viz.,

$$\frac{dv_0}{dt} = -\omega_0 v_0(t) - H_0 \omega_0 v_1(t) \qquad (1.85)$$

Using the summing integrator circuit of Fig. 1.10, we implement (1.85) with the circuit of Fig. 1.31*b*. We see, however, that this is really merely a block-diagram sketch of the circuit of Fig. 1.31*a* with

$$\omega_0 = \frac{1}{R_2 C} \qquad (1.86a)$$

$$H_0 \omega_0 = \frac{1}{R_1 C} \qquad (1.86b)$$

which agrees with (1.84).

Thus we have presented two related, but conceptually different, approaches to the implementation of (1.80), i.e.:

1 We can synthesize input and feedback *networks* for a circuit of the form of Fig. 1.15, 1.20, or 1.28, so that the overall transfer function is implemented (the approach of Fig. 1.31*a*).

2 Transfer functions may be implemented by *implementing their differential equation* with basic operations such as summing, inverting, integration, and amplification (the approach of Fig. 1.31*b*).

In this simple first-order example, we come up with the same design; in more complicated, higher-order cases, the resulting designs will in general differ in detail, depending on the approach used. The differential-equation-solving approach will generally lead to designs with more components. To see this better, let us consider a slightly different case.

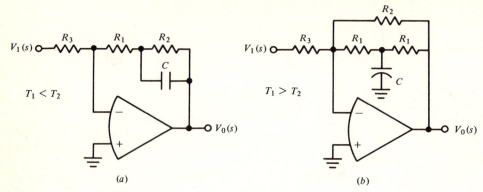

FIGURE 1.32

Two circuits for implementing one real zero and one real pole. (a) $T_1 < T_2$; (b) $T_1 > T_2$.

Implementing a Pole-Zero Pair

Suppose we want to implement the transfer function

$$\frac{V_0(s)}{V_1(s)} = \frac{H_0(1 + sT_1)}{1 + sT_2} = \frac{H_0 T_1(s + 1/T_1)}{T_2(s + 1/T_2)} \qquad (1.87)$$

This transfer function has the properties

$$\left|\frac{V_0(s)}{V_1(s)}\right|_{s=0} = H_0 = \text{dc gain} \qquad (1.88a)$$

$$\left|\frac{V_0(j\omega)}{V_1(j\omega)}\right|_{\omega=\infty} = \frac{H_0 T_1}{T_2} = \text{high-frequency gain} \qquad (1.88b)$$

pole at
$$s = -\frac{1}{T_2}$$

zero at
$$s = -\frac{1}{T_1}$$

Such a transfer function is commonly encountered in a variety of equalizer and compensator circuits for communication and control systems.

We will again take two approaches to implementing such a transfer function. First, in Fig. 1.32, we have one-amplifier circuits which implement (1.87). These circuits may be analyzed by finding the short-circuit transfer impedances of the feedback networks. In Fig. 1.32a, we can find that (assuming the structure of Fig. 1.15)

$$Z_0(s) = R_1 + \frac{R_2}{1 + sR_2 C} \qquad (1.89)$$

Thus

$$\frac{V_0(s)}{V_1(s)} = \frac{-Z_0(s)}{Z_1(s)} = \frac{-[R_1 + R_2/(1 + sR_2C)]}{R_3}$$

$$= \frac{-(R_1 + R_2 + sR_1R_2C)}{R_3(1 + sR_2C)}$$

$$= H_0 \frac{1 + sT_1}{1 + sT_2} \qquad (1.90a)$$

with

$$T_1 = \frac{R_1}{R_1 + R_2} R_2C = \frac{R_1}{R_1 + R_2} T_2 \qquad (1.90b)$$

$$T_2 = R_2C \qquad (1.90c)$$

and

$$H_0 = \frac{-(R_1 + R_2)}{R_3} \qquad (1.90d)$$

Also we note that for Fig. 1.32a,

$$T_1 < T_2 \qquad (1.90e)$$

This restriction can be overcome by the circuit of Fig. 1.32b. Here (see Prob. 1.28) we find that

$$\frac{V_0(s)}{V_1(s)} = H_0 \frac{1 + sT_1}{1 + sT_2} \qquad (1.91a)$$

with

$$T_1 = \frac{R_1C}{2} \qquad (1.91b)$$

$$T_2 = \frac{2R_1}{2R_1 + R_2} \frac{R_1C}{2} \qquad (1.91c)$$

and

$$H_0 = \frac{-2R_1R_2}{(2R_1 + R_2)R_3} \qquad (1.91d)$$

For Fig. 1.32b we note that now

$$T_1 > T_2$$

Thus the two circuits in Fig. 1.32 together permit us to implement transfer functions of the form (1.87) for either $T_1 < T_2$ (Fig. 1.32a) or $T_1 > T_2$ (Fig. 1.32b).

We should note that although we have implemented (1.87) with one amplifier, we have been limited to transfer functions which have a polarity inversion ($H_0 < 0$). Note that the circuits of Fig. 1.32 present a constant input impedance over the useful frequency range; this might not be true of alternative approaches.

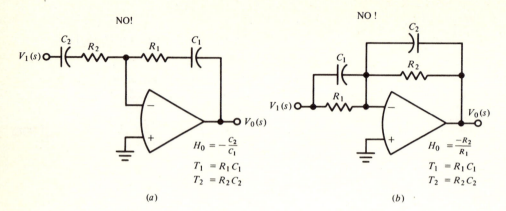

FIGURE 1.33
Alternative ways of implementing a pole-zero pair. (a) Avoid this (no dc feedback); (b) this works, but has capacitive input impedance.

Figure 1.33 shows two alternative, but generally less desirable, circuits for implementing (1.87). Figure 1.33a looks acceptable at first glance, but the *circuit of Fig. 1.33a* should normally *not be used*. The main reason for this recommendation is not apparent from the presentation of this chapter. Briefly, the dc offsets in the circuit, including those in the amplifier, will be integrated, i.e., will cause a charge to build up on capacitor C, eventually causing an amplifier overload. The circuit of Fig. 1.33a might be safe to use, however, as part of a larger system, provided there is some external closed-loop path around the circuit to provide feedback cancellation of long-term (dc) drifts. The circuit of Fig. 1.33b also implements (1.87). It will perform properly; however, we note that the *input impedance* is highly capacitive, that is,

$$Z_{in} = R_1 \, \| \, C_1$$

Thus this last circuit (Fig. 1.33b) may provide a poor load for the signal source driving it. In general, the circuits of Fig. 1.32 are preferred for implementing (1.87).

To illustrate the analog-computer or state-variable approach to implementing (1.87), let us write

$$\frac{V_0(s)}{V_1(s)} = \frac{H_0(1 + sT_1)}{1 + sT_2}$$

$$= \frac{T_1 H_0}{T_2} \frac{s + 1/T_1}{s + 1/T_2}$$

$$= \frac{T_1}{T_2} H_0 + \frac{(T_1 - T_2)H_0}{T_2} \frac{-1/T_2}{s + 1/T_2} \qquad (1.92)$$

or

$$V_0(s) = \frac{T_1 H_0}{T_2} V_1(s) + \frac{(T_1 - T_2)H_0}{T_2} V_2(s) \qquad (1.93a)$$

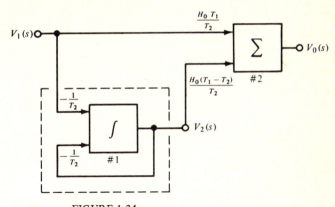

FIGURE 1.34
Two-amplifier circuit for implementing a pole-zero pair.

where

$$V_2(s) = \frac{-1/T_2}{s + 1/T_2} V_1(s) \qquad (1.93b)$$

We recognize (1.93b) as being of the form of (1.80), i.e., we can implement it by the circuit of Fig. 1.31, noting that

$$\frac{dv_2(t)}{dt} = \frac{-1}{T_2} [v_2(t) + v_1(t)] \qquad (1.94)$$

Thus (1.93) leads us to the block diagram of Fig. 1.34, which implies that *two amplifiers* are required (shown in dotted area); one implements (1.93b), and the other sums $V_2(s)$ and $V_1(s)$. Note that the polarity of the gains for the summing amplifier 2 depends upon the relative values of T_1 and T_2, as well as the polarity of H_0. Note that H_0 may be either positive or negative, whereas the circuits of Fig. 1.32 required $H_0 < 0$. The decomposition of (1.87) into (1.93a and b) may seem somewhat arbitrary, since the relative gains might have been split up differently. We chose the form of (1.93) so that the voltage gain from $V_1(s)$ to $V_2(s)$ is always less than or equal to 1 at all frequencies, thereby avoiding an overload at the output of amplifier 1. Students with prior analog-computer experience will recognize that this last approach we have taken has *amplitude-scaling problems* associated with the actual design implementation. We will not pursue the amplitude scaling any further, but the reader is cautioned to remember this consideration in a general problem.

EXAMPLE 1.13 Suppose we want the transfer function whose frequency response is sketched in Fig. 1.35a. From the corner or break frequencies we assume a transfer function of the form of (1.87), with

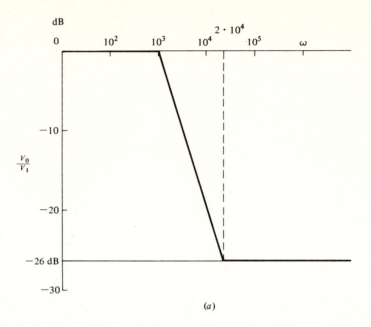

(a)

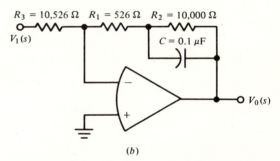

(b)

FIGURE 1.35
Circuit for Example 1.13. (a) Desired transfer function (gain magnitude only); (b) a one-amplifier implementation using (1.90).

$$T_1 = \frac{1}{2 \times 10^4} = 5 \times 10^{-5}$$

$$T_2 = \frac{1}{10^3} = 10^{-3}$$

and
$$H_0 = \pm 1$$

Let us assume that a polarity inversion is acceptable. Then, using the circuit of Fig. 1.32a $(T_1 < T_2)$, we obtain the design of Fig. 1.35b, where we have arbitrarily selected $R_2 = 10 \text{ k}\Omega$. Then from (1.90)

$$C = \frac{T_2}{R_2} = \frac{10^{-3}}{10^4}$$

$$= 0.1 \times 10^{-6} = 0.1 \ \mu F$$

Since

$$\frac{R_1}{R_1 + R_2} = \frac{T_1}{T_2} = 0.05$$

then

$$R_1 = 526 \ \Omega$$

and

$$R_3 = (R_1 + R_2) \times 1 = 10.526 \ k\Omega \qquad ////$$

1.10 MISCELLANEOUS LINEAR APPLICATIONS

This section is intended to present a variety of linear applications, discussed in terms of the ideal operational amplifier model, to broaden the reader's viewpoint about the potential range of operational amplifier applications. Since we are introducing these topics at this point, we cannot fully discuss the practical limitations imposed by the nonideal behavior of actual amplifiers. Later on, after we have described methods for treating nonideal behavior (Chap. 2), the reader will hopefully be able to more fully appreciate some of the practical limitations associated with the performance of the circuits we shall describe here.

Differentiation, High-Pass Filtering, DC Blocking

It is often desirable to implement the differentiator transfer function

$$\frac{V_0(s)}{V_1(s)} = H(s) = Ks \qquad (1.95)$$

at least approximately. Please note that this transfer function is not physically realizable (there is one more zero than pole in the transfer function); however, we may be satisfied with an approximation, such as the one shown in Fig. 1.36a; that is, we might try to implement either

$$H(s) = \frac{Ks\omega_0}{s + \omega_0} \qquad (1.96a)$$

which is a *high-pass filter* function, or, more realistically

$$H(s) = \frac{Ks\omega_0\omega_1}{(s + \omega_0)(s + \omega_1)} \qquad (1.96b)$$

Indeed, we never need the function of (1.95) in practice; rather, we want to perform the differentiation operation over some finite frequency range. The functions of (1.96) and Fig. 1.36 will block direct current, differentiate signals with spectral content well below $\omega = \omega_0$, and pass frequencies above ω_0 [and below ω_1 in the case of (1.96b)]. Differentiation of signals is a notoriously noisy operation, and a judicious

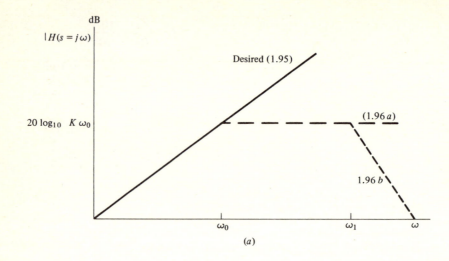

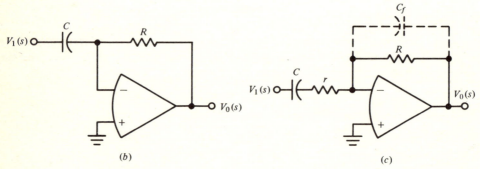

FIGURE 1.36
Differentiator and high-pass filter operators. (a) Magnitude plots of an ideal differentiator (1.95) and two high-pass filter approximations (1.96); (b) an ideal differentiator circuit; (c) a more practical circuit.

choice of ω_0 is important. Note that, in actuality, the transfer function of (1.96a) is still not physically realizable, since we do not have amplifiers of infinite bandwidth and there are always some stray capacitances which cause the gain of an amplifier circuit to eventually drop off at high frequencies. The transfer function of (1.96b) is more realistic. Here the second corner frequency $\omega = \omega_1$ may either result directly from the internal bandwidth limitations of the operational amplifier or be purposely introduced by the circuit designer to further attenuate unwanted high-frequency noise.

Figure 1.36b shows the basic differentiator circuit that *ideally* ought to implement (1.95). If we had an ideal operational amplifier, then, for Fig. 1.36b,

$$H(s) = -RCs \qquad (1.97)$$

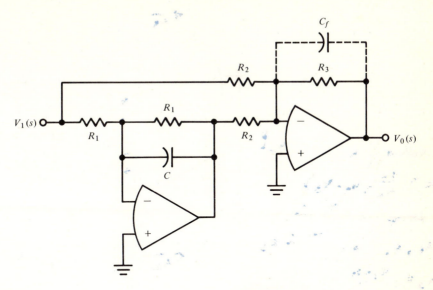

FIGURE 1.37
A two-amplifier high-pass filter circuit.

that is,

$$K = -RC$$

(see Prob. 1.37). There are, however, serious difficulties with the circuit of Fig. 1.36b:

1 The input impedance of the circuit is a pure capacitance, possibly of fairly high value; this is a poor load for most signal sources.
2 Although we have not shown it yet, most typical operational amplifiers, because of their high-frequency behavior, will cause the actual circuit to be unstable; viz., it will probably oscillate.
3 There is no control of the high-frequency response.

The *high-pass* circuit of Fig. 1.36c remedies all the above problems, if properly designed; the use of C_f is optional. With an ideal operational amplifier, but *without* C_f, we implement (1.96a), that is,

$$H(s) = \frac{-RCs}{1 + rCs} \qquad (1.98a)$$

with

$$K = -RC \qquad (1.98b)$$

$$\omega_0 = \frac{1}{rC} \qquad (1.98c)$$

and

$$|H(s = j\infty)| = K\omega_0 = \frac{R}{r} \qquad (1.98d)$$

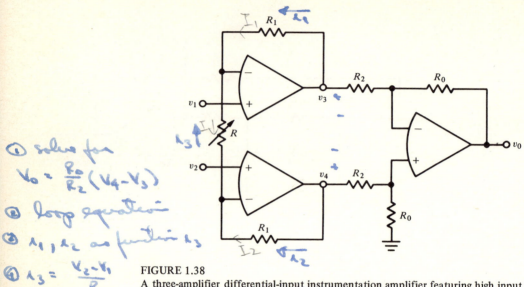

(handwritten notes in left margin:)

① solve for

$v_0 = \dfrac{R_0}{R_2}(v_4 - v_3)$

② loop equation

③ i_1, i_2 as function i_3

④ $i_3 = \dfrac{v_2 - v_1}{R}$

FIGURE 1.38
A three-amplifier differential-input instrumentation amplifier featuring high input impedance and easily adjustable gain.

The input impedance now has a resistive component, but is still highly capacitative. With C_f added to Fig. 1.36c, we obtain

$$H(s) = \frac{-RCs}{(1 + rCs)(1 + RC_f s)} \qquad (1.99a)$$

that is, we still have

$$K = -RC \qquad (1.99b)$$

and

$$\omega_0 = \frac{1}{rC}$$

but now

$$\omega_1 = \frac{1}{RC_f} \qquad (1.99c)$$

and the high-frequency gain falls off 20 dB/decade above ω_1 rad/s.

A somewhat more expensive approach is taken in Fig. 1.37. This is a two-amplifier circuit which features a *constant input impedance*, viz.,

$$Z_{\text{in}} = R_1 \| R_2 \qquad (1.100)$$

Note that Fig. 1.37 is a *high-pass filter* circuit; without C_f, we have

$$\frac{V_0(s)}{V_1(s)} = \frac{-R_1 Cs}{1 + R_1 Cs} \frac{R_3}{R_2} \qquad (1.101a)$$

that is,

$$K = -R_1 C \frac{R_3}{R_2} \qquad (1.101b)$$

$$\omega_0 = \frac{1}{R_1 C} \qquad (1.101c)$$

With C_f, we have

$$\frac{V_0(s)}{V_1(s)} = \frac{-R_1 Cs}{(1 + R_1 Cs)(1 + R_3 C_f s)} \frac{R_3}{R_2} \qquad (1.102a)$$

that is,

$$K = -R_1 C \frac{R_3}{R_2} \qquad (1.102b)$$

$$\omega_0 = \frac{1}{R_1 C} \qquad (1.102c)$$

$$\omega_1 = \frac{1}{R_3 C_f} \qquad (1.102d)$$

Note that Fig. 1.37 is really a special case of Fig. 1.34; that is, the numerator zero is now at the origin. Careful balancing of the resistors in the circuit is required to assure precise location of the zero, or there will be some dc feedthrough. This circuit permits implementing good dc blocking and low-frequency differentiation.

Instrumentation Amplifiers

The differential amplifier circuit of Fig. 1.25c has at least two fairly serious disadvantages:

1. The input impedance is $2R_1$; this may in some cases cause serious loading of high-impedance low-level signal sources (e.g., in biological instrumentation).
2. It is not easy to adjust the overall gain.

The circuit of Fig. 1.38 eliminates these two limitations by using three operational amplifiers. Typically, the complete circuit is packaged in one module as a general-purpose instrumentation amplifier (see Prob. 1.38). The *input impedance* of the circuit is *very high,* particularly if operational amplifiers with FET input stages are used. The overall differential gain expression is

$$v_0 = -\left(1 + \frac{2R_1}{R}\right) \frac{R_0}{R_2} (v_1 - v_2) \qquad (1.103)$$

assuming well-balanced resistors throughout. A wide range of gains may be implemented merely by adjusting R.

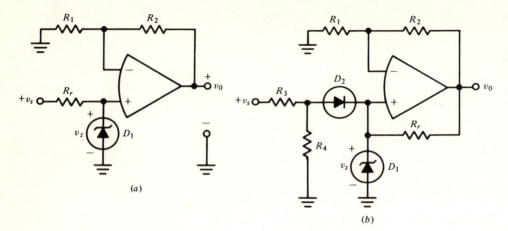

FIGURE 1.39
Voltage regulator circuits. (*a*) Simple noninverting regulator; (*b*) circuit with improved immunity to v_s.

Reference Voltage Sources and Voltage Regulators

Figure 1.39 presents circuits for providing a regulated voltage source, using an operational amplifier. In the circuit of Fig. 1.39*a*, the amplifier is used in a simple noninverting circuit, which derives a reference voltage from the Zener diode D_1.

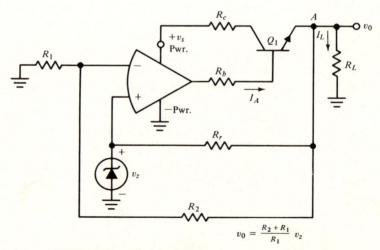

$$v_0 = \frac{R_2 + R_1}{R_1} v_z$$

FIGURE 1.40
Current-boosted regulator requiring only one power-supply potential.

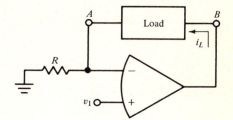

FIGURE 1.41
Simple voltage-controlled current source
(floating load).

$$v_0 = \frac{R_2 + R_1}{R_1} v_z \qquad (1.104)$$

In both circuits, v_s is a source of positive dc voltage; however, in Fig. 1.39b, improved immunity to variations in v_s is obtained by only using v_s to "start up" the circuit via R_3, R_4, and D_2; once the circuit is operating, the Zener diode D_1 obtains its "keep-alive" current from R_r, which is itself supplied by the well-regulated output voltage. Both of these circuits suffer at least two limitations:

1 The maximum output current obtainable is limited by the current-driving capability of the operational amplifier. The disadvantage can be alleviated by adding a booster output stage to the amplifier. Many manufacturers supply such a stage, or in many cases the output current may be boosted using a transistor emitter follower (see also Fig. 1.40 and Sec. 2.8).
2 The regulators of Fig. 1.39 require that the ± supply voltages be provided to the operational amplifier separately. This is often an undesirable nuisance.

The circuit of Fig. 1.40 eliminates both of the above disadvantages, and may be used to directly regulate the voltage supplied from power source v_s without requiring any additional power voltages. Here is a special occasion where the operational amplifier is operated from one supply voltage (that is, v_s), since both input terminals are at a positive potential v_z. Transistor Q_1 is a series regulator which is actually an output booster for the operational amplifier. Resistor R_c should be chosen to protect Q_1 from excessive current overload. Base drive resistor R_b is used to protect the operational amplifier from excessive output current demand. Of course, Q_1 must be chosen with sufficient current gain (usually called β) to provide current I_L from amplifier current I_A.

Voltage-Current Converters

It is sometimes useful to be able to generate a current proportional to a voltage. Typical applications include driving a deflection coil for a magnetically deflected cathode-ray tube. For floating loads (i.e., both terminals may be connected into the driving circuit), the circuit of Fig. 1.41 is useful.

It is easy to show that

$$i_L = \frac{v_1}{R} \qquad (1.105)$$

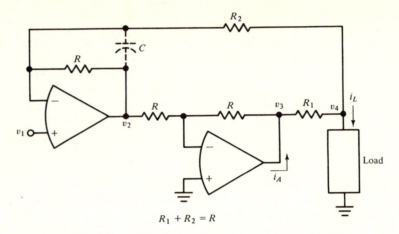

$$R_1 + R_2 = R$$

FIGURE 1.42
Voltage-controlled current source with high-input impedance and grounded load.

Note also that the circuit presents a high impedance to the driving signal source. Of course, the operational amplifier must be capable of providing the requisite driving current and the total load voltage at point B (point A is at a virtual ground potential). In some situations, it is necessary to drive a load which has one side already grounded externally. The circuit of Fig. 1.42, which requires two amplifiers, has a high input impedance and will drive a grounded load. Note also that the amplifier output swing must now be sufficient to provide the voltage drop associated with R_1 as well as the load. For Fig. 1.42, the current delivered to the load is (see Prob. 1.32)

$$i_L = \frac{-2v_1}{R_1} \qquad (1.106a)$$

where it is required that

$$R_2 = R - R_1 \qquad (1.106b)$$

This last circuit has the potential for presenting a negative output resistance (looking back into node v_4), and may cause oscillation if the load is highly inductive. A small stabilizing capacitor C may be required in some cases.

Other Applications

There are many other applications of operational amplifiers to linear amplifying and signal-processing circuits. In Chap. 4 we will discuss the general topic of active RC filter design. Chapter 6 of Ref. 1 also contains much useful information. Hopefully some of the problems at the end of this chapter will suggest other ways in which operational amplifiers may be used to precisely amplify, filter, or combine signals.

REFERENCES AND BIBLIOGRAPHY

1 TOBEY, G. E., J. G. GRAEME, and L. P. HUELSMAN: "Operational Amplifiers," McGraw-Hill, New York, 1971.

2 GIACOLETTO, L. J.: "Differential Amplifiers," Wiley, New York, 1970.

3 KORN, G. A., and T. M. KORN: "Electronic Analog and Hybrid Computers," 2d ed., McGraw-Hill, New York, 1972.

4 SMITH, JOHN I.: "Modern Operational Circuit Design," Wiley, New York, 1971.

5 *Electronics*: A biweekly journal for electronics engineers; in particular, see the Circuit Design features. McGraw-Hill, Inc.

6 *Electronic Design*: A biweekly journal for electronics engineers; in particular, see the articles in the Technology section. Hayden Publ. Co.

7 HUELSMAN, L. P.: "Circuit Theory with Digital Computations," Prentice-Hall, Englewood Cliffs, N.J., 1972.

8 HAUSNER, A.: "Analog and Analog/Hybrid Computer Programming," Prentice-Hall, Englewood Cliffs, N.J., 1971.

9 KOSTANTY, RAYMOND G.: Doubling Op Amp Summing Power, *Electronics,* Feb. 14, 1972, pp. 73–75.

10 MOSCHYTZ, G. S.: Inductorless Filters: A Survey, *IEEE Spectrum,* September 1970, pp. 63–75.

11 SHAH, M. J.: Stable Voltage Reference Uses Single Power Supply, *Electronics,* Mar. 13, 1972, p. 74.

12 CLAYTON, G. B.: "Operational Amplifiers," Butterworth, London, 1971.

PROBLEMS

For all problems in this chapter, assume an ideal operational amplifier model (see Sec. 1.2).

1.1 *(Sec. 1.3)* For both parts of Fig. P1.1, find v_0 in terms of v_1 and v_2.

1.2 *(Sec. 1.3)* Using one operational amplifier and resistors of 100 kΩ or less, but otherwise as large as possible, design an operational amplifier circuit to implement

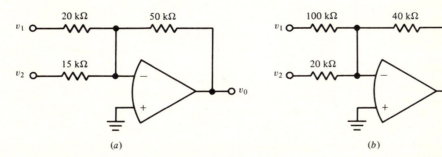

(a)   *(b)*

FIGURE P1.1

(a) $v_0 = -2v_1 - 4v_2$

(b) $v_0 = -3v_1 - 2v_2 - v_3$

(c) $v_0 = -3v_1 - 0.5v_2$

1.3 *(Sec. 1.3)* For the circuit of Fig. P1.3, make a table indicating which lettered terminals to connect to an input signal source in order to get all possible (different) amplification factors. Specify gains.

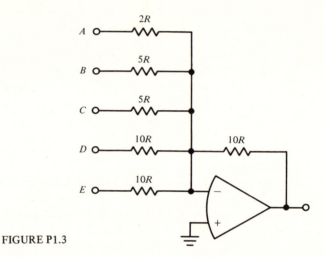

FIGURE P1.3

1.4 *(Sec. 1.3)* Given the circuit of Fig. P1.4, find *all* possible voltage gains that can be implemented by connecting the amplifier output to one or more lettered nodes. Assume we want to amplify only one signal (or possibly attenuate it). Make a table indicating which node(s) are to be connected to the output and which to the input.

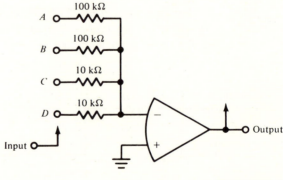

FIGURE P1.4

1.5 *(Sec. 1.4)* Refer to Figs. P1.5 and 1.10b. Draw a block-diagram symbol describing the nominal performance of the circuit; also write an expression similar to (1.18).

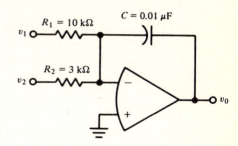

FIGURE P1.5

1.6 *(Sec. 1.4)* Using a single operational amplifier, one capacitor, and resistors of 100 kΩ or less, but otherwise as large as possible, design a circuit to implement:

(a) $v_0 = -100 \int_0^t v_1(t)\, dt$

(b) $v_0 = -10 \int_0^t v_1(t)\, dt - 2 \int_0^t v_2(t)\, dt$

Assume $v_0 = 0$ at $t = 0$

1.7 *(Sec. 1.4)* Given the situation of Fig. P1.7:

(a) Sketch v_0 and i_A versus time; assume no other output load on the amplifier.

(b) Repeat the above if an output load of 10 kΩ is connected from the amplifier to ground.

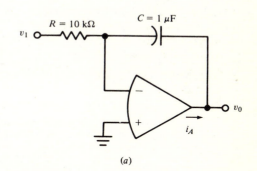

(a)

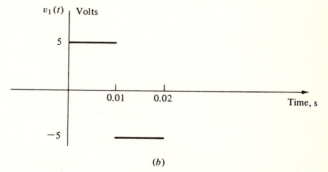

(b)

FIGURE P1.7

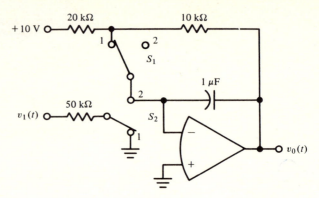

FIGURE P1.8

1.8 *(Sec. 1.4)* Refer to Fig. P1.8; at $t = 0$ the switches S_1 and S_2 are thrown from state 1 to state 2. Find $v_0(t)$ for $t \geqslant 0$, in terms of $v_1(t)$.

1.9 *(Sec. 1.4)* For the circuit of Fig. 1.13e, make separate sketches for each of the three modes; explain why the initial-condition setting will be faster than for the circuit of Fig. 1.13a.

1.10 *(Secs. 1.4 and 1.9)* For each of the circuits of Fig. P1.10:

    (a) Find the differential equation for v_0 in terms of v_1.

    (b) Find the Laplace transform transfer function

$$H(s) = \frac{V_0(s)}{V_1(s)}$$

    HINT: Draw block diagrams of the circuits.

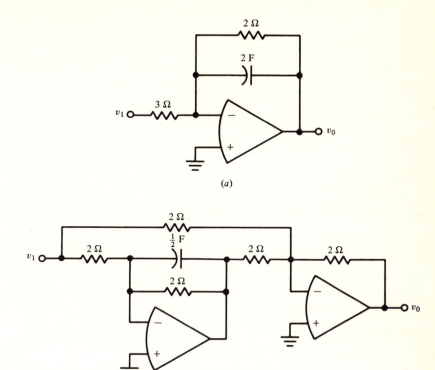

(a)

(b)

FIGURE P1.10

1.11 *(Sec. 1.4)* Given the circuit of Fig. P1.11 with

$$i_A < 20 \text{ mA}$$
$$v_0 < 10 \text{ V}$$
$$v_S = A \sin 2\pi f t$$

find the maximum value of A as a function of f.

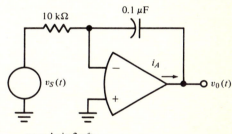

FIGURE P1.11 $v_S = A \sin 2\pi f t$

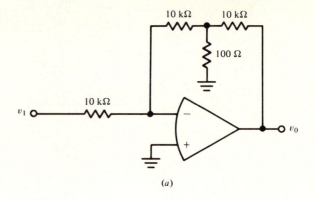

(a)

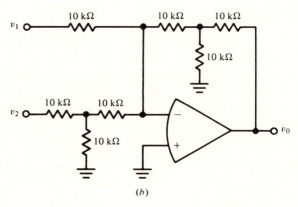

(b)

FIGURE P1.12

1.12 *(Sec. 1.5)* For both parts of Fig. P1.12, find v_0 in terms of the input variables.

1.13 *(Sec. 1.5)* In the circuit of Fig. P1.13, find the value of R required to obtain

$$v_0 = -50v_1$$

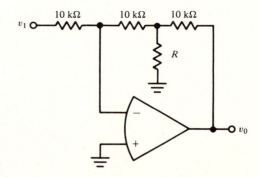

FIGURE P1.13

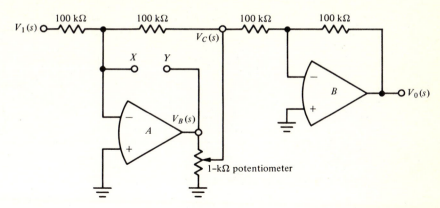

1.14 *(Secs. 1.5 and 1.9)* Given the circuit of Fig. P1.14:

(a) Find the Laplace transform transfer function

$$H(s) = \frac{V_0(s)}{V_1(s)}$$

in terms of the *loaded* potentiometer ratio α, i.e.,

$$\alpha = \frac{V_C(s)}{V_B(s)} \qquad 0 < \alpha \leqslant 1$$

(b) Repeat part a if a 0.001-μF capacitor is connected between points X and Y.

(c) Why is the output amplifier advisable? With the capacitor included, what is the convenient feature of this circuit? Comment on allowable signal level, particularly when α is small.

1.15 *(Sec. 1.6)* Using one operational amplifier and resistors of 100 kΩ or less, but otherwise as large as possible, design a circuit to implement each of the operations indicated:

(a) $v_0 = 3v_1$

(b) $v_0 = 0.75v_1$

1.16 *(Sec. 1.6)* For the circuit of Fig. P1.16 (see also Fig. 1.22):

(a) Find the Laplace transform transfer function

$$H(s) = \frac{V_0(s)}{V_1(s)}$$

in terms of general R's and C.

(b) Find the ncesssary condition on the resistances to obtain a *noninverting integrator*, i.e., to get

$$H(s) = \frac{+K}{s}$$

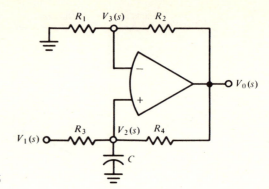

FIGURE P1.16

(c) What is the value of K?

(d) Comment on the importance of accurate resistor values for the accuracy with which the integrator pole is located at $s = 0$.

1.17 *(Secs. 1.6 and 1.8)* For both parts of Fig. P1.17, find v_0 in terms of v_1 and v_2.

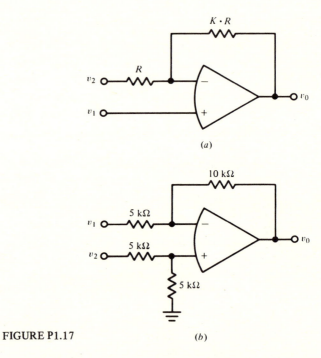

FIGURE P1.17

1.18 *(Sec. 1.7)* Design a differential amplifier circuit similar to Fig. 1.25b which will have a differential voltage gain of 20 and an input impedance (between input terminals) of 15 kΩ.

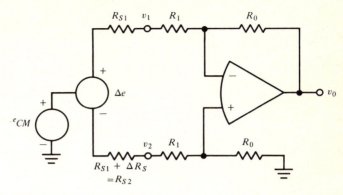

FIGURE P1.19

1.19 *(Sec. 1.7)* The differential amplifier in Fig. P1.19 is assumed to have an ideal operational amplifier and ideal, perfectly accurate resistors. However, the differential signal source is associated with unbalanced source resistances; that is,

$$R_{S2} = R_{S1} + \Delta R_S$$

(a) Find v_0 in terms of e_{cm}, Δe, R_0, R_1, R_{s1} and ΔR_S.

(b) Find the *differential gain*

$$G_D = \frac{v_0}{\Delta e}$$

assuming $\qquad\qquad \Delta R_S \ll R_{S1} \qquad R_{S1} \ll R_1$

(c) Find the *common-mode gain* (due to source resistance unbalance)

$$G_{cm} = \frac{v_0}{e_{cm}}$$

assuming $\qquad\qquad \Delta R_S \ll R_{S1} \qquad R_{S1} \ll R_1$

(d) Let us define a *common-mode rejection ratio* as

$$\text{CMRR} = \frac{G_D}{G_{cm}}$$

Find the CMRR, again assuming

$$\Delta R_S \ll R_{S1} \qquad R_{S1} \ll R_1$$

(e) Consider the circuit of Fig. 1.38 with respect to its immunity to the errors caused by R_{S1} and ΔR_S.

NOTE: In Sec. 2.4, we will discuss further errors due to internal unbalance in the operational amplifier itself. Overall differential amplifier performance must be analyzed considering all anticipated error sources.

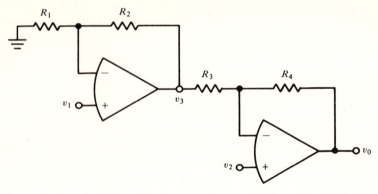

FIGURE P1.20

1.20 *(Secs. 1.7 and 1.8)* Given the circuit of Fig. P1.20, find the conditions on $R_1, R_2,$ $R_3,$ and R_4 so that

$$v_0 = K(v_2 - v_1)$$

and find K. Note that the circuit is a differential amplifier under these conditions. Can you think of an advantage of this circuit over that of Fig. 1.25b?

1.21 *(Sec. 1.8)* Using one operational amplifier and resistors of 100 kΩ or less, but otherwise as large as possible, design circuits to implement each of the operations indicated.

(a) $v_0 = 3v_1 - 3v_2$
(b) $v_0 = 2v_1 - v_2$
(c) $v_0 = 3v_1 + 4v_2$

1.22 *(Sec. 1.8)* Redesign the circuit of Example 1.12 to obtain

$$v_0 = -(3v_{I1} + 2v_{I2}) + 10v_{N1}$$

1.23 *(Sec. 1.8)* Given the circuit of Fig. P1.23, find v_0 in terms of $v_1, v_2,$ and v_3. Note that this implements combined summing and inverting as per (1.69), but $Z_A(s) \neq Z_B(s)$.

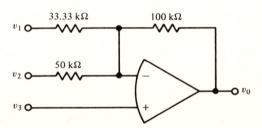

FIGURE P1.23

1.24 *(Sec. 1.9)* Design a circuit using a single operational amplifier to implement

$$V_0(s) = \frac{-[V_1(s) + 4V_2(s)]}{1 + s/10^5}$$

The circuit input impedances should be greater than 10 kΩ, and resistive; all resistors should be less than 100 kΩ, but otherwise as large as possible.

1.25 *(Sec. 1.9)* Suppose a signal from a single-ended transducer has the following properties:

Peak value: less than 0.1 V (open circuit)
Source impedance: less than 5 Ω, but otherwise not well known

Design a low-pass operational amplifier circuit that will amplify the signal 50 times (a polarity inversion is allowable); the circuit should pass all frequencies from direct current on up to around 1 kHz, and then exhibit a 6 dB/octave rolloff indefinitely above a -3-dB breakpoint of 1 kHz. All resistors should be no more than 100 kΩ; gain accuracy should be at least 1 percent. Use one operational amplifier.

1.26 *(Sec. 1.9)* For the circuit of Fig. P1.26:
(*a*) Find the Laplace transform transfer function

$$H(s) = \frac{V_0(s)}{V_1(s)}$$

for general R_0, R_1, R_2, and C.
(*b*) Make a Bode plot (magnitude of H in dB and phase of H in degrees, both versus log ω) of $H(s = j\omega)$ with

$$R_0 = 100 \text{ k}\Omega$$
$$R_1 = 10 \text{ k}\Omega$$
$$R_2 = 1 \text{ k}\Omega$$
$$C = 0.01 \text{ }\mu F$$

Show all critical frequencies on your plots.

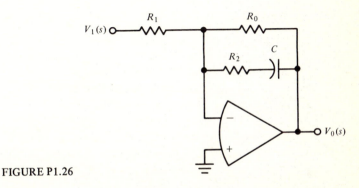

FIGURE P1.26

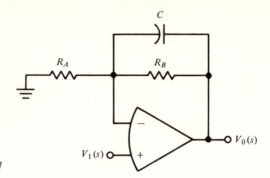

FIGURE P1.27

1.27 *(Sec. 1.9)* Given the circuit of Fig. P1.27, find:

(a) The Laplace transform transfer function

$$H(s) = \frac{V_0(s)}{V_1(s)}$$

(b) The magnitude of the circuit voltage gain at direct current; can you do this without doing all the work of part a?

(c) The magnitude of the circuit voltage gain at high frequencies, i.e., at angular frequencies satisfying

$$\omega \gg \frac{1}{R_B C} \quad \text{rad/s}$$

Again, can you do this without doing all the work of part a?

1.28 *(Sec. 1.9)* Prove that Eq. (1.91) is associated with Fig. 1.32b.

1.29 *(Sec. 1.9)*

(a) Design a frequency-compensated amplifier to have the following voltage gain characteristics:

Frequency, Hz	Voltage gain, dB
above 500	40
500	37
50	23
below 50	20

Assume that the signal source is single-ended with respect to ground, and that the signal source should see a load impedance of a constant 30 kΩ. A polarity inversion is permissible; use one operational amplifier.

(b) Add a 6 dB/octave high-frequency attentuation with a 2200-Hz break frequency.

1.30 *(Sec. 1.9)* Given the block diagram of Fig. P1.30, show that the resulting transfer function fits Eq. (1.87), and find H_0, T_1, and T_2 in terms of G_1, G_2, and K, where G_1 and G_2 may be positive or negative constants, but K is positive (that is, $-K$ is negative).

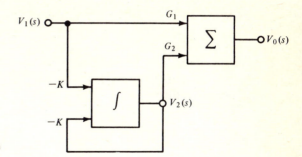

FIGURE P1.30

1.31 *(Sec. 1.9)* Referring to Fig. P1.31:

(*a*) Derive the Laplace transform transfer function

$$H(s) = \frac{V_0(s)}{V_1(s)}$$

(*b*) Find the circuit input impedance.

(*c*) For $R_1 = 100$ kΩ, $R_0 = 100$ kΩ, $R_2 = 10$ kΩ, $C_2 = 1000$ pF, $C_1 = 0.1$ μF, make a Bode plot of $H(s)$, i.e., magnitude in dB and angle in degrees versus log ω, $s = j\omega$.

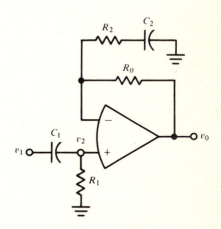

FIGURE P1.31

1.32 *(Sec. 1.10)* Refer to the circuit in Fig. 1.42.

(*a*) Derive Eqs. (1.106*a*) and (1.106*b*); i.e., show that

$$i_L = \frac{-2v_1}{R_1}$$

if

$$R = R_1 + R_2$$

(b) If $R = R_1 + R_2$ and the load is a pure resistance of value R_1, find expressions for v_2, v_3, i_A, and v_4.

(c) If $R = 10 \text{ k}\Omega, R_1 = R_2 = 5 \text{ k}\Omega$, find i_L in terms of v_1.

1.33 (Secs. 1.7 and 1.10) Derive Eq. (1.103), associated with Fig. 1.38.

1.34 (Sec. 1.10) Referring to Fig. P1.34, find the Laplace transform transfer functions (second-order filter):

$$H(s) \atop bp = \frac{V_2(s)}{V_1(s)}$$

and

$$H(s) \atop lp = \frac{V_3(s)}{V_1(s)}$$

HINT: Break the circuit up into basic operational blocks (analog-computer or state-variable approach for implementing transfer functions). (See also Chap. 4.)

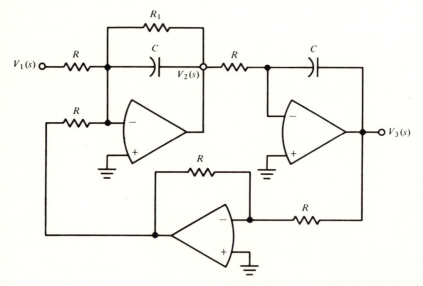

FIGURE P1.34

1.35 (Sec. 1.10) This problem deals with *bridge amplifiers*. Referring to Fig. P1.35, suppose we want to measure the *difference* between two resistances (e.g., two elements in a strain gage, or a thermistor bridge).

(a) In Fig. P1.35a, find v_0 in terms of δ, V, and the circuit resistances; assume $\delta \ll 1, R_0 \gg R$.

(b) In Fig. P1.35b, find v_0 in terms of δ, V, and the circuit resistances; do we have to assume δ is small? Note that the power supply has to be floating. Name an advantage of this circuit.

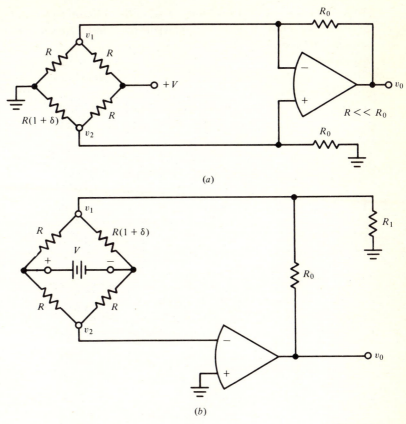

(a)

(b)

FIGURE P1.35

1.36 (Sec. 1.10) Show that the circuit of Fig. P1.36 "gyrates" a capacitance, i.e., show that $Z_{in}(s)$ looks like an inductance; find the value of the apparent inductance (Ref. 10).

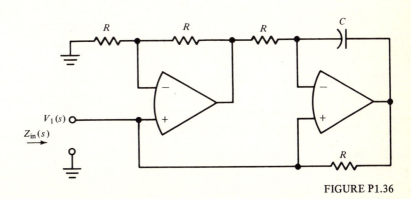

FIGURE P1.36

1.37 (Sec. 1.10) In Sec. 1.10 a number of differentiator and high-pass circuits are described. Suppose we want to implement the transfer function with the following sinusoidal frequency response function:

$$H(j\omega) = \frac{-0.01 j\omega}{1 + j\omega/10^4}$$

i.e., it differentiates well below 10^4 rad/s, with a differentiator gain of 0.01.

(a) Design a circuit based on Fig. 1.36c to implement this transfer function; make all resistors 100 kΩ or less.

(b) Add another pole to the transfer function to get a second high-frequency rolloff corner frequency at $\omega_1 = 10^4$ rad/s.

(c) Repeat parts a and b, but use the circuit of Fig. 1.37.

1.38 (Secs. 1.7 and 1.10) Refer to the instrumentation amplifier circuit of Fig. 1.38. Use this circuit to design a differential amplifier with a gain adjustable from 10 to 100 by the adjustment of a single resistor; give an equation for the value of the resistor required to get a gain K, $10 \leqslant K \leqslant 100$. List the required values of R for gains of 10, 20, 50, and 100.

1.39 (Sec. 1.10) The circuit of Fig. P1.39 was suggested by M. J. Shah (Ref. 11). It uses only one unregulated power supply to drive the operational amplifier. Derive an expression for v_0 in terms of v_Z, R_1, R_2, and R_3. Note that the Zener diode current is independent of the power-supply voltage.

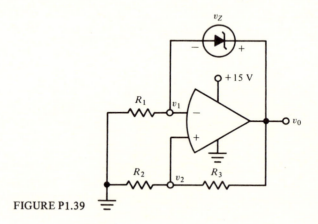

FIGURE P1.39

2

PERFORMANCE LIMITATIONS: THE NONIDEAL AMPLIFIER

In Chap. 1 we assumed an operational amplifier with the ideal characteristics of Sec. 1.2, viz., a VCVS with infinite voltage gain, zero output impedance, and infinite input impedance (Fig. 1.4b). Of course, such an amplifier cannot be perfectly realized. The implication thus far has been that an actual operational amplifier will not differ from the ideal model enough to significantly affect circuit performance. Table 2.1 lists typical specifications of currently available amplifiers.

This chapter discusses important ways in which modern solid-state operational amplifiers differ from the ideal model, and more importantly, how to estimate the effects of the various types of nonidealness on circuit performance. We will first take a look at a general linear model with no dc offsets, internal noise, or nonlinearities (Sec. 2.1). Using this model, we will study the effect of finite open-loop gain, finite bandwidth, finite input impedance, and nonzero output impedance on circuit behavior (Sec. 2.2). The effect of these factors on closed-loop stability is discussed in Sec. 2.3. Common-mode effects due to lack of symmetry in the amplifier's response to input at the inverting and noninverting terminals will be described, as they relate to differential amplifier performance (Sec. 2.4).

The effect of small, but ever-present, dc unbalances in the amplifier input stage (dc offsets) will be analyzed in Sec. 2.5. We use a similar model, involving random noise sources, in Sec. 2.6 to describe the effects of internal amplifier noise.

Large-signal output limitations, which are primarily of a nonlinear nature, are discussed in Sec. 2.7, including rated output voltage and current, slewing rate, full-power bandwidth, and settling time.

The use of output power boosting is discussed in Sec. 2.8. The chapter closes with two sections which describe peripheral practical considerations that are associated with the design of operational amplifier circuits: power supply and grounding systems (Sec. 2.9), and components (Sec. 2.10).

Table 2.1 TYPICAL OPERATIONAL AMPLIFIER SPECIFICATIONS

Specification	Discrete	Monolithic IC (all bipolar)
Open-loop gain, dc, no load, A_o	90–110 dB	100 dB
Rated output	±10 V; ±10 mA	±10 V, ±25 mA
Unity gain frequency f_u	1–100 MHz	1 MHz
Full-power response f_p	10 kHz–10 MHz	10–100 kHz
Slew rate S	1–1000 V/μs	1–10 V/μs
Input offset voltage e_{os}	0.1–1 mV bipolar 1–2 mV FET	1–5 mV
Input bias current i_1, i_2	50 nA bipolar 10 pA FET	10–500 nA
Input difference current i_{os}	2 nA bipolar 1–5 pA FET	10–30 nA
Differential input impedance Z_I	10^6 Ω bipolar 10^{11} Ω FET	10^5–10^7 Ω
Common-mode input impedance, Z_{cm1}, Z_{cm2}	10^9 Ω bipolar 10^{13} Ω FET	10^8–10^{10} Ω
Output impedance, open-loop, Z_o	1–2 kΩ	50–200 Ω
Common-mode rejection, CMRR	80–120 dB	80–100 dB

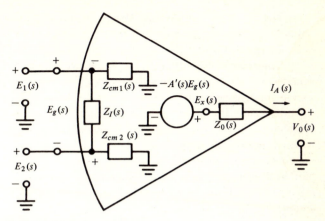

FIGURE 2.1
General linearized (small-signal) model of an operational amplifier.

2.1 THE GENERAL NONIDEAL LINEAR MODEL

Figure 2.1 shows the general *linearized (small-signal) model* of an operational amplifier (Fig. 2.1 is essentially Fig. 1.4, but here we imply that a complex frequency domain analysis will be employed). No internal dc offsets, noise sources, or lack of input symmetry are included; these factors will be considered later. Also we assume a perfectly linear model, in which all effects can be described in terms of the Laplace transform descriptions of the various model elements. So long as the amplifier *output voltage and current* and *output voltage rate* are relatively small compared to large-signal rated maximum values, our small-signal linear model will describe the amplifier's dynamics rather well. *Care must be taken,* however, *in applying* the results of an analysis based on the model of Fig. 2.1 *to large-signal situations.*

In Fig. 2.1, four properties of a nonideal amplifier are included:

1 The *open-loop* (voltage) *gain, $A'(s)$*
2 The (open-loop) *output impedance, $Z_0(s)$*
3 The *differential input impedance, $Z_I(s)$*
4 The two *common-mode input impedances, $Z_{cm1}(s)$ and $Z_{cm2}(s)$*

The common-mode input impedances are simply the impedances between each input terminal and ground. Usually these are *very* high, typically greater than 10^8 Ω. The differential impedance between terminals, $Z_I(s)$, is also high, but not always as high as the common-mode input impedances. Amplifiers with bipolar-transistor input stages will have a differential input impedance of perhaps 10^6 Ω; amplifiers with FET-input stages may have a very high differential input impedance, approaching the common-mode values (see Table 2.1). Usually these impedances are considered to be pure resistances, although at high frequencies one may need to consider small, but possibly important, common-mode and differential input capacitances.

The amplifier output impedance basically describes the small-signal incremental output resistance of the amplifier output stage. For most analyses, $Z_0(s)$ is treated as a pure resistance. It is perhaps surprising to note (Table 2.1) that typical values of amplifier output impedance are not really very low. We will see, however, that negative feedback, which is usually present around the amplifier, tends to suppress the effect of $Z_0(s)$ on overall performance.

Probably the most significant characteristic of the model of Fig. 2.1 is the open-loop gain function $A'(s)$. As the reader might expect, $A'(s)$ is not infinite, nor does it stay large for all frequencies. Rather, a typical variation of $A'(s)$ with frequency is sketched, somewhat idealized, in Fig. 2.2, where magnitude and phase of $A'(s = j\omega)$ are plotted separately versus log ω (Bode plot). Two important parameters of the plots are usually specified for a given amplifier:

1　The dc or low-frequency open-loop gain A_0, usually given in decibels.
2　The frequency at which the open-loop gain drops to 1 in magnitude. This is usually called the *unity gain bandwidth* f_u, typically specified in MHz. In Fig. 2.2, where ω is the independent variable,

$$\omega_u = 2\pi f_u \qquad (2.1)$$

ω_u is in rad/s and f_u in Hz.

For the present, we will treat A' as a complex-valued quantity, $A'(s)$ or $A'(s = j\omega)$, as needed.

As implied in Fig. 2.2, the magnitude of A' is usually constant (approximately equal to A_0) from direct current up to some fairly low frequency ω_1; then the magnitude of A' will drop at the rate of 20 dB/decade until beyond the unity gain frequency. Often a *simple one-pole model* is used for $A'(s)$, that is,

$$A'(s) \leqslant \frac{A_0}{1 + s/\omega_1} \qquad (2.2)$$

A_0, of course, is here expressed as a numeric, not in decibels. The -3-dB frequency ω_1 may thus be estimated from ω_u as follows:

$$|A'(j\omega_u)| = \frac{A_0}{|1 + j\omega_u/\omega_1|} \cong 1$$

assuming $A_0 \gg 1$,

$$\frac{\omega_u}{\omega_1} \cong A_0 \qquad A_0 \gg 1$$

or

$$\omega_1 \cong \frac{\omega_u}{A_0} = \frac{2\pi f_u}{A_0} \qquad \text{rad/s} \qquad (2.3a)$$

of course,

$$f_1 \cong \frac{f_u}{A_0} \qquad \text{Hz} \qquad (2.3b)$$

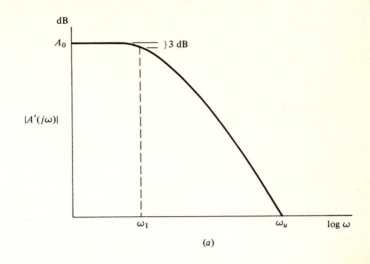

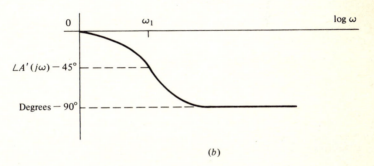

FIGURE 2.2
Typical variation of open-loop gain A' $(j\omega)$ with angular frequency ω (Bode plot).
(a) Gain in decibels; (b) phase angle in degrees.

EXAMPLE 2.1 Given an amplifier with $A_0 = 80$ dB and $f_u = 1$ MHz, find $A'(s)$, ω_1, and f_1.

First find A_0 as a numeric.

$$A_0 = \text{antilog}_{10} \frac{80}{20} = 10^4$$

Also,

$$\omega_u = 2\pi \times 10^6 \text{ rad/s}$$

Then, using (2.3),

$$\omega_1 = \frac{2\pi \times 10^6}{10^4} = 628 \text{ rad/s}$$

also

$$f_1 = 100 \text{ Hz}$$

and

$$A'(s) = \frac{10^4}{1 + s/628} \qquad ////$$

Note that in Fig. 2.1

$$V_0(s) = -A'(s)E_g(s) - I_A(s)Z_0(s) \qquad (2.4a)$$
$$= A'(s)[E_2(s) - E_1(s)] - I_A(s)Z_0(s) \qquad (2.4b)$$

Here we have included the polarity inversion from E_1 to V_0 separately from A'.

2.2 Closed-Loop Gain, Output Impedance, and Input Impedance

Figure 2.3 shows the general model of Fig. 2.1, as it applies to the inverting amplifier circuit. This circuit is basically that of Fig. 1.15, except we will assume that the external networks $Z_F(s)$, $Z_1(s)$, ..., $Z_n(s)$ are only *two-terminal networks*. The impedance $Z_g(s)$ is the total extraneous impedance between the inverting amplifier terminal and ground; typically,

$$Z_g(s) = Z_I(s)||Z_{cm1}(s) \qquad (2.5)$$

In some cases, it may also be necessary to include any shunt capacitance associated with this node. Often

$$|Z_I(s)| \ll |Z_{cm1}(s)|$$

at all frequencies of interest, and we can often say

$$Z_g(s) \cong Z_I(s) \qquad (2.6)$$

Note that the amplifier open-loop gain $A'(s)$ is the transfer function between $E_g(s)$ and $V_0(s)$ *under no-load conditions;* i.e., the amplifier output terminal has a Thévenin-equivalent voltage source

$$V_0(s) = E_x(s) = -A'(s)E_g(s) \qquad \text{no-load} \qquad (2.7)$$

associated with it. In general $V_0(s) \neq E_x(s)$, due to the voltage drop across $Z_0(s)$ created by the amplifier output current $I_A(s)$. (Note that we again use the symbol V for the external node voltages of the complete circuit.)

    CASE I: $Z_0(s) = 0$ This case serves as a starting point for more detailed analyses. It gives approximate results at low frequencies, when the amplifier output current is small. We can write

$$E_g(s) = K_1(s)V_1(s) + K_2(s)V_2(s) + \cdots + K_n(s)V_n(s) + \beta(s)V_0(s) \qquad (2.8)$$

Define

$$V_0(s) = -A(s)E_g(s)$$

In later developments, we will reserve the symbol $A(s)$ for the actual ratio:

$$A(s) = \frac{-V_0(s)}{E_g(s)} \qquad (2.9)$$

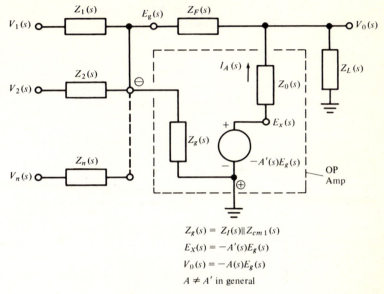

$$Z_g(s) = Z_I(s)\|Z_{cm\,1}(s)$$
$$E_X(s) = -A'(s)E_g(s)$$
$$V_0(s) = -A(s)E_g(s)$$
$$A \neq A' \text{ in general}$$

$\ominus$ and $\oplus$ are amplifier inverting and noninverting terminals

FIGURE 2.3
General linear model of an inverting amplifier circuit.

In the first case,

$$A(s) = A'(s) \qquad \text{if } Z_0(s) = 0 \qquad (2.10)$$

We also define

$$K_j(s) = \frac{1}{Z_j(s)} \left[\frac{1}{Z_F(s)} + \frac{1}{Z_1(s)} + \cdots + \frac{1}{Z_n(s)} + \frac{1}{Z_g(s)} \right]^{-1} \qquad j = 1, 2, \ldots, n$$

$$= \frac{Z_F(s)}{Z_j(s)} \beta(s) = \frac{Z_A(s)}{Z_j(s)} \qquad (2.11)$$

where

$$\beta(s) = \frac{1}{Z_F(s)} \left[\frac{1}{Z_F(s)} + \frac{1}{Z_1(s)} + \cdots + \frac{1}{Z_n(s)} + \frac{1}{Z_g(s)} \right]^{-1}$$

$$= \frac{Z_A(s)}{Z_F(s)} \qquad (2.12)$$

and

$$Z_A(s) = \left[\frac{1}{Z_F(s)} + \frac{1}{Z_1(s)} + \cdots + \frac{1}{Z_n(s)} + \frac{1}{Z_g(s)} \right]^{-1}$$

$$= \text{total parallel impedance of all branches terminating at } E_g(s) \qquad (2.13)$$

The ratio $\beta(s)$ will be important throughout this section; it is the actual *voltage feedback ratio* from the *amplifier output terminal* to the *inverting terminal*.

$$\beta(s) = \frac{E_g(s)}{V_o(s)} \qquad V_1(s) = V_2(s) = \cdots = V_n(s) = 0 \qquad (2.14)$$

Note also that from (2.12) and (2.13), $|\beta| < 1$; it may typically be a small fraction.

From (2.7), (2.8), (2.10), and (2.14) we obtain

$$V_o(s) = \frac{-A(s)}{1 + A(s)\beta(s)} \left[\sum_{j=1}^{n} K_j(s)V_j(s) \right]$$

or

$$V_o(s) = \frac{-A(s)\beta(s)}{1 + A(s)\beta(s)} \left[\frac{1}{\beta(s)} \sum_{j=1}^{n} K_j(s)V_j(s) \right] \qquad (2.15)$$

From (2.11),

$$\frac{K_j(s)}{\beta(s)} = \frac{Z_F(s)}{Z_j(s)} \qquad j = 1, 2, \ldots, n$$

Thus

$$V_o(s) = \frac{-A(s)\beta(s)}{1 + A(s)\beta(s)} \left[Z_F(s) \sum_{j=1}^{n} \frac{V_j(s)}{Z_j(s)} \right] \qquad (2.16)$$

or

$$V_o(s) = \frac{-1}{1 + 1/A(s)\beta(s)} \left[Z_F(s) \sum_{j=1}^{n} \frac{V_j(s)}{Z_j(s)} \right] \qquad (2.17)$$

These last two equations provide a lot of information. First, we note that the transfer function between any input and the output is given by

$$\frac{V_o(s)}{V_j(s)} = \frac{A(s)\beta(s)}{1 + A(s)\beta(s)} \left[\frac{-Z_F(s)}{Z_j(s)} \right] \qquad (2.18a)$$

$$= \frac{-Z_F(s)/Z_j(s)}{1 + 1/A(s)\beta(s)} \qquad (2.18b)$$

Thus each transfer function has unwanted poles at the roots of

$$1 + A(s)\beta(s)$$

Hopefully these roots will have large negative values, and will correspond to unimportant high-frequency poles. The possibility of instability is, however, suggested. By using the techniques of feedback system analysis (Bode plots, Nyquist plots, root locus, etc.), we can predict the stability of the closed-loop system. Normally Bode plots are the most useful.

We see from (2.18b) that

$$\frac{V_o(s)}{V_j(s)} \cong \frac{-Z_F(s)}{Z_j(s)} \qquad \text{as } |A(s)| \to \infty \qquad (2.19)$$

and for $|A(s)\beta(s)| \gg 1$, the circuit behaves in the manner predicted by the ideal-amplifier equations developed in Chap. 1!

Also (2.16) may be rewritten

$$V_0(s) = -\left[1 - \frac{1}{1 + A(s)\beta(s)}\right] Z_F(s) \sum_{j=1}^{n} \frac{V_j(s)}{Z_j(s)} \qquad (2.20)$$

From this, we can see that the fractional *error* in $V_0(s)$ [i.e., the amount by which $V_0(s)$ differs from the ideal of (2.19)] may be estimated by

$$\text{Fractional error} = \epsilon(s) = \frac{1}{1 + A(s)\beta(s)} \qquad (2.21)$$

That is, if $V_{0D}(s)$ is the desired output,

$$\epsilon(s) = \frac{V_{0D}(s) - V_0(s)}{V_{0D}(s)}$$

Typically $A(s)$ is a large positive quantity and $|A(s)\beta(s)| \gg 1$. Thus

$$\epsilon(s) \cong \frac{1}{A(s)\beta(s)} \qquad (2.22)$$

Throughout this section we will see that the quantity $A(s)\beta(s)$ is of primary importance in estimating the effects of finite $A(s)$.

CASE II: *Effect of Nonzero* $Z_0(s)$ If $Z_0(s)$ is not negligibly small, then $A(s) \neq A'(s)$. If we can find $A(s)$ in terms of $A'(s)$, $Z_0(s)$, and other quantities, the above results still apply. It is possible to show that

$$A(s) = A'(s)\left[a_I(s) - \frac{a_T(s)}{A'(s)}\right] \qquad (2.23a)$$

$$\cong A'(s)a_I(s) \qquad (2.23b)$$

where $a_I(s)$ is an attenuation ratio caused by the voltage drop through $Z_0(s)$.

$$a_I(s) = \frac{Z_F(s)Z_L(s)}{Z_0(s)Z_L(s) + Z_0(s)Z_F(s) + Z_F(s)Z_L(s)}$$

$$= \frac{Z_B(s)}{Z_0(s)} \qquad (2.24)$$

where

$$Z_B(s) = Z_F(s) \| Z_0(s) \| Z_L(s) \qquad (2.25)$$

The ratio $a_T(s)$ is a feedforward effect representing the effect of any voltage at $E_g(s)$ on $V_0(s)$ in the absence of the amplifier itself.

$$a_T(s) = \frac{Z_0(s)Z_L(s)}{Z_0(s)Z_F(s) + Z_0(s)Z_L(s) + Z_F(s)Z_L(s)}$$

$$= \frac{Z_B(s)}{Z_F(s)} \qquad (2.26)$$

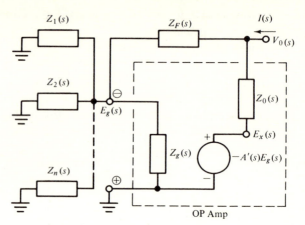

FIGURE 2.4
Circuit for finding the closed-loop output impedance.

Note that if $Z_0(s) = 0$, as in case I, then $a_T = 0$, $a_I = 1$, and

$$A(s) = A'(s) \qquad Z_0(s) = 0$$

The factor $a_T(s)$ measures the undesirable lack of isolation between operational amplifier input and output. Normally this factor may be neglected, except possibly at very high frequencies, and (2.23b) is used to estimate $A(s)$.

Equation (2.23b) may be used in (2.18) to (2.22) to estimate the effect of $Z_0(s)$ and $Z_L(s)$ on overall performance. Note that since $A(s)$ is not now equal to $A'(s)$, the closed-loop poles at the roots of $[1 + A(s)\beta(s)]$ depend upon $a_I(s)$, and *thus $Z_L(s)$ can affect stability.* In particular, if $Z_L(s)$ includes a large capacitive component at high frequencies, it is possible for the circuit to become unstable, even though it is stable without $Z_L(s)$.

Closed-Loop Output Impedance

In Fig. 2.4 we apply a test current $I(s)$ to the amplifier output terminal; the resulting value of $V_0(s)$ allows us to find the closed-loop output impedance (all inputs grounded). Using the symbols from the previous development, noting that

$$E_x(s) = -A'(s)E_g(s) \qquad (2.27a)$$

and

$$E_g(s) = \beta(s)V_0(s) \qquad (2.27b)$$

we write a nodal equation

$$I(s) = \frac{V_0(s) - E_x(s)}{Z_0(s)} + \frac{V_0(s) - E_g(s)}{Z_F(s)} \qquad (2.28)$$

then

$$E_x(s) = -A'(s)E_g(s) = -A'(s)\beta(s)V_0(s) \qquad (2.29)$$

and
$$I(s) = \frac{V_0(s)\,[1 + A'(s)\beta(s)]}{Z_0(s)} + \frac{V_0(s)\,[1 - \beta(s)]}{Z_F(s)} \qquad (2.30)$$

or
$$Y_{out}(s) = \frac{I(s)}{V_0(s)} = \frac{1 + A'(s)\beta(s)}{Z_0(s)} + \frac{1 - \beta(s)}{Z_F(s)} \qquad (2.31)$$

Thus we see that the output impedance is the parallel combination of two impedances, that is,

$$Z_{out}(s) = \frac{Z_0(s)}{1 + A'(s)\beta(s)} \;\|\; \frac{Z_F(s)}{1 - \beta(s)} \qquad (2.32)$$

Since the magnitude of β is always less than 1, and $Z_0(s)$ is usually relatively small, then

$$Z_{out}(s) \cong \frac{Z_0(s)}{1 + A'(s)\beta(s)} \qquad (2.33)$$

Over the useful frequency range of the complete operational amplifier circuit, we expect

$$|A'\beta| \gg 1$$

and we have

$$Z_{out}(s) \cong \frac{Z_0(s)}{A'(s)\beta(s)} \qquad (2.34)$$

Note that from our definition of $A'(s)$, it is a large positive quantity at direct current, and the closed-loop output impedance is usually a very small fraction of $Z_0(s)$. Note also that it is the value of $A'(s)$, *not* $A(s)$, that is of interest in (2.34).

EXAMPLE 2.2 Given an operational amplifier with the following static (dc) characteristics:

Open-loop, unloaded voltage gain $\quad A_0 = A_{dc} = 10^6$ (or 120 dB)
Output resistance $\qquad\qquad\qquad\quad R_0 = 10^3\ \Omega$
Differential input impedance $\qquad\quad R_I = 10^6\ \Omega$

Find the dc gain error and circuit output impedance.
Assume common-mode input impedances are very high (over $10^8\ \Omega$).
In the simple resistive inverter circuit of Fig. 2.5,

$$R_F = 10^5\ \Omega$$
$$R_1 = 10^4\ \Omega$$

Equation (1.5) predicts (for an ideal amplifier) that

$$\frac{V_0}{V_1} = -\frac{R_F}{R_1} = -10$$

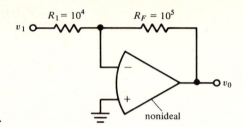

FIGURE 2.5
Simple inverter circuit of Example 2.2.

More precisely, using the approach of Sec. 2.2 ($Z_L = \infty$),

$$\beta = \frac{10^4 \| 10^5 \| 10^6}{10^5} = 0.0909$$

Note that here the effect of $Z_g \cong Z_i = 10^6 \ \Omega$ could be ignored without any serious loss of accuracy. Also,

$$a_I = \frac{10^3 \| 10^5}{10^3} = 0.9901$$

and

$$A = a_I A' = 0.9901 \times 10^6$$

Thus at direct current,

$$A\beta \cong 9 \times 10^4$$

or a more precise estimate of the gain is by (2.18):

$$\frac{V_0}{V_1} = -\frac{A\beta}{1 + A\beta} \times 10$$

$$= -0.99989 \times 10$$

$$= -9.9989$$

The fractional error from (2.22) is

$$\epsilon = \frac{1}{A\beta} = \frac{1}{9 \times 10^4}$$

which agrees. Note that our ideal amplifier model of Chap. 1 is not so bad!
The output impedance from (2.34) is

$$Z_{\text{out}}(s) = \frac{10^3}{A'\beta}$$

$$= 0.011 \ \Omega$$

i.e., very low.
Suppose we load the circuit of Fig. 2.5 with a 1000-Ω load. This changes a_I, that is,

$$a_I = \frac{10^3 \| 10^3 \| 10^5}{10^3} \cong 0.5$$

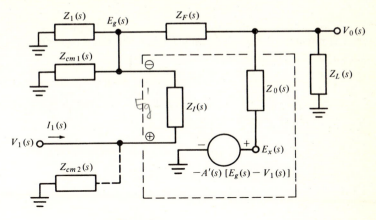

FIGURE 2.6
General linear model for a noninverting amplifier circuit.

Now

$$A = 0.5 \times 10^6$$
$$A\beta \cong 4.5 \times 10^4$$

and the fractional gain error is twice as large (but still very small).   ////

Before leaving the inverting amplifier circuit for a while, we should note that the *input impedance* at each input node V_j may still normally be assumed to equal the associated input summing impedance Z_j. Thus in the last example, the circuit input impedance is assumed to be 10^4 Ω. At very high frequencies, the input impedance will be Z_j in series with some other impedance, due to the fact that the summing junction is no longer a virtual ground.

The Noninverting Amplifier Circuit

In Fig. 2.6 we have the linear nonideal amplifier model embedded in a basic noninverting amplifier circuit. The analysis of this circuit is quite similar to the inverting amplifier, and we will present only the final results. We first define the following:

$$E_x(s) = -A'(s)[E_g(s) - V_1(s)] \qquad (2.35)$$
$$V_0(s) = -A(s)[E_g(s) - V_1(s)] \qquad (2.36)$$
$$Z_1'(s) = Z_1(s) \| Z_{cm1}(s) \qquad (2.37)$$
$$Z_A(s) = Z_1'(s) \| Z_F(s) \| Z_I(s) \qquad (2.38)$$
$$Z_B(s) = Z_0(s) \| Z_F(s) \| Z_L(s) \qquad (2.39)$$

Note that typically $Z_{cm1}(s) \gg Z_1(s)$; thus

$$Z_1'(s) \cong Z_1(s) \qquad (2.40)$$

except possibly *at high frequencies,* when the *shunt capacity component* of Z_{cm1} may be significant. In words,

Z_A is the total shunt impedance at node E_g

Z_B is the total shunt impedance at node V_0

An analysis of the circuit of Fig. 2.6 yields

$$\frac{V_0(s)}{V_1(s)} = \frac{[Z_1'(s) + Z_F(s)]}{Z_1'(s)} \frac{A(s)\beta(s)}{1 + A(s)\beta(s)} \qquad (2.41)$$

where again, $\beta(s)$ is a *voltage feedback ratio*

$$\beta(s) = \frac{Z_A(s)}{Z_F(s)} \qquad (2.42)$$

Also, if we neglect feedforward effects around the amplifier, i.e., assume that

$$Z_0(s) \ll |Z_F(s)A'(s)| \qquad (2.43)$$

then again

$$A(s) = a_I(s)A'(s) \qquad (2.44)$$

where

$$a_I(s) = \frac{Z_B(s)}{Z_0(s)} \qquad (2.45)$$

These last results are similar to those for the inverting amplifier. The *loop gain* $A(s)\beta(s)$ affects the closed-loop transfer function in the same way as in the inverting circuit, and the actual gain is equal to the desired gain multiplied by

$$\frac{A(s)\beta(s)}{1 + A(s)\beta(s)} = \frac{1}{1 + 1/A(s)\beta(s)} \qquad (2.46)$$

Thus the fractional error is again

$$\epsilon(s) \cong \frac{1}{A(s)\beta(s)} \qquad A(s)\beta(s) \gg 1 \qquad (2.47)$$

Again, an examination of stability and gain errors can be made from an analysis of the loop gain $A(s)\beta(s)$.

Input Impedance of the Noninverting Amplifier

The input impedance of the noninverting amplifier is (Fig. 2.6)

$$Z_{in}(s) = \frac{V_1(s)}{I_1(s)} \qquad (2.48)$$

but

$$I_1(s) = \frac{V_1(s) - E_g(s)}{Z_I(s)} + \frac{V_1(s)}{Z_{cm2}(s)} \qquad (2.49)$$

From (2.36) we have

$$I_1(s) = \frac{V_1(s) + V_0(s)/A(s) - V_1(s)}{Z_I(s)} + \frac{V_1(s)}{Z_{cm2}(s)}$$

$$= \frac{V_0(s)}{Z_I(s)A(s)} + \frac{V_1(s)}{Z_{cm2}(s)}$$

but if $A(s)\beta(s) \gg 1$,

$$V_0(s) \cong \frac{Z_1'(s) + Z_F(s)}{Z_1'(s)} V_1(s)$$

and

$$I_1(s) \cong \left[\frac{Z_1'(s) + Z_F(s)}{Z_1'(s)} \frac{1}{Z_I(s)A(s)} + \frac{1}{Z_{cm2}(s)} \right] V_1(s)$$

Thus

$$Z_{in}(s) \cong \left[Z_I(s)A(s)\beta(s) \middle\| Z_{cm2}(s) \right] \qquad (2.50)$$

where

$$\frac{Z_1'(s) + Z_F(s)}{Z_1'(s)} \cong \text{design closed-loop gain} \cong \frac{1}{\beta(s)} \qquad (2.51)$$

We note that the input impedance is thus normally very large compared to $Z_I(s)$, since typically $A(s) \gg G(s)$ and $Z_{cm2} \gg Z_I$.

Output Impedance of the Noninverting Amplifier

With respect to output impedance, the noninverting circuit of Fig. 2.6 is essentially the same as the inverting amplifier of Fig. 2.4. That is, the output impedance may be found from the results of (2.32), (2.33), and (2.34). In summary,

$$Z_{out}(s) \cong \frac{Z_0(s)}{A'(s)\beta(s)}$$

The Unity Gain Follower

Note that the inequality of (2.43) does not apply to the unity gain follower circuit of Fig. 2.7. Let us make a special analysis of this case. Here we see that the feedback ratio is also approximately unity, that is,

$$\beta(s) \cong 1 \qquad (2.52)$$

The nodal equation at $V_0(s)$ is

$$\frac{A'(s)}{Z_0(s)} [V_1(s) - V_0(s)] = \frac{V_0(s)}{Z_x(s)} - \frac{V_1(s)}{Z_I(s)} \qquad (2.53)?$$

where

$$Z_x(s) = Z_0(s) \| Z_L(s) \| Z_I(s) \| Z_{cm1}(s)$$

$$\cong Z_0(s) \| Z_L(s) \qquad |Z_I|, |Z_{cm1}| \gg |Z_0| \qquad (2.54)$$

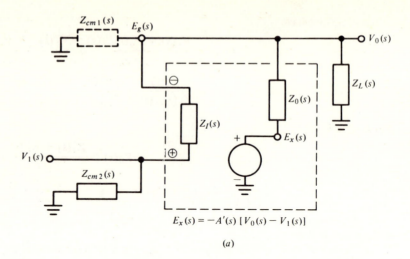

$$E_x(s) = -A'(s)[V_0(s) - V_1(s)]$$

(a)

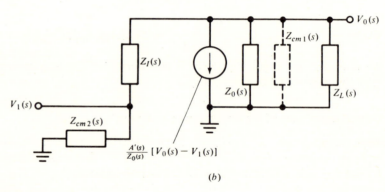

(b)

FIGURE 2.7
A general linear model for the unity gain follower circuit. (a) Adaptation from Fig.
2.6; (b) a Norton equivalent.

or

$$\frac{V_0(s)}{V_1(s)} = \frac{1 + Z_0(s)/A'(s)Z_I(s)}{1 + Z_0(s)/A'(s)Z_x(s)} \qquad (2.55)$$

Typically $A'(s)Z_i(s) \gg Z_0(s)$ and $Z_I(s) \gg Z_x(s)$, and we have

$$\frac{V_0(s)}{V_1(s)} \cong \frac{1}{1 + Z_0(s)/A'(s)Z_x(s)} = \frac{1}{1 + 1/A(s)}$$

$$= \frac{A(s)}{1 + A(s)} \qquad (2.56)$$

where we define

$$A(s) = A'(s)\frac{Z_x(s)}{Z_0(s)} \qquad (2.57)$$

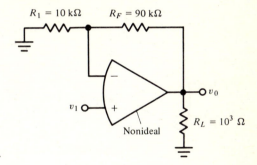

FIGURE 2.8
Circuit of Example 2.3.

Thus our general expression, (2.41), still holds for this special case where $\beta(s) = 1$ and $A(s)$ is defined by (2.57).

The input impedance is easily found from (2.50) to be

$$Z_{in} \cong A(s)Z_I(s)||Z_{cm2}(s) \qquad (2.58)$$

The output impedance is also found from (2.32), but with $\beta(s) = 1$, so that now

$$Z_{out}(s) = \frac{Z_0(s)}{1 + A'(s)} \qquad (2.59a)$$

or

$$Z_{out}(s) \cong \frac{Z_0(s)}{A'(s)} \qquad A'(s) \gg 1 \qquad (2.59b)$$

EXAMPLE 2.3 Consider the noninverting amplifier circuit of Fig. 2.8; note that the values of R_F and R_1 are chosen to obtain a noninverting gain of 10. Otherwise, the circuit is very similar to that of Example 2.2. If we assume the same nonideal amplifier model as in Example 2.2, viz.,

$$A_0 = 10^6$$

$$R_0 = 10^3 \ \Omega$$

$$R_I = 10^6 \ \Omega \qquad (\text{neglect } R_{cm})$$

then we can estimate the fractional gain error by (2.47). In this case

$$\beta = \frac{10^4||9 \times 10^4||10^6}{9 \times 10^4} \cong 0.1$$

Including the 10^3-Ω output load,

$$a_I = \frac{10^3||10^3||9 \times 10^4}{10^3}$$

$$\cong 0.5$$

Thus

$$A \cong 0.5 \times 10^6$$

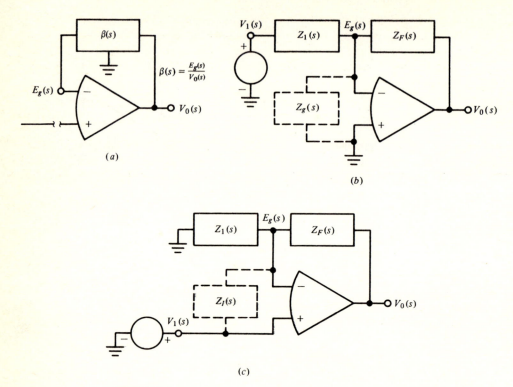

FIGURE 2.9
Finding $\beta(s)$ in typical circuits. (a) General two-port feedback network; (b) estimating $\beta(s)$ in a simple inverting amplifier circuit; (c) estimating $\beta(s)$ for a noninverting amplifier.

at dc, and the fractional error is

$$\frac{1}{A\beta} = \frac{1}{0.5 \times 10^5} = 2 \times 10^{-5}$$

The circuit input impedance may be estimated from (2.50) as follows:

$$Z_{\text{in}} \cong \frac{AR_I}{G} \bigg\| R_{cm2}$$

where

$$G = 10 = \frac{R_1 + R_F}{R_1}$$

Thus

$$Z_{\text{in}} \cong \frac{0.5 \times 10^6 \times 10^6}{10}$$

$$= 0.5 \times 10^{11} \ \Omega$$

in parallel with R_{cm2}.   ////

We note that in many simple amplifier configurations, the feedback ratio $\beta(s)$ is relatively easy to calculate. Essentially we are trying to find the voltage feedback ratio from the amplifier output terminal to the inverting input terminal. This ratio is in general established by some three-terminal network (Fig. 2.9a) connecting these two points, that is,

$$\beta(s) = \frac{E_g(s)}{V_0(s)} \qquad (2.60)$$

For specific circuits, the resulting value of $\beta(s)$ may be found; for many approximate analyses, it is all right to neglect the effect of $Z_I(s)$ and the common-mode impedances; however, particularly for accurate high-frequency analysis, the effects of the capacitive components of these impedances may need to be included.

For both the single-input inverting amplifier circuit (Fig. 2.9b), and the noninverting amplifier circuit (Fig. 2.9c), we find that

$$\beta(s) = \frac{Z_1(s)}{Z_1(s) + Z_F(s)} \qquad (2.61)$$

where, if needed, the effects of the amplifier input impedance may be included by including $Z_g = Z_I \parallel Z_{cm1}$ in the calculation of Z_1.

It is often useful to be aware of the fact that for the *single-input* inverting amplifier (Fig. 2.9b),

$$\beta(s) = \frac{1}{1 - G_I(s)} \qquad (2.62)$$

where $G_I(s)$ is the desired or design value of the *closed-loop gain,* that is,

$$G_I(s) = \frac{-Z_F(s)}{Z_1(s)} \qquad (2.63)$$

[Note the inclusion of the minus sign with $G_I(s)$.] Thus an inverting amplifier with a design gain of magnitude 10 ($G_I = -10$) at some frequency will have a corresponding value of

$$|\beta| = \frac{1}{1 - (-10)} = \frac{1}{11}$$

For an amplifier with a high open-loop gain, the value of $|\beta|$ is *approximately* the reciprocal of the magnitude of the design gain, since

$$|\beta| = \frac{1}{|1 - G_I|} \cong \frac{1}{|G_I|} \qquad |G_I| \gg 1$$

For the *noninverting* amplifier (Fig. 2.9c), we still have

$$\beta(s) = \frac{Z_1(s)}{Z_1(s) + Z_F(s)}$$

where again the effects of $Z_g(s)$ may warrant inclusion. Here, however, if we define $G_N(s)$ as the design value of (noninverting) closed-loop gain, that is,

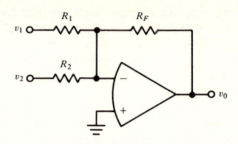

FIGURE 2.10
Circuit for Example 2.4, a two-input
resistive inverter.

$$G_N(s) = \frac{Z_1(s) + Z_F(s)}{Z_1(s)} \qquad (2.64)$$

then for the noninverting amplifier,

$$\beta(s) = \frac{1}{G_N(s)} \qquad (2.65)$$

Thus a noninverting amplifier with a design gain of magnitude 10 will have $|\beta| = 0.1$.

Many articles on amplifier design will say that the magnitude of β is equal to the reciprocal of the magnitude of the desired closed-loop gain. This is true for the noninverting amplifier, and also approximately true for the single-input inverting circuit *if* the closed-loop gain is large compared to 1. Note, however, the following example.

EXAMPLE 2.4 Consider the simple two-input inverting amplifier of Fig. 2.10. An estimation of β follows directly from (2.12), that is,

$$\beta = \frac{R_1 \| R_2 \| R_F}{R_F}$$

$$= \frac{R_1 R_2}{R_2 R_F + R_1 R_F + R_1 R_2}$$

Defining

$$G_1 = \frac{-R_F}{R_1}$$

and

$$G_2 = \frac{-R_F}{R_2}$$

we can show that

$$\beta = \frac{1}{|G_1| + |G_2| + 1} \cong \frac{1}{|G_1| + |G_2|}$$

which is an extension of (2.62) to the two-input case.

$////$

2.3 CLOSED-LOOP STABILITY; BODE PLOT METHODS; COMPENSATION

In the last section, we saw [e.g., in Eq. (2.15)] that the closed-loop transfer function of a linear operational amplifier circuit has unwanted poles at roots of

$$1 + A(s)\beta(s) = 0 \qquad (2.66)$$

It is important to determine where the roots of (2.66) lie in the s plane. Specifically, for the circuit to be stable, they must have negative real parts; moreover, these spurious roots should have a large magnitude, and relatively small imaginary parts, if unwanted oscillatory transient components in the circuit's response are to be minimized. The well-known Bode plot methods of estimating stability apply very nicely to the analysis of closed-loop operational amplifier circuit behavior; a discussion of this method appears in most control system texts (see, e.g., Ref. 2, Sec. 3.4, or Ref. 3, Sec. 7.9). We summarize the important points here.

Given a linear system transfer function $H(s)$, the Bode plots consist of:

1 Magnitude of $H(s = j\omega)$ in the form $20 \log_{10} |H(j\omega)|$ versus $\log \omega$, that is, dB versus $\log \omega$.
2 Phase of $H(j\omega) = \angle H(j\omega)$ versus $\log \omega$, usually in degrees, but possibly in radians.

Reference 3, Sec. 7.9, summarizes the technique for making approximate plots from a knowledge of critical frequencies of $H(s)$. In the analysis of operational amplifier circuits, we are interested in Bode plots of $H(s) = A(s)\beta(s)$.

To determine closed-loop stability via a Bode plot, we are primarily interested in two frequencies (Fig. 2.11):

1 The *unity gain crossover frequency* ω_0; that is, the frequency where the magnitude curve first crosses the 0-dB point
2 The *phase crossover frequency* ω_p, where the phase curve first crosses $-180°$ (or $-\pi$ radians)

The criterion *for stability* may be stated in various ways:

1 If

$$\omega_0 < \omega_p \qquad (2.67)$$

that is, if the unity gain crossover frequency is less than the phase crossover frequency.
2 If the phase is less negative than $-180°$ at the unity gain crossover frequency; the difference between $-180°$ and the phase at ω_0 is termed the *phase margin* (ϕ_m, Fig. 2.11). Thus the phase margin

$$\phi_m = \angle A(j\omega_0)\beta(j\omega_0) + 180° > 0° \qquad (2.68)$$

3 If the gain is less than 0 dB at the phase crossover frequency. This difference is termed the *gain margin* (G_m, Fig. 2.11). Thus the gain margin

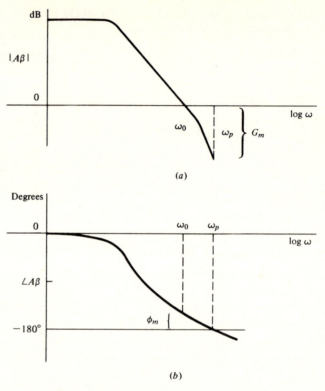

FIGURE 2.11
Typical Bode plots of loop gain $A\beta$. (a) Magnitude in decibels; (b) phase angle in degrees.

$$G_m = -20 \log_{10}|A(j\omega_p)\beta(j\omega_p)| > 0 \text{ dB} \qquad (2.69)$$

Note that $\angle A\beta$ is usually negative at all frequencies, hence the form of (2.68). Also $|A\beta| > 0$ dB below ω_0; however, G_m is usually defined as positive for stability, hence the form of (2.69).

The above criteria merely ensure stability. Going beyond this, we are usually interested in whether the roots of $A(s)\beta(s) + 1 = 0$ influence the desired closed-loop response; e.g., the frequency response may exhibit peaking, and the transient response will have some form of oscillatory behavior. To ensure against this, the phase margin should be reasonably high; a common rule of thumb that is often applied is that, to ensure good transient behavior,

$$\phi_m > 45°$$

that is, the phase of $A\beta$ should be about $-135°$ at ω_0. This means that typically the slope of the magnitude curve (dB versus log ω) will be less than or equal to 20 dB/decade near ω_0.

It is common practice to analyze operational amplifier circuits by plotting the magnitude curves of $A(j\omega)$ and $1/\beta(j\omega)$ on the same graph. Thus at the point of intersection of the two curves, $\omega = \omega_0$, $|A\beta| = 1$ (0 dB), and the *rate of closure* of the two curves indicates the approximate phase margin. If the rate of closure is less than 20 dB/decade, we have a good phase margin. If it is greater than or equal to 40 dB/decade, we have instability. In between, we have stability, but probably poor phase margin (Fig. 2.12).

If the above criteria are not met, then normally oscillation will result, since there will be roots of

$$A(s)\beta(s) + 1 = 0 \qquad (2.70)$$

with positive real parts.

EXAMPLE 2.5   In Fig. 2.12, we have a simple gain-of-10 inverting amplifier (Fig. 2.12c). For this circuit

$$\beta = \frac{1}{11}$$

and $1/\beta$ in decibels is merely a constant.

$$\left|\frac{1}{\beta}\right| = 20 \log_{10} 11 = 20.8 \text{ dB}$$

The solid curve (Fig. 2.12a) representing a typical $|A|$ shows a rate of closure of 20 dB/decade, indicating stability, with probably a good phase margin. The dotted curve, indicating an extra pole in $A(s)$ which is going to add more phase shift to $A\beta$, indicates a higher rate of closure. If the rate of closure is less than 40 dB/decade, then the circuit is presumably closed-loop stable, but the phase margin may not be good. A further analysis of the phase curves for A and β may thus be warranted. Figure 2.12b shows corresponding plots of $\angle A\beta$.   ////

EXAMPLE 2.6   In Fig. 2.13, we have a Bode plot of a typical $A(j\omega)$, indicating two first-order poles. Two different values of β are shown (assumed constant, i.e., for some simple amplifier with resistances for Z_1 and Z_F). Case 1 would presumably be for an amplifier with a relatively high closed-loop gain, case 2 for a lower closed-loop gain. The rate of closure for the two situations indicates that case 1 would lead to good closed-loop stability with good phase margin. Case 2 has a poor phase margin. If β were much larger (lower closed-loop gain), then instability might occur.   ////

At this point, it is useful to note that the simple unity gain follower circuit, which has a β of 1, provides an acid test for the closed-loop stability of an operational amplifier. Here the amplifier output is connected directly to the inverting terminal (Fig. 2.7). In this case, the phase margin is found by merely examining the phase angle of A at

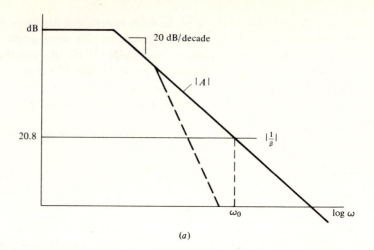

(a)

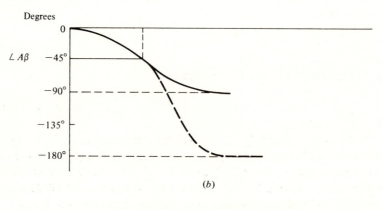

(b)

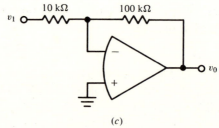

(c)

FIGURE 2.12
Refer to Example 2.5. (a) Two possible magnitude curves for A; (b) associated phase curves; (c) simple gain-of-10 inverter, with $\beta = \frac{1}{11}$.

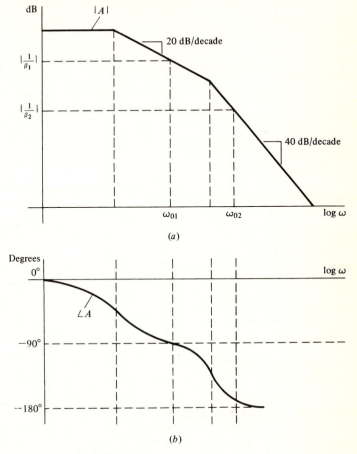

FIGURE 2.13
Refer to Example 2.6. (*a*) Magnitude plots; (*b*) phase plots.

$\omega_0 = \omega_u$ = the unity gain crossover frequency of the loaded amplifier! Amplifiers which are stable under this condition are termed *short-circuit stable*; they must have $\angle A(j\omega) > -180°$, and preferably in the neighborhood of $-135°$ at ω_u.

It is common practice in the design of integrated-circuit operational amplifiers to provide for the connection of *compensation networks*. These often take the form of a simple series RC network that controls the first break frequency in $A'(s)$, for example, ω_1 in Fig. 2.2. By adjusting this break frequency, it is possible to trade bandwidth for closed-loop gain, within the limits of the overall gain-bandwidth product. (See Prob. 2.12 to gain some insight into how this works.)

Figure 2.14 shows some typical compensation schemes that may be used with an amplifier that has no provision for internal compensation; optimum component values are often best determined empirically. (See also Ref. 1, Secs. 3-18 to 3-20.)

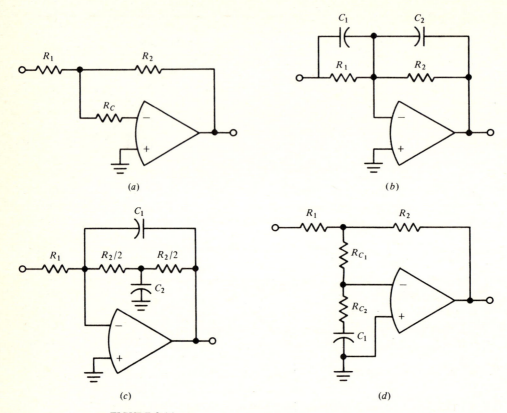

FIGURE 2.14
Typical compensation schemes that may be used to improve the stability and/or frequency response of a simple inverting amplifier. (a) Resistor R_C in series with the inverting terminal improves stability, but will generally degrade bandwidth; (b) compensating capacitors can be used to correct closed-loop phase margin and also improve bandwidth; C_1 and C_2 are normally adjusted empirically, C_1 often omitted; (c) the tee network is easier to adjust and eliminates the capacitive loading of the input source; (d) this network has been successfully used in D/A converters to improve the step response by phase-shift correction.

2.4 COMMON–MODE EFFECTS

In the differential amplifier circuit shown below (Fig. 2.15), an *ideal* operational amplifier provides an output

$$e_0 = -\frac{R_0}{R_1}\Delta e \quad \text{(ideal)} \quad (2.71)$$

where Δe is the differential input signal.

The input *common-mode signal* e_{cm} is ideally rejected. This type of situation is common in many instrumentation systems, where Δe originates from some low-level

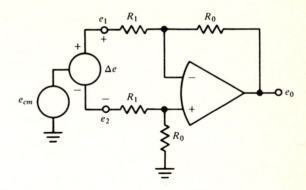

FIGURE 2.15
A differential amplifier circuit, with the input resolved into a differential component Δe and a common-mode component e_{cm}.

transducer; e_{cm} may be a dc offset or some form of noise (60-cycle hum or crosstalk from nearby electrical systems), i.e., an *unwanted* signal to be ignored by the amplifier.

Unfortunately, real operational amplifiers exhibit a *common-mode gain* (CMG), which may be thought of as the ratio of output signal to input signal, if both inputs are tied together and excited simultaneously; hence in Fig. 2.16a,

$$\text{CMG} = \left| \frac{e_0}{e_i} \right| \qquad (2.72)$$

The common-mode rejection ratio (CMRR or CMR) is the ratio

$$\text{CMRR} = \frac{|A_0|}{|\text{CMG}|} \qquad \text{(polarity not specified)} \qquad (2.73)$$

where A_0 is the dc or low-frequency value of the normal open-loop differential input gain. *CMRR is often expressed in decibels.*

To determine the effect of the common-mode gain on the differential amplifier performance, consider the circuit of Fig. 2.16b. The usual model of the amplifier is shown in Fig. 2.16c, or equivalently, in Fig. 2.16d, which implies

$$e_0 = -Ae_x = -A \left(e_s - \frac{e_A}{\text{CMRR}} \right) \qquad (2.74)$$

Also

$$e_{cm} = \frac{e_1 + e_2}{2} = \text{common-mode input signal} \qquad (2.75a)$$

$$\Delta e = e_1 - e_2 = \text{differential input signal} \qquad (2.75b)$$

$$e_s = e_B - e_A \qquad (2.76)$$

From Fig. 2.16d we have

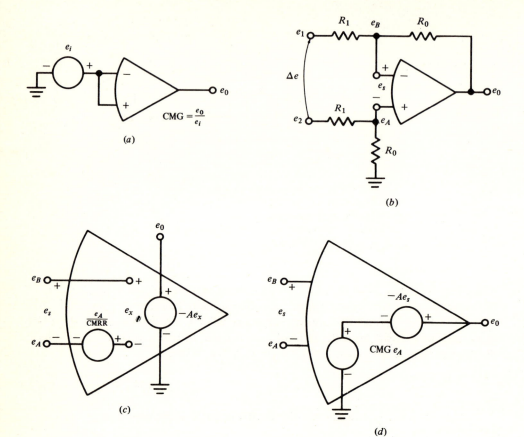

FIGURE 2.16
Analysis of common-mode gain effect. (a) Circuit for measuring common-mode gain; (b) amplifier circuit; (c) operational amplifier mode with common-mode effect in input circuit; (d) referral of common-mode gain to the output circuit.

$$e_0 = -(Ae_s - \text{CMG}e_A) = -A\left(e_s - \frac{e_A}{\text{CMRR}}\right) \qquad (2.77)$$

$$e_B = \frac{e_0 R_1}{R_1 + R_0} + \frac{e_1 R_0}{R_1 + R_0} \qquad (2.78)$$

$$e_A = e_2 \frac{R_0}{R_1 + R_0} \qquad (2.79)$$

or

$$e_s = \frac{e_0 R_1}{R_1 + R_0} + \frac{\Delta e R_0}{R_1 + R_0} \qquad (2.80)$$

From (2.77) and (2.80),

$$e_0 = -A\left(\frac{e_0 R_1}{R_1 + R_0} + \frac{\Delta e R_0}{R_1 + R_0}\right) + \frac{A e_A}{\text{CMRR}} \qquad (2.81)$$

or

$$e_0\left(1 + \frac{A R_1}{R_1 + R_0}\right) = \frac{-A \Delta e R_0}{R_1 + R_0} + \frac{A e_A}{\text{CMRR}}$$

If we let

$$\beta = \frac{R_1}{R_1 + R_0} \qquad (2.82)$$

then

$$e_0 = \frac{-A}{1 + A\beta}\left(\frac{R_0}{R_1 + R_0}\Delta e - \frac{e_A}{\text{CMRR}}\right) \qquad (2.83)$$

For $|A\beta| \gg 1$,

$$e_0 \cong \frac{-1}{\beta}\frac{R_0}{R_1 + R_0}\Delta e + \frac{1}{\beta}\frac{e_A}{\text{CMRR}}$$

$$\cong -\frac{R_0}{R_1}\Delta e + \frac{R_1 + R_0}{R_1}\frac{e_A}{\text{CMRR}} \qquad (2.84)$$

From (2.79) and (2.84),

$$e_0 = \frac{-R_0}{R_1}\Delta e + \frac{R_0}{R_1}\frac{e_2}{\text{CMRR}} \qquad (2.85)$$

Note that we usually know only the magnitude of CMRR, not its polarity. But also,

$$e_2 = e_{cm} - \frac{\Delta e}{2} \qquad (2.86)$$

Thus

$$e_0 = \frac{-R_0}{R_1}\Delta e + \frac{R_0}{R_1}\frac{1}{\text{CMRR}}\left(e_{cm} - \frac{\Delta e}{2}\right) \qquad (2.87)$$

Typically

$$|\Delta e| \ll |e_{cm}|$$

and

$$e_0 \cong \frac{-R_0}{R_1}\Delta e + \frac{R_0}{R_1}\frac{1}{\text{CMRR}}e_{cm} \qquad (2.88)$$

From Eq. (2.88) we see that the CMRR indicates the relative effect of the differential and common-mode signals on the output.

Noninverting Amplifiers

From Fig. 2.17 it is fairly easy to show that if

$$\beta = \frac{R_1}{R_1 + R_0} \qquad (2.89)$$

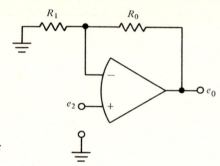

FIGURE 2.17
A simple resistive noninverting amplifier
circuit (see Sec. 2.4).

then

$$\frac{e_0}{e_2} \cong \frac{R_1 + R_0}{R_1} \frac{A\beta}{1 + A\beta} \left(1 + \frac{1}{\text{CMRR}}\right) \qquad (2.90a)$$

$$\cong \frac{R_1 + R_0}{R_1} \left(1 + \frac{1}{\text{CMRR}}\right) \qquad |A\beta| \gg 1 \qquad (2.90b)$$

In (2.90) we see an instance where the CMRR can affect the gain expression in a simple noninverting amplifier, leading to another source of gain error in addition to that predicted by the analysis in Sec. 2.2. Indeed, common-mode gain effects may be a principal source of gain inaccuracy, even in single-ended amplifier circuits.

EXAMPLE 2.7 In Fig. 2.16*b*, let

$$R_0 = 10^5 \ \Omega$$
$$R_1 = 10^4 \ \Omega$$
$$\text{CMRR} = 100 \ \text{dB}$$

The circuit has a differential gain

$$\frac{e_0}{\Delta e} = \frac{-10^5}{10^4} = -10$$

and a common-mode gain

$$\frac{e_0}{e_{cm}} = \frac{R_0}{R_1} \frac{1}{\text{CMRR}}$$

As a ratio

$$\text{CMRR} = 10^5$$

or

$$\frac{e_0}{e_{cm}} = 10^{-4}$$

(may be either polarity).

////

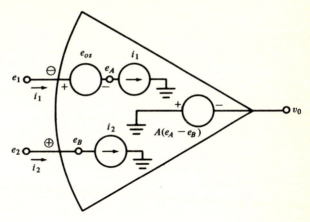

FIGURE 2.18
An equivalent circuit for representing operational amplifier internal dc offsets.

2.5 DC OFFSETS AND TEMPERATURE DRIFT

Thus far we have been assuming that there were no internal dc offsets in the operational amplifier. In fact, there are always offsets, hopefully small, inherent in any operational amplifier, as a result of lack of symmetry in the differential stages, lack of precise bias, etc. Moreover, even though the dc offsets may be small at normal room temperature, at elevated temperatures they may be enhanced appreciably. Power-supply voltage drifts may also create offsets. Thus any practical operational amplifier circuit must be designed with a view toward the possible effects of dc offsets.

Figure 2.18 illustrates a fairly general amplifier equivalent circuit for use in treating dc offsets. Offset effects are usually described by the following parameters:

e_{os} = Differential dc offset voltage (referred to the amplifier input), usually called the *input offset voltage*. It is the differential dc input voltage required to provide zero output voltage, with no other input signal and zero resistance in either input terminal path to ground.

i_1, i_2 = *Input bias currents*. These are residual input currents required by the input stage transistors (base current if bipolar, gate current if FET). They are the direct biasing currents required at either input to produce zero output voltage with no input offset voltage.

In most operational amplifiers, i_1 and i_2 will be similar in magnitude and polarity, and will tend to track each other with temperature or power-supply voltage changes. That is, one may define an *input offset current*

$$i_{os} = i_1 - i_2 \qquad (2.91)$$

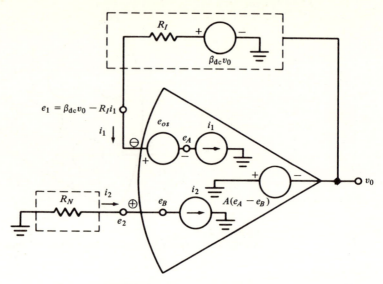

FIGURE 2.19
A general equivalent circuit for analyzing dc offset effects.

i_{os} is typically a small fraction of either i_1 or i_2. Thus, it is possible to minimize the effects of the input bias currents by using compensation techniques to be described below.

Note that we usually have no way of predicting the polarity of the offset components; hence, our estimates of dc offset effects will be worst-case bounds, which is a sensible approach. The designer should be careful to note whether the specified offsets are guaranteed maximum values or merely typical!

Figure 2.19 shows the typical environment where an operational amplifier may be operating. The resistances R_I and R_N represent the *Thévenin-equivalent dc resistances* associated with the input terminals. The feedback network provides a dc path between v_0 and e_1 which has associated with it a *dc feedback ratio* β_{dc}. That is, the Thévenin-equivalent voltage in the inverting amplifier branch has the value $\beta_{dc}v_0$. As an example, Fig. 2.20 shows a simple inverting amplifier circuit with input resistance R_1, feedback resistance R_F, and possibly an offset-compensating resistance R_2 as the noninverting terminal connection to ground. R_2 is used here solely to permit some compensation for input offset currents, as we shall see later. In Fig. 2.20, it is fairly easy to show that

$$\beta_{dc} = \frac{R_1}{R_1 + R_F} \tag{2.92}$$

$$R_I = R_1 \| R_F = \frac{R_1 R_F}{R_1 + R_F} \tag{2.93}$$

and
$$R_N = R_2 \tag{2.94}$$

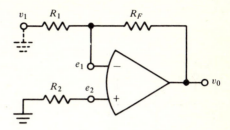

FIGURE 2.20
A simple resistive inverting amplifier circuit; note the compensation resistor in the noninverting terminal.

In a more general circuit, one would have to find the dc values of β, R_I, and R_N for the specific circuit configuration. It is emphasized that only the dc values are of interest, so in finding values for β, R_I, and R_N, all capacitors should be replaced by open circuits and inductors by short circuits.

Basic Error Equations

Referring to Fig. 2.19, Kirchhoff's voltage law gives

$$\beta v_0 - i_1 R_I = e_{os} - R_N i_2 \qquad (2.95)$$

The above assumes that the amplifier open-loop gain is very high, and therefore $|e_A - e_B| \cong 0$. We are here assuming that the amplifier is otherwise ideal, and are not including such effects as finite open-loop gain and finite input impedance. Our results will be normally quite satisfactory, since we do not know the magnitude of the offsets with any great precision anyway.

From (2.95) we find

$$v_0 = \frac{1}{\beta}(e_{os} + i_1 R_I - i_2 R_N) \qquad (2.96)$$

From (2.91), we also have

$$v_0 = \frac{1}{\beta}[e_{os} + i_1(R_I - R_N) + i_{os}R_N] \qquad (2.97)$$

Usually we have no idea about the polarity of the offset components, but we can estimate the worst-case offset as follows:

$$|v_0| \leqslant \frac{1}{\beta}[|e_{os}| + |i_1|(R_I - R_N) + |i_{os}|R_N] \qquad (2.98)$$

There are two important special cases based upon (2.98):

CASE I: $R_N = 0$ If $R_N = 0$, (2.98) becomes

$$|v_0| \leqslant \frac{1}{\beta}(|e_{os}| + |i_1|R_I) \qquad (2.99)$$

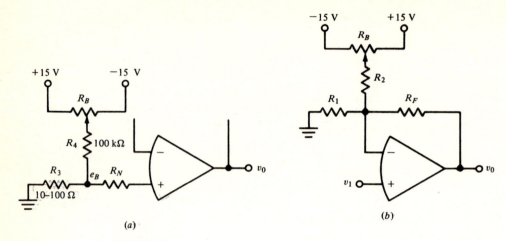

FIGURE 2.21
Typical dc offset-adjusting networks. (*a*) Offset network for the noninverting terminal; (*b*) a similar network for the inverting terminal.

CASE II: $R_N = R_I$ We can usually minimize the total dc offset by setting $R_N = R_I$; in this case

$$|v_0| \leqslant \frac{1}{\beta}(|e_{os}| + |i_{os}|R_I) \qquad (2.100)$$

For the circuit of Fig. 2.20 we would then find from (2.100) that

$$|v_0| \leqslant \frac{R_1 + R_F}{R_1}(|e_{os}| + R_I|i_{os}|) \qquad (2.101a)$$

with

$$R_2 = R_N = R_I = \frac{R_1 R_F}{R_1 + R_F} \qquad (2.101b)$$

In (2.98) to (2.101) we note that the offset effects are aggravated by making β smaller. This means, in general, that operational *amplifier circuits* with a *high dc gain* have *more offset*. Also, as indicated by (2.98) and (2.101*a*), the *offset is*, in general, *increased* as we go to circuit designs with a *higher dc impedance* level. The compensation scheme of case II above permits us to reduce the effects of current offsets, and it is generally recommended. (Note, however, that we will find in Sec. 2.6 that adding R_2 to the circuit in Fig. 2.20 will often enhance the random noise in the circuit—another design tradeoff.)

DC Offset Balancing

For a given pair of power-supply voltages and a given ambient temperature, it is possible to introduce a dc offset balancing signal into the operational amplifier circuit, so that the output voltage is zero in the absence of any normal inputs. Figure 2.21*a* illustrates one simple method. R_N is chosen so that the conditions of case II above are met, that is,

$$R_N = R_I$$

An adjustable offset voltage is established at node e_B. Note that the impedance at node e_B is quite low and that adjustments of potentiometer R_B merely introduce a dc offset counterbalance without appreciably disturbing the dc resistance in the noninverting terminal path to ground. R_3 is typically 10 to 100 Ω; R_4 is chosen to permit e_B to range over the anticipated offsets. Specifically, we must be able to set e_B such that

$$|e_B| > |e_{os}|_{\max} + |i_{os}|_{\max} R_N \qquad (2.102)$$

Of course, the circuit of Fig. 2.21a will not work if the noninverting terminal must be used for other purposes, e.g., in a differential amplifier or a noninverting circuit. In that case, other alternatives are possible. It is often possible to inject an offset current into the inverting terminal as well, as shown in Fig. 2.21b. Here we have a noninverting circuit. By properly choosing R_2 and potentiometer R_B, one can inject sufficient direct current to counteract the amplifier internal offset components. Usually R_2 is chosen to be large compared to the parallel combination of R_1 and R_F. For differential amplifier circuits, it may be preferable to use an operational amplifier that has separate terminals for the addition of a dc balance adjustment network.

We should point out that recent improvements in operational amplifier fabrication have led to the development of amplifiers with very small offsets *at room temperature,* so that the dc offset problem may not be important. However, even these newer units may have significant offsets at elevated temperatures (for example, 150°F).

Offset Drifts

As the ambient temperature in the vicinity of the operational amplifier changes, the dc offset components will change. Even though the dc offsets have been balanced out for a specific set of operating conditions, changes in temperature or power-supply voltage will generally cause changes in e_{os}, i_1, i_2, and i_{os}. One may then need to estimate the effect of these changes. Assuming no dc offset at some point, if the offset components undergo a change, the resulting output dc offset in a circuit fitting the general model of Fig. 2.19 will be

$$|\Delta v_0| \leqslant \frac{1}{\beta_{dc}} [|\Delta e_{os}| + |\Delta i_1|(R_I - R_N) + |\Delta i_{os}| R_N] \qquad (2.103)$$

Again, the compensation scheme of setting $R_N = R_I$ yields

$$|\Delta v_0| \leqslant \frac{1}{\beta_{dc}} (|\Delta e_{os}| + |\Delta i_{os}| R_I) \qquad (2.104)$$

Table 2.2 shows some typical dc offset parameters for today's operational amplifiers. Special units with low offset current are available; however, they often have reduced bandwidth, so that a design compromise may be involved. It is safe to say that almost daily we will see announcements of new amplifiers with improved offset drift characteristics. Note that, for amplifiers with bipolar transistors in the input stage, an offset-current drift-current coefficient in amperes per degree Celsius is often stated, which may be fairly valid over a reasonable temperature range. For amplifiers with FET input stages, more typically the offset-current components will double for a 10°C change. Both types will typically exhibit voltage offset drifts of 0.5–20 $\mu V/°C$.

EXAMPLE 2.8 Refer to the inverting amplifier circuit of Fig. 2.20. Suppose we want to design a gain-of-10 circuit. Use the specifications for the poor-quality IC amplifier given below. Pick $R_F = 100$ kΩ and estimate the inherent offset with and without R_N. Here $\beta = 1/11$, $R_F = 10^5$ Ω, $R_1 = 10^4$ Ω, and thus $R_I = 9.09 \times 10^3$ Ω. Let

$$e_{os} = \pm \ 5 \text{ mV}$$
$$i_1, i_2 = \pm 500 \text{ nA}$$
$$i_{os} = \pm \ 50 \text{ nA}$$

First, with $R_N = 0$, from (2.99),

$$|v_0| \leqslant 11(5 \times 10^{-3} + 5 \times 10^{-7} \times 9.09 \times 10^3)$$
$$\leqslant 0.105 \text{ V}$$

Picking $R_N = R_I = 9.09 \times 10^3$ Ω,

$$|v_0| \leqslant 11(5 \times 10^{-3} + 5 \times 10^{-8} \times 9.09 \times 10^3)$$
$$\leqslant 0.06 \text{ V}$$

The compensation scheme *is* worthwhile here.

If we make $R_F = 10^4$ Ω, then $R_I = 9.09 \times 10^2$ Ω; with $R_N = 0$,

$$|v_0| \leqslant 11(5 \times 10^{-3} + 5 \times 10^{-7} \times 9.09 \times 10^2)$$
$$\leqslant 0.060 \text{ V}$$

Table 2.2 TYPICAL AMPLIFIER DC OFFSET SPECIFICATIONS

	Discrete bipolar	Discrete, FET input	Monolithic IC (low-cost)	Monolithic IC, FET input
Offset voltage e_{os}*	±0.2 mV	±1 mV	±5 mV	±3.5 mV
Input bias current i_1 or i_2 *	±20 nA	−20 pA†	50 nA†	−10 pA†
Input offset current i_{os}*	±2 nA	±10 pA	±5 nA†	±5 pA
Temperature drift of voltage offset $\partial e_{os}/\partial T$	±0.5 $\mu V/°C$	±20 $\mu V/°C$	±5 $\mu V/°C$	±10 $\mu V/°C$
Temperature drift of input current $\partial i_1/\partial T$	±0.2 nA/°C	‡	±0.5 nA/°C	‡
Temperature drift of input offset current $\partial i_{os}/\partial T$	±0.02 nA/°C	‡	±0.05 nA/°C	‡

*At 25°C.
†Often i_1 and i_2 have known polarity, but the polarity of i_{os} is not known.
‡Doubles over 10°C change.

and with $R_N = R_I$,

$$|v_0| \leqslant 0.055 \text{ V}$$

that is, R_N does little good.

If the circuit was otherwise balanced so that $v_0 = 0$ at, say, 25°C, then at 50°C,

$$\Delta e_{os} = \pm 125 \ \mu\text{V}$$
$$\Delta i_1 = \Delta i_2 = \pm \ 25 \text{ nA}$$
$$\Delta i_{os} = 12.5 \text{ nA}$$

With $R_F = 100 \text{ k}\Omega$ and $R_N = 0$, the output voltage drift can be estimated from (2.103).

$$|\Delta v_0| \leqslant 11 \ (125 \times 10^{-6} + 9.09 \times 10^3 \times 25 \times 10^{-9})$$
$$\leqslant 3.88 \text{ mV}$$

The addition of $R_N = R_I$ reduces the drift to

$$|\Delta v_0| \leqslant 11(125 \times 10^{-6} + 9.09 \times 10^3 \times 125 \times 10^{-9})$$
$$\leqslant 2.62 \text{ mV}$$

a modest improvement.   ////

Chopper-Stabilized and Varactor Diode Amplifiers

Since the early days of vacuum-tube operational amplifiers, the problem of dc offset and offset drift has always been present. At present, one can often use the types of low-offset units described in Table 2.2 with satisfactory results. There are still situations, however, where wide temperature variations or extreme requirements for dc offset reduction make it necessary to employ special amplifiers. These usually use modulated carrier techniques to transmit the dc and very-low-frequency components, while the high-frequency components may be transmitted through the amplifier with ac-coupled techniques (see Ref. 4, Sec. 4.4). Two principal types of units are readily available in discrete form: those using a vibrating reed or solid-state *chopper* channel for dc offset sensing and stabilization, and *varactor* input amplifiers. These units are relatively expensive; however, integrated-circuit amplifiers with chopper stabilization are just now becoming available. This should make it easier to employ these techniques where required.

2.6 AMPLIFIER INTERNAL NOISE EFFECTS

In addition to the dc offsets discussed in the previous section (which could be considered a form of unwanted signal, or noise), an amplifier's output will always contain additional spurious components, which we collectively call noise. In many cases, much of this noise comes from *external sources.* In this category are spurious signals coupled into the input circuit or ground return (60-Hz hum or other electromagnetically coupled induced noise), thermally generated Johnson noise in circuit resistances, or noise introduced from the power supply (e.g., due to excessive ripple). We shall have more to say about techniques

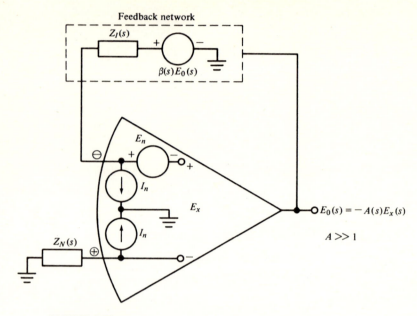

FIGURE 2.22
A general equivalent circuit for analyzing operational amplifier noise effects.

for eliminating these forms of external interference in Sec. 2.9. In this section we are concerned with *internal amplifier noise* produced by random phenomena, primarily in the amplifier input stage. We will not describe the noise mechanisms in detail; Refs. 4 and 5 may be consulted for more information on the nature of the noise sources. We will merely categorize the noise sources as follows:

1 Broadband white noise with an essentially constant noise spectrum, primarily due to Schottky and Johnson noise in the input transistor stage.
2 Low-frequency ($1/f$) noise with a power spectrum that is inversely proportional to frequency. Typically this noise is not important above about 10 Hz, but it can be important at very low frequencies in special applications.
3 Burst or popcorn noise with a power spectrum of the form

$$\frac{K}{1 + (f/f_b)^2}$$

Typically this noise is not important above 200 to 300 Hz. It is not always apparent, but has been found in some silicon planar-diffused bipolar integrated-circuit transistors (see Ref. 10).

Figure 2.22 shows a useful model for representing internal amplifier noise effects. The noise phenomena are represented in terms of equivalent input noise voltage and current sources e_n and i_n. Also important is β, the feedback ratio between the output

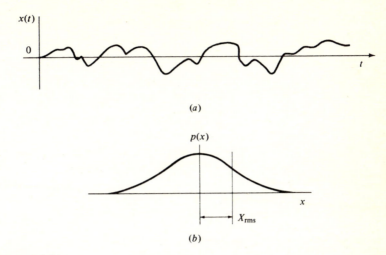

FIGURE 2.23
A typical noise signal. (*a*) A sample waveform; (*b*) probability density of the amplitude.

terminal and the inverting input terminal. (Now we must in general consider the frequency-dependent nature of β.) The Thévenin-equivalent impedances in the inverting and noninverting terminal branches, Z_I and Z_N, are also important. Note that the symbol Z_I has a different meaning here than in Secs. 2.1 to 2.3, where Z_I stood for the differential input impedance of the operational amplifier.

Spectral Description of Noise

At this point we must explore the means by which noise is specified. We assume that the noise signals E_n and I_n (Fig. 2.22) are *gaussian stationary random processes,* which may be *specified in terms* of either their *correlation function* or their *power spectral density* (see, for example, one of the many good texts on random process theory, such as Ref. 9). We cannot elaborate on these concepts here, but will attempt to present principal design equations. Figure 2.23*a* shows a typical sample of a random noise signal. Here we assume that the noise is stationary (i.e., the statistical properties do not change with time); thus we can describe the general nature of the noise signal by its *spectral* properties. We also assume that the signals can be described as having a gaussian amplitude distribution with zero average or mean value (Fig. 2.23*b*); hence we are able to describe the overall noise intensity rather well by specifying its rms value (i.e., the standard deviation of the noise amplitude). Actually, the rms value that is observed for a given noise signal depends upon the bandwidth (frequency range) observed. For the present discussion, let us describe the noise by a *one-sided power spectral density*.

Let $G(f)$ be the one-sided power spectral density of a stationary random signal $x(f)$ with zero average value. If we observe the process with a unity gain "brickwall" filter

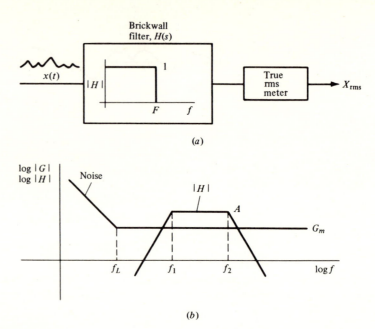

FIGURE 2.24
Effect of filtering on noise. (*a*) Use of a brickwall filter to measure power spectral density, Eqs. (2.105) and (2.106); (*b*) rejection of noise outside the passband f_1 to f_2, Eq. (2.109).

which passes frequencies from zero up to some upper frequency F, then we will observe (Fig. 2.24*a*) the rms value

$$X_{rms} = \sqrt{\int_0^F G(f)\,df} \qquad (2.105)$$

or conversely,

$$G(f) = \frac{d}{df} X_{rms}^2 \qquad (2.106)$$

Usually we are interested in the *apparent rms value* of a random signal, as observed over a *finite frequency range*, viz., from f_1 to f_2, since no physical system can respond to all frequencies. In this case, the apparent rms value of the signal is given by

$$X_{rms} = \sqrt{\int_{f_1}^{f_2} G(f)\,df} \qquad (2.107)$$

Figure 2.24*b* shows a typical power spectral density, with a low-frequency component that increases below some frequency f_L (typically 1 to 10 Hz) and a "white noise" component at higher frequencies. If a filter with transfer function

$$H(j\omega) = H(j2\pi f)$$

is used to observe the noise, then the apparent rms value of the noise will be dependent on the filter bandwidth. In this case, if the filter output is y, then

$$y_{\rm rms} = \sqrt{\int_{f_1}^{f_2} |H(j2\pi f)|^2 G(f)\, df} \qquad (2.108)$$

We can often approximately evaluate this last interval from a knowledge of the frequency range where the major contribution of noise is present. For example, in Fig. 2.24b, the filter function H passes primarily frequencies in the range from f_1 to f_2, with gain A and $f_1 > f_L$. Thus,

$$y_{\rm rms} \cong \sqrt{\int_{f_1}^{f_2} |H|^2 G\, df} \qquad (2.109a)$$

$$\cong \sqrt{(f_2 - f_1)A^2 G_m} \cong A\sqrt{f_2 G_m} \qquad (2.109b)$$

where A and G_m are as indicated in Fig. 2.24b. This last estimate will be somewhat in error, but often we are looking for an order of magnitude anyway.

General Design Equations

Using the equivalent noise model of Fig. 2.22, we can show that the rms noise voltage at the amplifier output is

$$E_{0,\rm rms} = \sqrt{\int_{f_1}^{f_2} A_n^2 [G_e + G_i(|Z_I|^2 + |Z_N|^2)]\, df} \qquad (2.110)$$

where G_e = input noise voltage power spectral density associated with generator E_n; units are volts squared per hertz

G_i = input noise current power spectral density associated with I_n; units are amperes squared per hertz

A_n = noise gain of circuit in volts per volt

The factor

$$G_e + G_i(|Z_I|^2 + |Z_N|^2) \qquad (2.111)$$

is an *equivalent input noise voltage power spectral density* which takes into account both the voltage and the current components of the input noise. We see from (2.111) that the amplifier output noise will be affected by the impedance level of the associated circuit, and that the noise can be reduced by keeping

$$G_i(|Z_I|^2 + |Z_N|^2) < G_e \qquad (2.112)$$

The noise gain A_n is in general

$$A_n = \frac{1}{\beta}\frac{A\beta}{1 + A\beta} \qquad (2.113)$$

In many cases, the effective bandwidth of the noise gain is small compared to the operational amplifier bandwidth itself, and we can assume $A \gg 1$ throughout the frequency range of interest. If this is the case,

$$A_n \cong \frac{1}{\beta} \qquad A \gg 1 \qquad (2.114)$$

As we have seen before, $1/\beta$ is approximately proportional to the circuit voltage gain [see, e.g., (2.61), (2.65), and Example 2.4]. Hence amplifier circuits with a high voltage gain will have a greater output noise level, as expected. It is obvious from (2.110) that one should shape the amplifier closed-loop transfer function to transmit only those signals that really have important spectral content, and to reject noise outside of the frequency range of interest.

One should not normally attempt to evaluate an integral of the form of (2.110) with $f_1 = 0$, since that implies that an infinitely long time of observation is involved. In particular, if the noise has a $1/f$ component, then since

$$\int_0^F \frac{K}{f} \, df = \infty$$

we will get an estimate of infinite rms noise!

A more reasonable approach is to use a fairly small value of f_1, for example, 0.001 or 0.01 Hz, in evaluating integrals of the (2.110) type. Lower values of f_1 imply very long observation times, while other types of drifts due to aging, temperature changes, etc., are probably more important in creating very-low-frequency error sources, and a dc offset analysis is more appropriate.

Also one need not use a value of f_2 greater than the frequency where A_n becomes small. At high frequencies, one should limit f_2 to the frequency where $|A\beta|$ finally drops below 1.

Amplifier Noise Specifications

There are a variety of ways in which manufacturers specify the noise properties of their amplifiers. Most of them will give separate voltage and current parameters. Some will give details of the spectral density and its variations with frequency. More commonly, however, one finds specifications for the effective rms value of the noise over a range of frequencies. In addition, one will find separate figures for a low-frequency range (for example, 0.01 to 10 Hz), where the noise is presumably of the $1/f$ variety, and a mid-frequency range (for example, 10 Hz to 10 kHz), where the noise spectral density is typically white, i.e., of constant intensity. We will here attempt to interpret some typical specifications in terms of the terminology of this section.

Table 2.3 lists a typical set of manufacturer's specifications for amplifier input noise. Implied in these specifications is a noise voltage power spectral density of the form

$$G_e(f) = \frac{K_{e1}{}^2}{f} + K_{e2}{}^2 \qquad (2.115a)$$

and a current power spectral density

$$G_i(f) = \frac{K_{i1}^2}{f} + K_{i2}^2 \quad (2.115b)$$

that is, there is a $1/f$ component implied in the low-frequency specification and a constant component implied in the mid-frequency specification. This may only be approximately true, but this is about all we can infer from the given specifications. Note that it is common to give the low-frequency specification in terms of a peak-to-peak value, but the mid-frequency specification is usually given in terms of an rms value. Lacking any other information, we may assume that the peak-to-peak value is 6 to 7 times the rms value; let us use 6.5 times here. (Peak-to-peak value is usually the range wherein the signal lies P percent of the time, where P is typically between 99 and 99.9 percent.) If we assume a $1/f$ variation for the low-frequency spectrum and a white spectrum for the mid-frequency spectrum, then we can find the constants in (2.115) as follows:

Let E_{n1} and I_{n1} be the low-frequency range (f_1 to f_2) peak-to-peak specifications of noise voltage and current, and let E_{n2} and I_{n2} be the mid-frequency range (f_3 to f_4) rms specifications; then

$$E_{n1} = \sqrt{\int_{f_1}^{f_2} \frac{K_{e1}^2}{f} 6.5 \, df} \quad \text{(peak to peak)}$$

or

$$K_{e1} = \frac{E_{n1}}{6.5\sqrt{\ln (f_2/f_1)}} \quad (2.116a)$$

Similarly,

$$K_{i1} = \frac{I_{n1}}{6.5\sqrt{\ln (f_2/f_1)}} \quad (2.116b)$$

also,

$$E_{n2} = \sqrt{\int_{f_3}^{f_4} K_{e2}^2 \, df} \quad \text{(rms)}$$

or

$$K_{e2} = \frac{E_{n2}}{\sqrt{f_4 - f_3}} \cong \frac{E_{n2}}{\sqrt{f_4}} \quad (2.117a)$$

and

$$K_{i2} = \frac{I_{n2}}{\sqrt{f_4 - f_3}} \cong \frac{I_{n2}}{\sqrt{f_4}} \quad (2.117b)$$

Table 2.3   TYPICAL AMPLIFIER NOISE SPECIFICATIONS (BURR-BROWN 3357/15)*

Frequency range	Voltage, μV	Current, pA
0.01 Hz to 10 Hz	0.5 p–p	20 p–p
10 Hz to 10 kHz	2 rms	2 rms

*Low-frequency specification is typically given in terms of peak-to-peak value, mid-frequency in terms of rms value.

EXAMPLE 2.9 Given the specifications of Table 2.3, estimate the power spectral density of the amplifier noise voltage and current. Here we want to take the original specifications and convert them to the constants of (2.115). The low-frequency noise constants K_{e1} and K_{i1} are found using (2.116); that is,

$$K_{e1} = \frac{0.5 \times 10^{-6}}{6.5\sqrt{\ln{(10/0.01)}}}$$

$$= \frac{0.6 \times 10^{-6}}{6.5 \times 2.63}$$

$$= 2.92 \times 10^{-8} \text{ V}$$

and

$$K_{i1} = \frac{20 \times 10^{-12}}{6.5 \times 2.63}$$

$$= 1.17 \times 10^{-12} \text{ A}$$

From (2.117) we estimate the mid-frequency parameters K_{e2} and K_{i2}:

$$K_{e2} = \frac{2 \times 10^{-6}}{\sqrt{10^4 - 10}} = 2 \times 10^{-8} \text{ V}/\sqrt{\text{Hz}}$$

and

$$K_{i2} = \frac{2 \times 10^{-12}}{\sqrt{10^4 - 10}} = 2 \times 10^{-14} \text{ A}/\sqrt{\text{Hz}}$$

Note the units of the above parameters. The low-frequency parameters ($1/f$ noise) are correctly specified in volts or amperes; the mid-frequency (white noise) parameters are correctly specified in volts per root-hertz or amperes per root-hertz.

The power spectral density functions are then obtained from (2.115):

$$G_e = \frac{8.4 \times 10^{-16}}{f} + 4 \times 10^{-16} \text{ V}^2/\text{Hz} \qquad (2.118a)$$

$$G_i = \frac{1.4 \times 10^{-24}}{f} + 4 \times 10^{-28} \text{ A}^2/\text{Hz} \qquad (2.118b)$$

These functions may be used in (2.110) to estimate the rms output noise of a particular circuit. $////$

We should note here that a model for noise spectral density such as is implied by (2.115) may be considerably in error. In particular, we have not treated burst or popcorn noise in the simple model of Eq. (2.115). Its spectrum is usually of the form

$$\frac{K_{e3}}{1 + (f/f_b)^2}$$

where f_b is typically 200 to 300 Hz (see Ref. 10). In critical low-noise applications, proper selection procedures should be instituted to ensure that units with significant burst noise are rejected.

To make realistic estimates of the noise effects in a specific circuit with a specific amplifier, one should have detailed plots of the input noise voltage and current spectra (usually given in volts per root-hertz and amperes per root-hertz, i.e., plots of $\sqrt{G_e}$ and $\sqrt{G_i}$ versus f. Without such detailed information, one can make only very rough estimates. Reference 5 presents an excellent description of the use of graphical methods for estimating effects, given the details of the input noise spectral density; we will not attempt to present these techniques here, since this type of detailed information is often unavailable. We will, however, present one example of a typical analysis, based upon equations of the form of (2.110) and (2.115).

EXAMPLE 2.10   Suppose we have the simple inverting low-pass filter circuit of Fig. 2.25a, to be used with the amplifier having the noise specifications of Table 2.3. The circuit impedance level will be determined by the value of R; C is then chosen to give a 10-kHz cutoff frequency. Let us estimate the output rms noise, assuming a noise spectrum as given by the results of Example 2.9, i.e., in (2.118). To be realistic, let us assume that the amplifier has a unity gain bandwidth of 1 MHz and a dc open-loop gain of 10^4. Figure 2.25b shows a plot of the noise gain (heavy curve). Note that the amplifier open-loop gain affects the noise gain curve at high frequencies.

In order to *simplify* the estimation of the noise output, we will approximate the noise gain as follows:

$$A_n \cong \begin{cases} 11 & 10^{-3} < f < 10^4 \\ 1 & 10^4 < f < 10^6 \\ 0 & \text{otherwise} \end{cases} \quad (2.119)$$

This seemingly crude approximation will be acceptable here, since we only have a rough idea of the input noise spectrum to begin with.

We will now proceed to evaluate (2.110) for this circuit. To facilitate this, we will use the following relationships:

$$\int_{f_a}^{f_b} \frac{K^2 \, df}{f} = K^2 \ln (f_b/f_a) \quad (2.120a)$$

or if

$$f_b = 10Nf_a$$

$$\int_{f_a}^{10Nf_a} \frac{K^2 \, df}{f} = 2.3K^2 \ln N \quad (2.120b)$$

also,

$$\int_{f_a}^{f_b} K^2 \, df = (f_b - f_a)K^2 \quad (2.120c)$$

and

$$\int_{f_a}^{f_b} K^2 \, df \cong f_b K^2 \qquad f_b > 10f_a \quad (2.120d)$$

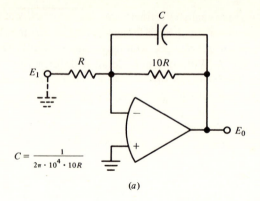

$$C = \frac{1}{2\pi \cdot 10^4 \cdot 10R}$$

(a)

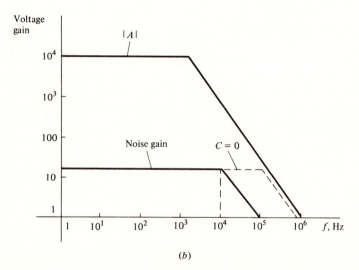

(b)

FIGURE 2.25
Refer to Example 2.10. (a) Simple inverting low-pass filter with a 10-kHz cutoff frequency; (b) plots of noise gain and amplifier open-loop gain (magnitude only).

The equivalent series impedance in the inverting terminal branch is approximately

$$Z_I \cong \begin{cases} \frac{10}{11}R & f < 10^4 \\ 0 & 10^4 < f \end{cases} \qquad (2.121)$$

Again, we are using a fairly crude representation, since we are seeking a rough estimate.

Using (2.118) through (2.121) in (2.110), we have

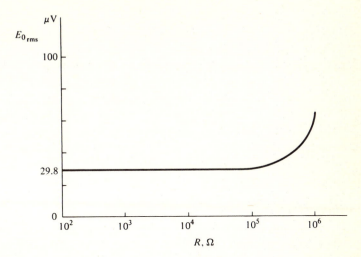

FIGURE 2.26
Effect of impedance level on the output noise produced by the circuit of Fig. 2.25.

$$E_0{}^2 \cong \left(11^2 \ln \frac{10^4}{10^{-3}} + 1^2 \ln \frac{10^6}{10^4}\right) \times 8.4 \times 10^{-16}$$

$$+ \left(11^2 \ln \frac{10^4}{10^{-3}}\right) \times 1.4 \times 10^{-24} \left(\frac{10}{11}\right)^2 R^2$$

$$+ [11^2(10^4 - 10^{-3}) + 1(10^6 - 10^4)] \times 4 \times 10^{-16}$$

$$+ 11^2(10^4 - 10^{-3}) \times 4 \times 10^{-28} \left(\frac{10}{11}\right)^2 R^2$$

$$\cong (121 \times 7 \times 2.3 + 1 \times 2 \times 2.3) \times 8.4 \times 10^{-16}$$

$$+ (121 \times 7 \times 2.3) \times 1.16 \times 10^{-24} R^2$$

$$+ (1.21 \times 10^6 + 10^6) \times 4 \times 10^{-16}$$

$$+ 1.21 \times 10^6 \times 3.31 \times 10^{-28} R^2$$

or

$$E_0{}^2 \cong 1.64 \times 10^{-12} + 2.26 \times 10^{-21} R^2 + 8.84 \times 10^{-10} + 4 \times 10^{-22} R^2$$

Thus
$$E_0{}^2 \cong 8.86 \times 10^{-10} + 0.266 \times 10^{-20} R^2 \qquad V^2 \qquad (2.122)$$

Figure 2.26 shows a plot of E_0 versus R. We note that for values of R less than about 10^5 Ω, the output noise is 29.8 μV rms. As R is increased above the value, the effect of the input *current noise component becomes important.*   ////

From Example 2.10 we see that the choice of impedance level affects the noise level. Note also that if we introduce a dc offset compensating resistor $(10/11R)$ in the noninverting terminal, we increase the noise level, i.e., we replace R^2 by $2R^2$ in (2.122).

If the circuit of Fig. 2.25 must be built with large resistance values, then it would be advisable to use an amplifier with a lower noise current. Amplifiers with an FET or varactor input stage are usually preferred for this case. Note also that if the capacitor is omitted in Fig. 2.25, the output noise will be greater (see Prob. 2.21). It is important to restrict the bandwidth of amplifiers to only the frequency range where useful signal components lie in order to reduce the effects of amplifier noise.

2.7 LARGE-SIGNAL LIMITATIONS

In this section we will discuss a variety of nonideal amplifier performance measures that relate to the amplifier large signal. These include output limitations such as slewing rate, full-power bandwidth, settling time, overload recovery, and rated common-mode input voltage. Normally, the specifications are derived from large-signal testing of typical units, and they may not apply with high accuracy to all circuit configurations.

Rated Output

We mentioned the basic limitations on amplifier output load-driving capabilities as early as Chap. 1; they seldom can be overlooked. These ratings specify the *maximum* allowable output *voltage swing* (at low frequencies). The rating is usually given under the assumption that an output load is connected which causes the *maximum* or peak *output current* to flow at the rated peak output voltage. Often the output voltage swing is a result of the output current-limiting effects, and the two output signal ratings are interdependent. Complete specifications should include a graph of peak voltage versus peak load current. In many cases, however, the manufacturer will only specify the rated output current (assumed ±) over a maximum nominal output voltage range, for example, ±20 mA and ±10 V. The ±10-V range is typical of most contemporary operational amplifiers, but the current rating will vary greatly, depending upon the intended range of applications (and cost) of the unit.

If the rated output is exceeded, the amplifier will go into an *overload* condition. If the amplifier output stage is properly designed (usually called short-circuit overload protection), then no permanent damage will occur, but there will, of course, be a temporary high distortion of the output signal, and there may be an appreciable overload recovery time. Note that if an amplifier is allowed to remain in an overload condition for a long time, internal heat buildup may cause the input stage to develop significant internal dc offsets, which may take a long time to subside.

Slewing Rate and Full-Power Response

These limitations are primarily associated with internal rate limitations in the amplifier. Typically the amplifier input stage can only supply a certain maximum amount of current

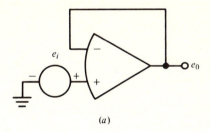

(a)

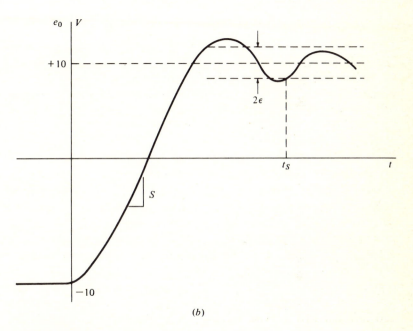

(b)

FIGURE 2.27
Circuit for making large-signal tests. (a) Circuit for measuring slewing rate and
full-power response (settling-time measurement); (b) typical output response.

to charge associated stray capacitances. The general result is that the large-signal dynamic
behavior of the amplifier is degraded, as compared with the small-signal response. The
circuit of Fig. 2.27 may be used to measure this effect; test results are usually
summarized by two specifications: slewing rate and full-power bandwidth.

If the circuit of Fig. 2.27a is used with a ±10-V square wave (with fast rise time),
we get a rigorous test for slewing rate. As shown in Fig. 2.27b, the amplifier output
waveform will exhibit a rate limitation. The maximum rate of change of the output
voltage under these conditions is often taken as a measure of slewing rate, that is,

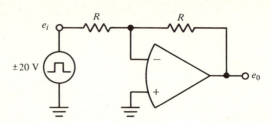

FIGURE 2.28
Measuring overload recovery; input is driven to produce 100 percent overload.

$$S = \left| \frac{de_0}{dt} \right|_{max} \qquad \text{V/s} \qquad (2.123)$$

Usually the positive and negative maximum rates will differ, in which case the lower slewing rate should be specified.

Full-power bandwidth may also be measured with the circuit of Fig. 2.27a, but using a variable-frequency sine-wave input and adjusting the input amplitude to produce a ±10-V output waveform. As the frequency is increased, one will observe a point where distortion appears in the output waveform, as a result of rate limiting. The frequency at which this distortion becomes apparent is usually taken as a measure of the *full-power bandwidth* f_p. The slewing rate provides an *upper bound* on this figure, that is,

$$f_p \leqslant \frac{S}{2\pi E_{op}} \qquad \text{Hz} \qquad (2.124)$$

where E_{op} is the peak output swing (for example, 10 V or less). Hence, (2.124) may be used to estimate the maximum large-signal frequency that may be accommodated at a given output voltage level.

Note that (2.124) is an inequality; in many amplifiers approximate equality holds, while in others which have been specially designed for fast step response, f_p may be somewhat less than the value implied by (2.124).

EXAMPLE 2.11 Given a ±10-V amplifier with a slewing rate of 10^6 V/s, estimate the full-power bandwidth. Also estimate the maximum frequency at which one can output a 2-V peak sine wave.

From (2.124), we find f_p with $E_{op} = 10$ V:

$$f_p = \frac{10^6}{2\pi \times 10} = 15.9 \text{ kHz}$$

Similarly, with $E_{op} = 2$ V,

$$f_p = 80 \text{ kHz} \qquad ////$$

Settling Time and Overload Recovery

In the slewing rate test of Fig. 2.27b, we will often observe a small oscillatory transient associated with the large-signal step response of the amplifier. The *settling time* t_s is the total time t_s required for the output to settle to within a specified percentage of its final value. Thus, in Fig. 2.27b, if the desired percentage is 0.1 percent (of 10 V), then ϵ is 0.01 V. Settling time is an important specification for amplifiers which are to be used in a switching mode (e.g., in D/A converters, sample-hold circuits, etc.).

Additional useful information is conveyed by a specification of overload recovery time. This is the time the amplifier takes to return to linear operation after being driven into overload. For example, if the pulse response of the circuit of Fig. 2.28 is measured with the input increased so that overdrive is produced, then additional delay due to overload recovery will be observed in the output response. Typically, the overload recovery is measured using 100 percent overdrive.

Maximum Common-Mode Input Voltage

This is a specification of the maximum magnitude of voltage that may be applied to either input terminal of a differential-input operational amplifier. In Fig. 2.16a, this would be the maximum value of e_i that still yielded reasonably linear operation. Typical values will be ±10 to 11 V.

2.8 OUTPUT POWER BOOSTING

Occasionally one encounters a situation where an otherwise suitable operational amplifier cannot provide the required output load driving. That is, either the available output current is insufficient or the allowable output swing will not accommodate the signal levels anticipated at the output. Frequently this problem can be resolved by using an output power booster (Fig. 2.29).

Here the booster amplifier is a single-ended amplifier with the following general properties:

1 An output voltage driving capability exceeding that required by the load $v_{0,\max}$.
2 An output current driving capability exceeding that required by the load, $i_{0,\max}$.
3 A full-power bandwidth (see Sec. 2.7) consistent with the anticipated load and signal frequency content (or else a slew rate that meets anticipated requirements).
4 An input impedance that can be accommodated by the driving operational amplifier.
5 A forward-voltage gain of at least G_v from direct current up to some high frequency well beyond the anticipated range. Here G_v must satisfy

$$G_v > \frac{v_{0,\max}}{v_{A,\max}}$$

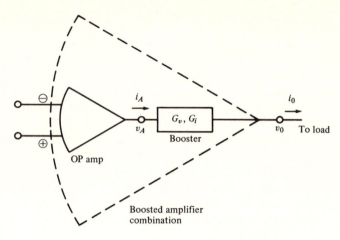

FIGURE 2.29
Boosting amplifier output.

where $v_{A,\max}$ is the maximum output voltage of the driving operational amplifier.

6 Implicit in *4*, above, is a required booster current gain G_i such that

$$G_i > \frac{i_{0,\max}}{i_{A,\max}}$$

where $i_{A,\max}$ is the maximum rated output current of the driving operational amplifier.

7 The small-signal bandwidth of the booster must be broad enough to prevent significant extra phase shift in the overall loop gain $AG_v\beta$ when the combined operational amplifier–booster combination is placed in an appropriate feedback network; i.e., the overall stability of the final circuit must be preserved.

Note that so long as these general conditions are met, the booster amplifier can be rather sloppy in performance since it is in the feedback loop. In Fig. 2.30*a*, the feedback will suppress the dc offset inherent in the emitter follower being used as a current booster.

Often only current gain is required of the booster, i.e., the necessary output voltage swing is no more than that of the driving operational amplifier, for example, ±10 V. To meet this situation, many commercial operational amplifier manufacturers provide prepackaged booster amplifiers with only unity voltage gain, but with a fairly high current gain, for example, 10 to 100, and a husky output current, for example, 100 to 500 mA. These units may be used to boost a typical operational amplifier's current driving capability.

Note that if the required voltage swing exceeds that of the operational amplifier, this normally means that the booster amplifier will have to be powered by a new power system with large supply voltages. It is, nevertheless, feasible to use the boosting scheme

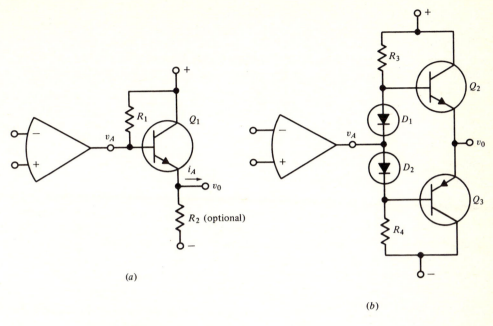

FIGURE 2.30
Some simple booster circuits. (*a*) A transistor emitter follower; (*b*) a complementary symmetry output booster improves bipolar operation.

of Fig. 2.29 if the feedback network properly restricts the terminal potentials of the driving operational amplifier. Thus, for example, a ±10-V operational amplifier could conceivably drive a ±100-V power booster amplifier driving some sort of servomotor.

Figure 2.30 shows two fairly simple boosting circuits that often will do the job. (Design details are not given, since they depend on the specific application.) The emitter-follower circuit of Fig. 2.30*a* is particularly handy. It may be used whenever the output current is essentially of one polarity. The *n-p-n* transistor shown would be used when i_A is positive. Note that v_0 could be bipolar. The emitter load resistor R_2 would be optional; it serves to "keep the circuit alive" under no-load conditions. Base resistor R_1 is chosen to set the base current so as to keep transistor Q_1 in the active region.

In some situations, we may need a booster that can both deliver and absorb current. In this event, one can often use some form of complementary-symmetry amplifier stage. Figure 2.30*b* shows the general form of such a stage with a unity voltage gain. If the required voltage gain is greater than 1, something more complex is required. The two simple boosters of Fig. 2.30 provide a current gain of the order of the transistor current gain, usually called β_F or h_{FE}. Higher current gains can be achieved using Darlington-connected transistors or some form of multistage circuit. Commercial manufacturers currently make hybrid current-booster circuits (see, e.g., National Semiconductor Corp. NH0002).

2.9 POWER SUPPLIES, GROUNDING, AND SHIELDING

The topics in this section could well have an entire book devoted to them. We shall attempt here to present only some fundamental points that should be considered in selecting a power supply and laying out grounds and shielding. References 1 and 6 to 8 provide considerable detail.

Power Supplies for Operational Amplifiers

We have already indicated that most operational amplifiers normally must be powered by a pair of symmetrical dc voltages, for example, ±15 V dc; a common ground point should be available (Fig. 1.2). We will have more to say about grounding below. The power-supply voltages should be equal in magnitude (matched to 1 percent, for example); more important, the voltages should be well regulated over a *wide range of frequencies*. Long-term drift of the power-supply potentials will create dc offset drifts in the amplifiers. Many manufacturers will specify *power-supply rejection ratios* indicating how much the input offset components e_{os} and i_{os} (Sec. 2.6) will drift with supply voltage. If the two voltages tend to track each other, i.e., if the magnitudes stay equal, but change, then the drift effects will be reduced.

Good high-frequency regulation is very important. For broadband operational amplifiers, the power-supply internal impedance should be very low (e.g., less than 0.1 Ω) over the entire frequency range where the amplifiers have any appreciable gain. Most operational amplifiers have class B output stages; i.e., they draw a low power-supply current when their output current is zero, but they will draw a fairly large current when their output current is high. Further, the current demand will not be symmetrical. In Fig. 2.31, we see a sketch of a typical situation. If the amplifier load current is a bipolar sinusoid (Fig. 2.31*b*), then alternating pulses of current will be drawn from the power supplies (Fig. 2.31*c* and *d*). The output waveform can conceivably have a frequency in the 10- to 100-kHz range, even at high signal levels; hence the power-supply current waveforms will contain relatively strong high-frequency components. Any power-supply internal impedance will cause a high-frequency ripple to be induced in the power-supply buses.

From the above discussion, we see that it is advisable to minimize the high-frequency impedance of power-supply connections. Figure 2.32 shows a recommended technique. A separate heavy (e.g., No. 10) copper bus is run from a common tie point near the power supply to each part of a system where a collection of amplifiers is located. That is, we do not string a long "party line" with a large number of amplifiers all being fed from the same bus. Typically each group of three to six amplifiers would have its own power bus. In this way we reduce the common paths for power-supply currents. In addition, it is advisable to place bypass capacitors at each amplifier location to shunt high-frequency currents around the busing system. The capacitors should provide a good RF bypass, e.g., a 0.01-μF ceramic type would be typical (Fig. 2.32*b*).

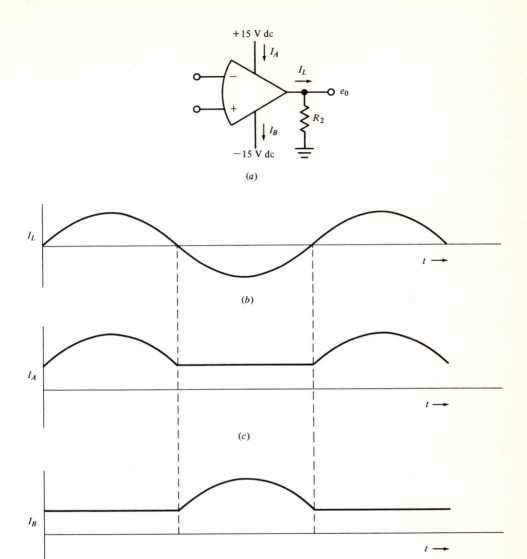

FIGURE 2.31
Power-supply demands. (*a*) Output circuit of an operational amplifier; R_L is total output load. (*b*) Typical amplifier output current waveform; (*c*) and (*d*) are associated power bus currents.

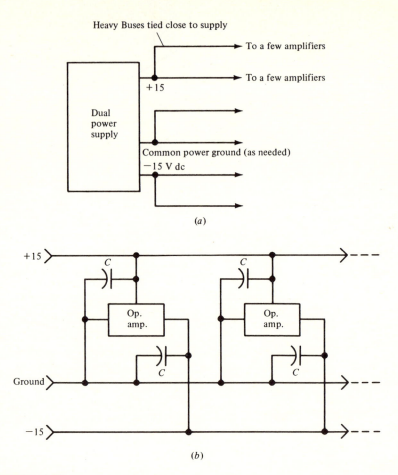

(a)

(b)

FIGURE 2.32
Technique for minimizing impedance of power-supply connections. (a) Use of separate buses for each group of amplifiers; (b) decoupling at each amplifier location via capacitors.

Shielding

The subject of shielding for long low-level signal lines is a complex one. If possible, it is best to use a differential system with both signal lines floating symmetrically with respect to signal ground, e.g., as in Fig. 2.33a. Twisted pairs make fairly noise-immune signal paths. Better yet, it is helpful to enclose the signal lines in a coaxial shield, which is then grounded *only at one point*, to avoid ground loops. It is better if the shield itself carries no signal current. Thus, even though we are transmitting single-ended signals, as in Fig. 2.33b, it is still better to use a separate signal ground return inside of the shield.

It is important to ground the shield properly. Often the best ground point must be determined experimentally, since the point of grounding can strongly influence the

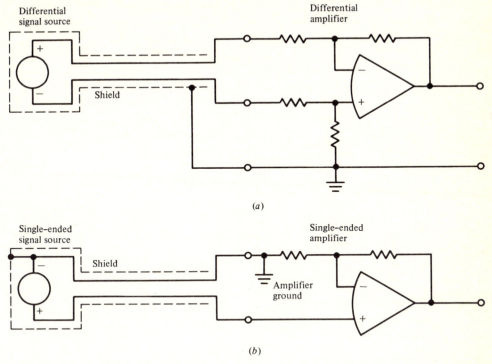

(a)

(b)

FIGURE 2.33
Shielding against externally induced noise. (*a*) Differential amplifier and floating
signal source; note shield is grounded at only one point; (*b*) single-ended signal
source; note use of separate ground return for signal.

common-mode voltage induced from external fields. Normally the shield should be
connected to the signal source ground for single-ended sources (Fig. 2.33*b*).

Ground Systems

Figure 2.34 shows a typical plan for an effective grounding system which has many
desirable features. It contains an *earth ground* (use the power-line ground, if nothing else
is available), preferably shielded until it enters the earth to keep it from being an antenna!
A *common tie point* is used where all ground lines are brought together in a fanlike
array, to avoid ground loops. Note that the chassis ground is separate from all other
grounds except at this point. Shield grounds are brought in individually, as are signal
grounds and power bus grounds. Figure 2.34 represents perhaps an extreme case, but the
general technique is useful in avoiding crosstalk problems in complex signal-processing
systems. Much misery may be avoided if an adequate grounding system is planned as part
of the initial design effort.

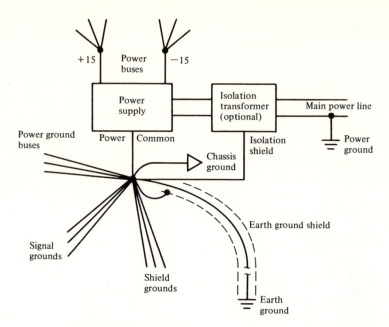

FIGURE 2.34
General layout of a grounding system to eliminate ground loops.

2.10 COMPONENTS: RESISTORS AND CAPACITORS

Operational amplifiers may be incorporated into a wide variety of circuit designs. Discrete and hybrid (thin-film/thick-film and IC) packages are more commonly used for precision circuits, although all-monolithic circuits are feasible.

Resistors

The three most common types are carbon, metal film, and wire-wound. Carbon-composition resistors are used in low-precision discrete circuits. They have a poor temperature coefficient and seldom have the tolerance and aging characteristics required for precision circuits. Metal-film resistors have good temperature coefficients (+10 ppm/°C) and can usually be obtained with stable, accurate values (for example, 0.1 to 1 percent) at modest cost. For very-high-accuracy resistors intended for use in precision instruments and analog computers, wire-wound (noninductive) resistors often must be employed, even though they are considerably more expensive.

For high-frequency circuits (for example, 10 kHz and above) one must worry about stray reactive components. Here the metal-film units are preferred; typically they will have little parasitic inductance (0.01 μH), and a parallel end-to-end capacitance of perhaps 0.5 pF for a 1/2-W unit.

In microcircuit applications, resistors may be fabricated by standard monolithic IC methods (diffused and pinch types) or by thin- or thick-film methods. Here one cannot usually achieve high absolute-value accuracy, so that the clever designer will attempt to take advantage of the fact that resistance *ratios* can usually be fairly well matched. Table 2.4 shows some typical parameters of microcircuit resistors. Note that, when required, thin-film resistors may be given precision values by temperature annealing and laser trimming techniques.

Capacitors

Accurate, low-loss capacitors are fairly expensive, particularly in the larger values. One must look carefully into specifications of dissipation coefficient (or power factor) and temperature coefficient. Large-value (over 0.1 μF) low-loss precision units preferably have polystyrene dielectric, although Mylar and Teflon are often acceptable. In the smaller sizes, one may choose from Mylar, mica, NPO ceramic, glass, and polycarbonate units. Cheap ceramics are probably not suitable for any precision work. Electrolytic capacitors are seldom useful, since they are unipolar.

The dissipation coefficient, or power factor, is a measure of the effective resistance component in the capacitor. If the coefficient is D, then the effective series resistance at f Hz is

$$R_s = \frac{D}{2\pi f C}$$

The corresponding effective parallel resistance is

$$R_p = \frac{1}{2\pi f C D}$$

Dissipation factors may typically range from 1×10^{-4} for Teflon or polystyrene to 50×10^{-4} for ceramic and polycarbonate dielectrics (see Ref. 1, Table 3.1, and Ref. 4, Table 8.1).

Table 2.4   TYPICAL MICROCIRCUIT RESISTOR PARAMETERS*

Type	Range, Ω	Tolerance, %	Matching, %
Base-diffused	500–30 kΩ	±10	±0.5
Emitter-diffused	20–500	±15	±0.5
Base pinch	10–200 kΩ	±50	±5
Collector pinch	10–500 kΩ	±50	±5
Thin-film	500–100 kΩ	± 2	±0.5†
Thick-film	1–10 MΩ	±20‡	±5

*These figures were provided by Professor D. J. Hamilton, University of Arizona, Department of Electrical Engineering.
†May be trimmed to better than 0.05 percent.
‡May be trimmed to 0.5 percent.

REFERENCES AND BIBLIOGRAPHY

1 KORN, G. A., and T. M. KORN: "Electronic Analog and Hybrid Computers," 2d ed., McGraw-Hill, New York, 1972.

2 GUPTA, S. L., and L. HASDORFF: "Fundamentals of Automatic Control," Wiley, New York, 1966.

3 HUELSMAN, L. P.: "Basic Circuit Theory with Digital Computations," Prentice-Hall, Englewood Cliffs, N.J., 1972.

4 TOBEY, G. E., T. G. GRAEME, and L. P. HUELSMAN: "Operational Amplifiers," McGraw-Hill, New York, 1971.

5 Noise and Operational Amplifier Circuits, *Analog Dialog*, vol. 3, no. 1, Analog Devices, Inc., March 1969.

6 MORRISON, R.: "Grounding and Shielding Techniques in Instrumentation," Wiley, New York, 1972.

7 STEWART, E. L.: Grounds, Grounds, and More Grounds, *Simulation*, August 1965, pp. 121–128.

8 ———: Noise Reduction on Interconnect Lines, *Simulation*, September 1965, pp. 149–155.

9 PAPOULIS, A.: "Probability, Random Variables and Stochastic Processes," McGraw-Hill, New York, 1965.

10 BRODERSON, A. J., E. R. CHENETTE, and R. C. JAEGER: Noise in Integrated-Circuit Transistors, *IEEE Journal of Solid-State Circuits*, vol. SC-5, no. 2, pp. 63–66, April 1970.

PROBLEMS

2.1 *(Sec. 2.1)* Given the following operational amplifier specifications:

Open-loop gain at dc   100 dB
Unity gain frequency   2 MHz

find $A'(s)$ using Eq. (2.2).

2.2 *(Sec. 2.2)* Suppose a simple unity gain follower is constructed using an operational amplifier with these specifications:

Open-loop gain at dc   100 dB
Unity gain frequency   2 MHz

Assume zero output impedance and a negligibly large differential input impedance; use a simple one-pole model of the amplifier. Make a Bode plot of the magnitude of the circuit closed-loop gain; what is the −3-dB frequency of the circuit?

2.3 *(Sec. 2.2)* For this problem, and the two to follow, assume that the following operational amplifier characteristics are important:

Unity gain frequency f_u 2 MHz
Open-loop gain at dc A_0 100 dB
Differential input impedance Z_I 10 MΩ
Output impedance Z_0 1 kΩ

in terms of A_0, f_u, and G. What do you get as $A_0 \to \infty$?

(b) Find the -3-dB closed-loop frequency of the circuit in terms of A_0, f_u, and G; in particular, what do you get if $A_0 \to \infty$?

2.7 *(Sec. 2.2)* Refer to the circuit of Fig. P2.7; here we assume that the input impedance of the operational amplifier is negligibly high below 100 kHz. This circuit can then be used to estimate the open-loop gain of the operational amplifier. We will also assume that the output loading is negligible. The following test results are obtained:

1 V dc input:

v_0, V dc	v_x, V dc
-0.997	0.000997

1 V ac input:

Frequency, Hz	v_0, V rms	v_x, V rms
10	0.997	0.00111
20	0.997	0.00141
50	0.997	0.00268
100	0.997	0.00508
200	0.997	0.0100
500	0.994	0.0249
1,000	0.986	0.0493
2,000	0.955	0.0955
5,000	0.798	0.200
10,000	0.554	0.277
20,000	0.316	0.316
50,000	0.132	0.330
100,000	0.0665	0.333

(a) Estimate the open-loop gain of the amplifier, i.e., estimate the dc value of the open-loop gain A_0, and the unity gain frequency f_u.

(b) Note that it is difficult to measure the voltages in some frequency ranges. At low frequencies, v_x is small in magnitude; at high frequencies, v_0 is small. Careful shielding is required to make meaningful measurements. Suggest

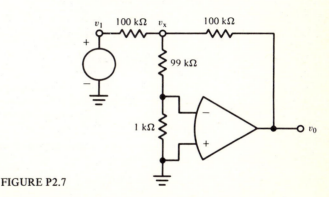

FIGURE P2.7

alternative circuits that could be used at very low and very high frequencies to improve the signal levels.

2.8 *(Sec. 2.3)* Show that if the rate of closure of Bode magnitude plots of A and $1/\beta$ is exactly 6 dB/octave, and the curves intersect at $\omega = \omega_0$, and there are no nearby corner frequencies, then the closed-loop gain of the associated circuit is 3 dB less than the desired gain at $\omega = \omega_0$ (refer to Fig. P2.8).

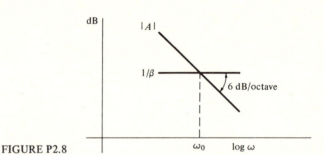

FIGURE P2.8

2.9 *(Sec. 2.3)* Design an operational amplifier circuit to implement

$$v_0 = -(v_1 + 10v_2)$$

i.e., a two-input summer-inverter. Assume perfect resistors, but all less than 100 kΩ in value. Assume the operational amplifier has zero output impedance and negligibly large differential input impedance.

(*a*) Specify the minimum value of dc open-loop gain required to achieve an overall gain accuracy of 0.1 percent at direct current.

(*b*) Determine the minimum unity gain frequency in hertz for the operational amplifier required to ensure that the -3-dB frequency of the actual circuit is at least 100 kHz (assume a simple one-pole model for A).

(*c*) Using the values of A_0 and f_u obtained above, at what frequency (in hertz) is the circuit accurate to 1 percent?

2.10 *(Sec. 2.3)* Refer to Fig. P2.10. Assume the operational amplifier has a negligibly large differential input impedance, but with:

Open-loop gain at dc	100 dB
Unity gain frequency	1 MHz

(*a*) Find the fractional error in the voltage gains at dc, compared to the ideal amplifier case; express the answer as a percentage.

(*b*) Make a Bode plot (magnitude and phase) of $1/\beta$.

(*c*) Make a Bode plot of A on the same graphs as part *b*.

(*d*) What are the gain and phase margins?

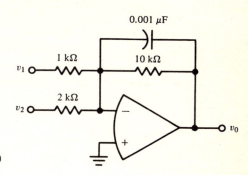

FIGURE P2.10

(e) At what frequencies in hertz are the magnitudes of the circuit gains in error by 1 percent? At what frequency are they in error by 10 percent? (Give approximate answers.)

(f) Does the fact that the operational amplifier is nonideal greatly affect the performance of this circuit as a summer, amplifier, and low-pass filter?

2.11 *(Sec. 2.3)* Refer to the integrator circuit of Fig. P2.11, with

$$R = 1 \text{ k}\Omega$$

$$C = 1 \text{ }\mu\text{F}$$

Assume that we have an operational amplifier with the following specifications:

Unity gain frequency f_u (that is, $\omega_u = 10^7$ rad/s)	1.59 MHz
Open-loop gain at dc	100 dB
Differential input impedance	Negligibly large
Output impedance	0 Ω

(a) Make a Bode plot of A and $1/\beta$; put both magnitude curves on one plot, both phase curves on another.

(b) Make a separate Bode plot for $A\beta$.

(c) Estimate the frequency range where the integrator gain accuracy is 1 percent or better (be sure to look at both low and high frequencies).

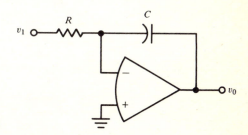

FIGURE P2.11

(d) Find the overall circuit transfer function

$$H(s) = \frac{V_0(s)}{V_1(s)}$$

2.12 *(Sec. 2.3)* Refer to Fig. P2.12; assume an ideal operational amplifier, except that

$$A(s) = A'(s) = \frac{10^4}{(1 + s/\omega_1)(1 + s/\omega_2)(1 + s/\omega_3)}$$

where $\omega_1 = 10$ rad/s

$\omega_2 = \omega_3 = 10^5$ rad/s

(a) Use Bode plot methods to find the phase margin of this circuit if it is designed to have a gain of 1; repeat for gains of 10 and 100. Are any of these cases unstable? Be sure to plot carefully; in particular, plot $|A|$ fairly precisely at the corner frequencies, rather than using a simple asymptotic corner plot. In each case, what is the approximate bandwidth of the circuit (for the stable cases)?

(b) Some operational amplifiers (notably low-cost IC varieties) permit the user to adjust the frequency response. Typically, ω_1 can be made smaller by adding an extra RC network. Repeat part a if ω_1 is moved down to 1 rad/s. Comment on advantages and disadvantages of the *compensation*.

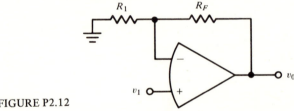

FIGURE P2.12

2.13 *(Sec. 2.3)* Refer to the differentiator circuit of Fig. 1.36b. Discuss the potential problems of using this circuit without modification. Use Bode plots to describe potential conditions for instability. Discuss the reasons why the modified circuit of Fig. 1.36c improves the stability situation.

2.14 *(Sec. 2.4)* Given an operational amplifier with the following specifications:

Common-mode rejection ratio   80 dB

Common-mode input impedance   10 MΩ

(a) Design a differential amplifier circuit (one operational amplifier) with a differential gain of 20. Pick the resistors so that the largest value is 100 kΩ.

(b) Neglecting the common-mode input impedances, estimate the common-mode gain of the circuit.

(c) If the common-mode input impedances are both equal to 10 MΩ, will this affect the differential and common-mode gains of the circuit, and if so, how much?

(HINT: Refer to Sec. 1.8.)

(d) If one of the common-mode input impedances is 10 MΩ and the other is 15 MΩ, repeat part c.

2.15 *(Sec. 2.5)* Refer to Table 2.2. Estimate the maximum values of e_{os} and i_{os} at 100°C for all four types of amplifiers. For the FET-input amplifiers, be sure to observe footnote ‡; that is, the magnitudes of the currents double every 10°C.

2.16 *(Sec. 2.5)* Refer to Fig. P2.16; modify the design to improve the sensitivity of the circuit to dc current offsets. Also add circuitry to permit balancing out up to 10 mV of input offset, in accordance with Eq. (2.102). Use standard RTMA 10 percent resistor values in your design, and assume ±15 V dc is available.

2.17 *(Sec. 2.5)* In the circuit of Fig. 2.20,

$$R_F = 100 \text{ k}\Omega$$
$$R_1 = 10 \text{ k}\Omega$$

(a) Pick R_2 for minimum dc offset.

(b) Make a table of estimated output dc offsets for this circuit, using the typical parameters of Table 2.2. Consider all four amplifier types. Consider the case of $R_2 = 0$ and R_2 equal to the optimum value from part a.

(c) Assume that some form of offset adjustment is available, so that at 25°C the offset is zero. For each amplifier of Table 2.2, find the drift that will occur if the ambient temperature changes 10°C. Consider cases where $R_2 = 0$ and R_2 equals the value from part a.

(d) Repeat all the above, but with

$$R_F = 1 \text{ M}\Omega$$
$$R_1 = 100 \text{ k}\Omega$$

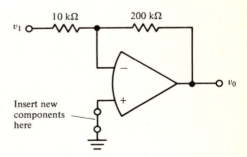

10 kΩ   200 kΩ

v_1

v_0

Insert new
components
here

FIGURE P2.16

2.18 *(Sec. 2.5)*

(a) Using the monolithic (low-cost) IC amplifier of Table 2.2, design a gain-of-20 inverting amplifier. Use the largest resistance values that yield an output dc offset drift of 0.5 mV/°C.

(b) Repeat the above design using the FET-input IC amplifier of Table 2.2.

2.19 *(Sec. 2.5)* Refer to the integrator mode control circuit in Fig. 1.13a; we want to have one input with a gain of 10 s^{-1} for integrating. The circuit uses ideal switches, resistors, and capacitors; however, the operational amplifier has the following specifications:

Dc offset voltage	$\pm 1 \text{ mV}$
Dc input bias current	$\pm 1 \ \mu\text{A}$
Dc input offset current	$\pm 0.1 \ \mu\text{A}$

(a) Using all resistors of 100 kΩ, design the circuit and estimate the effects of the dc offset currents on the three modes of operation; i.e., how much dc offset and/or drift do they create?

(b) Can you improve the situation with regard to dc offset effects by changing the values of the components to achieve a different impedance level?

2.20 *(Sec. 2.5)* Refer to Fig. P2.20; here we are interested in the effects of operational amplifier dc offset on the performance of an integrator. Basically, we find that in the absence of an external input (v_1 grounded), the output will drift with a certain maximum rate of change. Assume that the operational amplifier has dc offsets e_{os}, i_1, i_2, and i_{os}, as defined in Sec. 2.5.

(a) Derive an expression for the maximum rate of change of v_0 with time, if $v_1 = 0$.

(b) Would the addition of a resistance in the noninverting input branch of the amplifier improve this drift rate? If so, what should its value be, and how much is the drift with the resistor added?

(c) How would you add an offset adjustment network to the circuit?

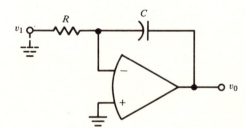

FIGURE P2.20

2.21 *(Sec. 2.6)* Using the circuit of Example 2.10, find the rms output noise for the following cases; compare your answers to the result of the example.

(a) $R = 10^4$, but C is adjusted to give a closed-loop bandwidth of 1 kHz.

(b) $R = 10^4$, $C = 0$.

2.22 *(Sec. 2.6)* Given an operational amplifier with the following noise specifications, estimate the power spectral density functions G_e and G_i.

Frequency range	Voltage, μV	Current, pA
0.001 Hz to 5 Hz	1 p-p	10 p-p
5 Hz to 5 kHz	2 rms	2 rms

2.23 *(Sec. 2.7)* Given that a particular operational amplifier has a minimum slew rate of 25×10^6 V/s, and is rated at an output of ±10 V, estimate the full-power bandwidth of the amplifier, in hertz.

2.24 *(Sec. 2.7)* Refer to Fig. P2.24. Given that a particular operational amplifier has a rated full-power bandwidth of 50 kHz at ±10 V, sketch the approximate response of the circuit to an input ideal step of +2 V.

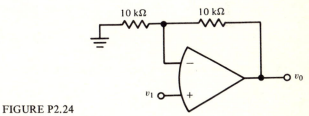

FIGURE P2.24

2.25 *(Sec. 2.7)* Refer to Fig. P2.25. Given that a particular operational amplifier has the ratings

Peak output voltage	±10 V
Peak output current	±15 mA

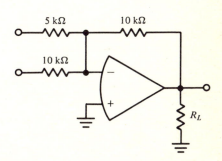

FIGURE P2.25

(a) Find the minimum value of R_L that can be driven over the full output $\pm$ voltage range; be sure to include all current drains on the amplifier.

(b) Repeat part a if R_L is connected to -15 V dc instead of to ground.

2.26 *(Sec. 2.9)* Refer to Fig. P2.26. Suppose we have an operational amplifier rated at ± 20 mA and ± 10-V output. The required currents drawn from the power buses depend on the output load current. In this case we know that

$$I_+ = \begin{cases} 0.005 + I_L & I_L > 0 \\ 0.005 & I_L < 0 \end{cases}$$

and similarly we know I_- (currents in amperes). If the parasitic inductance in the power supply bus is 0.1 μH, and the output of the amplifier is a 10-V peak sinusoid of frequency 100 kHz, find the peak ripple voltage induced in the parasitic inductances if the load R_L is drawing maximum rated output current.

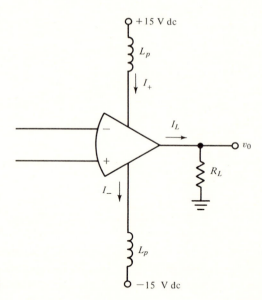

FIGURE P2.26

3

NONLINEAR CIRCUIT APPLICATIONS

In the previous two chapters, we have attempted to present a detailed picture of how an operational amplifier may be used as a general circuit element, and how nonideal effects in the amplifier may affect circuit performance. To this point, our examples have all been associated with so-called linear applications, in which simple RC networks were used to provide the desired performance of a complete circuit. If this were all we ever did with operational amplifiers, they would still be eminently useful components. In this chapter we will go a step further and describe some of the ways in which operational amplifiers may be combined with not only linear but also nonlinear circuit elements (primarily diodes and transistors) to obtain an extremely broad range of circuit functions.

In this realm of application, the clever circuit designer can use principles of feedback to suppress undesirable properties of a nonlinear element, and at the same time take advantage of desirable properties. The operational amplifier is used to do the things it does best, e.g., sensing current magnitude and direction without loading a node, and providing a low-impedance output circuit.

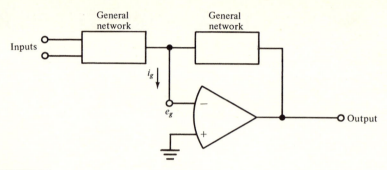

FIGURE 3.1
General nonlinear operational amplifier circuit (grounded noninverting terminal).

The number of possibilities available to the designer is boundless. We will not attempt to provide a complete catalog of all special or trick circuits that have been devised. We will, however, present a number of well-known nonlinear circuits and describe their function and design. Our hope is to thereby whet the reader's appetite and convince him that almost any analog signal-processing operation he can think of can be implemented.

Section 3.1 will describe the general nonlinear design approach, based primarily on ideal models for the operational amplifier and the associated nonlinear devices; nonideal properties will be discussed throughout the chapter, where appropriate. In Sec. 3.2, we will discuss some important first examples, the feedback limiter and the comparator. Section 3.3 discusses the general design of piecewise-linear-function generators, and also includes examples of absolute-value circuits and peak-detection and holding operators. In Sec. 3.4, the design of logarithmic operators introduces the use of the exponential law of *p-n* junction behavior. The last two sections deal with well-known applications of nonlinear operational amplifier circuits. Section 3.5 describes how nonlinear function generators may be used to implement analog squaring, multiplying, and dividing operations. Section 3.6 discusses techniques that may be used for precision waveform generation.

Again, our aim is not to provide a comprehensive presentation of every circuit ever designed; manufacturer's applications notes are often very helpful, as are the many popular industrial journals on electronic design (e.g., Refs. 1 and 2). References 3 and 4 also provide many examples and design variations.

3.1   GENERAL NONLINEAR DESIGN APPROACH

For most of our initial discussions, we still assume an ideal operational amplifier (Chap. 1, Sec. 1).

We embed such an amplifier in general (nonlinear) networks (Fig. 3.1), and assume the following:

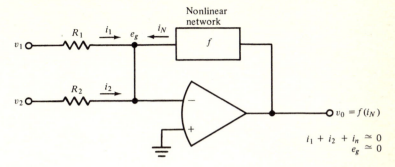

FIGURE 3.2
Refer to Example 3.1; input-output characteristic is controlled by the short-circuit conductance f.

1 The current flowing into the amplifier terminals is very small (essentially zero); $i_g \cong 0$.
2 The voltage between the amplifier input terminals is very small; typically (but not always) the noninverting terminal is grounded, and $e_g \cong 0$.

The above conditions are assumed to be true at every instant of time; the amplifier output will attempt to act upon the accompanying network to *force* the above conditions to be true. If it *cannot*, due to output current, voltage, or slewing rate limitations, then the amplifier will *overload*. This condition is normally not permitted even momentarily, since most operational amplifiers have a relatively slow overload recovery time.

EXAMPLE 3.1 The circuit of Fig. 3.2 is a general nonlinear operator which also performs a summing of two or more inputs if desired. The circuit behavior is dependent on the voltage-current characteristic

$$v_0 = f(i_N) \qquad (3.1)$$

where v_0 is the amplifier output terminal voltage and i_N is the short-circuit current that flows as indicated in Fig. 3.2. The amplifier forces $e_g = 0$, and thus

$$i_1 + i_2 + i_N = 0 \qquad (3.2)$$

Also, since $e_g = 0$,

$$i_1 = \frac{v_1}{R_1}$$

$$i_2 = \frac{v_2}{R_2}$$

and thus

$$v_0 = f\left(-\frac{v_1}{R_1} - \frac{v_2}{R_2}\right) \qquad (3.3) \qquad ////$$

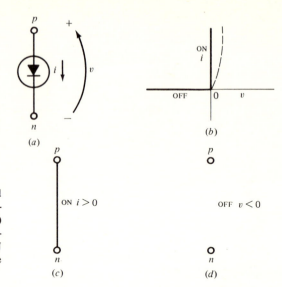

FIGURE 3.3
Ideal diode model. (*a*) Diode symbol and terminal conventions; (*b*) ideal "demon-with-a-switch" characteristic (solid line) and more realistic characteristic including forward drop (dotted line); (*c*) ON state equivalent circuit; (*d*) OFF state equivalent circuit.

Nonlinear Device Models

In many parts of this chapter, we use some form of circuit model for nonlinear devices, primarily diodes and transistors. To make our analyses tractable, we use for the most part the so-called ideal diode model. Figure 3.3 illustrates what we mean by this for a simple *p-n* junction diode. This model is often known as the "demon-with-a-switch," for fairly obvious reasons. This model describes the diode as follows:

1 Zero forward resistance if $i > 0$; diode ON.
2 Infinite reverse resistance if $v < 0$; diode OFF.

(Current and voltage polarities are as shown in Fig. 3.3*a*.) Hence the ideal model assumes that the diode is either a perfect short circuit or a perfect open circuit, depending upon the direction of current flow that the rest of the circuit attempts to establish. Note that v *is never positive,* and i *is never negative.*

This ideal model is, of course, only approximately true. An actual diode will have a forward voltage of 0.3 to 0.6 V if any appreciable current is flowing (dotted line, Fig. 3.3*b*). There will be a small reverse current, although at room temperature this can often be neglected. Nevertheless, the ideal diode model is extremely useful for making a first-cut analysis of how a circuit will operate. Once the general regions where diodes are ON or OFF are established, then more careful models may be employed, e.g., to estimate the effects of forward drop and leakage current. In this chapter we will also neglect the dynamics of the switching of a diode from one state to another. In practice one may have to examine the effects of diode storage time and junction capacitance on circuit speed. Reference 5 provides a good discussion of diode and transistor circuit properties, or see any modern electronics text. It is also encouraging to note that, by means of the high amount of negative feedback normally provided by the operational amplifier, many of the nonideal properties of diodes and transistors are suppressed!

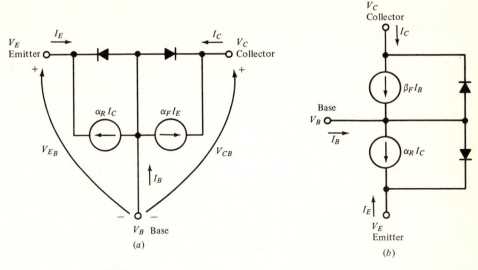

FIGURE 3.4
Ideal bipolar transistor equivalent circuits. (*a*) Common-base model with emitter and collector diodes and current generators; (*b*) common-emitter equivalent.

Figure 3.4 shows similar ideal diode models for an *n-p-n* bipolar transistor; for a *p-n-p* type, the diode polarities are reversed. Again, we assume ideal diodes and neglect the diode forward drop and reverse leakage. Even with these omissions, we still retain a model that is generally very useful for analyzing the behavior of the transistor when it is embedded in a complex circuit. Four basic modes of operation (corresponding to the four possible ON and OFF states of diodes) are apparent (Fig. 3.5):

1. *Normal or active:* This state is the one normally used for amplification. It occurs when the emitter-base junction is forward-biased and the collector-base junction is reverse-biased, best stated as:

n-p-n	*p-n-p*
$V_{CB} > 0$	$V_{CB} < 0$
$I_E < 0$	$I_E > 0$

   In this mode, there is an effective current generator in the collector circuit, that is, $I_C \cong -\alpha_F I_E = \beta_F I_B$. Here $\alpha_F < 1$ and $\beta_F = \alpha_F/(1 - \alpha_F)$.

2. *Cutoff:* In this state, which is one of the two switching modes of a bipolar transistor, both diodes are reverse-biased, that is:

n-p-n	*p-n-p*
$V_{EB}, V_{CB} > 0$	$V_{EB}, V_{CB} < 0$

The actual boundary of this region is where I_E tends to reverse polarity from the active region, which requires at least some reverse biasing of the emitter-base

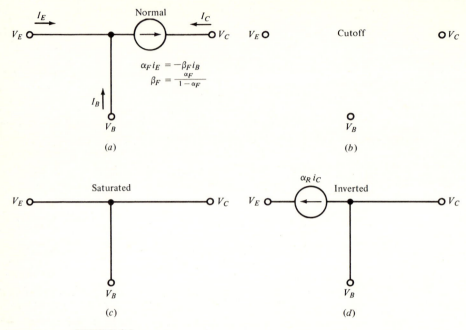

FIGURE 3.5
The four major bipolar transistor states. (*a*) Normal or active mode; (*b*) cutoff mode; (*c*) saturated or ON state; (*d*) inverted mode.

junction (for example, 0.1 to 0.2 V). In the cutoff mode the three transistor terminals have very little current flowing into them.

3 *Saturated* or ON: This is the second switching mode of a bipolar transistor. In this state, both diodes are forward-biased, and V_{CE} is smaller than either V_{EB} or V_{BC} in magnitude. The state is usually described by:

n-p-n	*p-n-p*
$V_{CB} < 0$	$V_{CB} > 0$
$I_E < 0$	$I_E > 0$

In this mode V_{CE} may be reduced to less than 0.1 V if the base current satisfies

$$|I_B| \gg \frac{|I_C|}{\beta_F}$$

4 *Inverted:* This mode is not frequently encountered, but it does occur in many nonlinear and switching applications. It occurs when the collector-base junction is forward-biased and the emitter-base junction is reverse-biased. The reverse-current gain is usually very poor, compared with the normal mode. In this state,

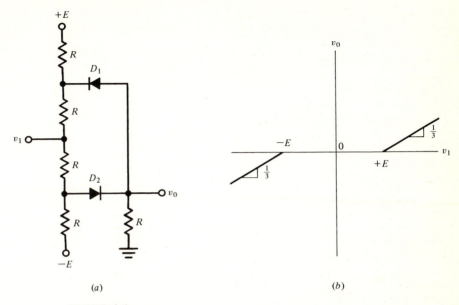

FIGURE 3.6
Refer to Example 3.2. (*a*) Simple passive dead-space circuit; (*b*) transfer characteristic.

n-p-n	p-n-p
$V_{EB} > 0$	$V_{EB} < 0$
$I_C < 0$	$I_C > 0$

One commonly overlooked problem in both the cutoff and inverted modes is that typically the emitter-base reverse breakdown voltage of most high-speed transistors is relatively low, for example, 2 to 3 V.

For most of our discussions in this chapter, we will use the ideal models of the diode and the bipolar transistor described above. Where the effects of junction forward drops and reverse leakage currents are worth noting, we will attempt to do so in at least a qualitative way. More will be said about both bipolar and field-effect transistors (FETs) in Chap. 5, when we discuss precision switching of analog signals. In later sections of this chapter, we will discuss a representative collection of nonlinear operational amplifier circuits that may be analyzed with considerable confidence using the ideal diode models of the junction diode or the bipolar transistor. The key to success in analyzing these circuits will be to test all possible diode ON or OFF states, and to deduce the points where a transition is made from one state to another.

EXAMPLE 3.2 Consider the simple diode dead-space circuit of Fig. 3.6. We want to know the transfer characteristic, that is, v_0 versus v_1. We can analyze the circuit as follows:

1 D_1 and D_2 OFF

$$v_0 = 0 \qquad (3.4a)$$

2 D_1 ON, D_2 OFF

$$v_0 = \frac{v_1 + E}{3} \qquad (3.4b)$$

3 D_2 ON, D_1 OFF

$$v_0 = \frac{v_1 - E}{3} \qquad (3.4c)$$

4 The diodes are never both ON at the same time (by inspection).

Figure 3.6*b* shows a sketch of the transfer characteristic obtained from Eqs. (3.4), where we observe that the segments of the three equations are chosen to obtain a continuous, single-valued curve. Note that if the circuit were not symmetrical, the analysis would be more involved but qualitatively the same.   ////

EXAMPLE 3.3   Figure 3.7*a* shows a simple transistor inverter circuit; suppose we want to estimate the transfer characteristic, v_0 versus v_1. Again, we can analyze the circuit with a simple model (Fig. 3.7*b*).

1 D_1, D_2 OFF

$$v_0 = E \qquad (3.5a)$$

2 D_2 ON, D_1 OFF

$$v_0 = E - \frac{v_1 R_C \beta_F}{R_B} \qquad (3.5b)$$

3 D_1 and D_2 both ON

$$v_0 = 0 \qquad (3.5c)$$

Figure 3.7*c* shows a sketch of the desired input-output transfer characteristic, which is that of a grounded-emitter inverter.   ////

EXAMPLE 3.4   Figure 3.8 shows a very useful four-diode bridge circuit (note that the output load R_0 could be the input resistor of an inverting amplifier circuit). All output load current must be provided from the outside legs of the bridge; the input signal controls the way in which current branches. Using the ideal diode model, one could look at all 16 possible diode states; such detail is not usually required. Some thought will reveal that only three possibilities need be considered in detail:

1   All four diodes ON (Fig. 3.8*c*)

$$v_0 = v_X \qquad (3.6a)$$

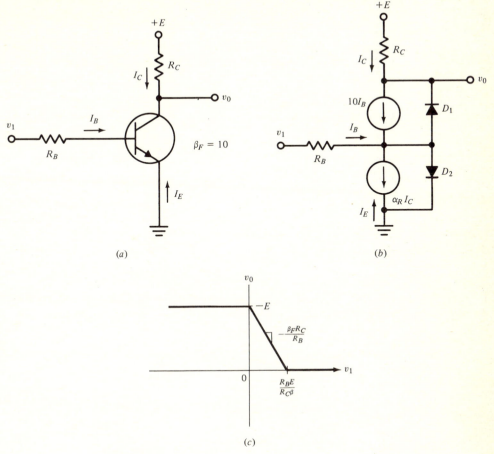

FIGURE 3.7
Refer to Example 3.3. (*a*) Simple transistor inverter; (*b*) ideal model, based upon modes of Fig. 3.5; (*c*) transfer characteristic predicted by model of part *b*.

2 D_2, D_3 ON; D_1, D_4 OFF (Fig. 3.8*d*)

$$v_0 = \frac{E_1 R_0}{R_1 + R_0} = E_A \qquad (3.6b)$$

3 D_1, D_4 ON; D_2, D_3 OFF

$$v_0 = \frac{-E_2 R_0}{R_2 + R_0} = E_B \qquad (3.6c)$$

The resulting sketch which combines these three cases is shown in Fig. 3.8*b*. Note that we have a limiter action. The limiting levels are given by (3.6*b* and *c*). The four-diode bridge circuit is one well worth remembering. Besides being a fairly precise limiter, it can be used

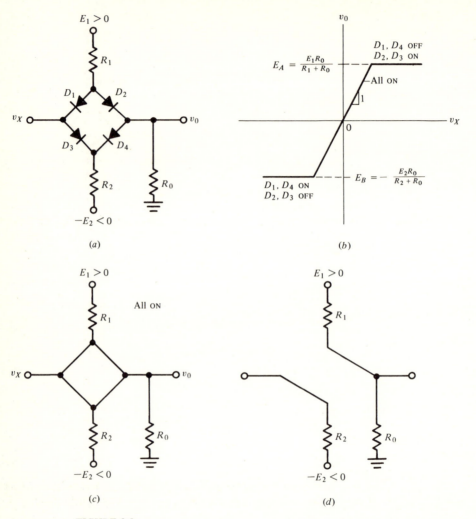

FIGURE 3.8
Refer to Example 3.4. (a) A four-diode bridge circuit; (b) transfer characteristic predicted by ideal diode model; (c) equivalent circuit with all four diodes ON; (d) equivalent circuit with D_2 and D_3 ON, D_1 and D_4 OFF.

as an analog switch; that is, levels E_1 and E_2 can be reversed in polarity by a switching waveform so as to completely block transmission. Note also that the limiting levels are dependent on the load R_0.   ////

3.2 FEEDBACK LIMITERS AND COMPARATORS

Limiting is a very important nonlinear operation. Figure 3.9 describes the *limiter operator* with a possible block symbol (Fig. 3.9a) and a sketch of the transfer characteristic (Fig. 3.9b). In general we want the ideal characteristic:

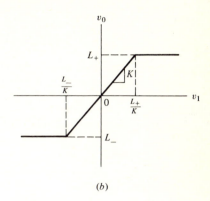

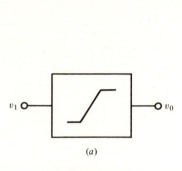

(a)

(b)

FIGURE 3.9

The limiter operator. (a) A typical block-diagram symbol; (b) important parameters; K is linear gain, and L_+ and L_- are the positive and negative limiting levels, respectively.

$$v_0 = Kv_1 \quad \begin{cases} L_- \leqslant v_0 \leqslant L_+ \\ \dfrac{L_-}{K} \leqslant v_1 \leqslant \dfrac{L_+}{K} \end{cases} \quad (3.7a)$$

$$v_0 = L_+ \qquad v_1 > \dfrac{L_+}{K} \qquad (3.7b)$$

$$v_0 = L_- \qquad v_1 < \dfrac{L_-}{K} \qquad (3.7c)$$

Of course,

$$L_- \leqslant L_+ \qquad (3.7d)$$

Such an operation is ideally a zero-memory operation, i.e., (3.7) is satisfied regardless of the frequency content of the input. In actuality such an operation can be implemented with reasonable accuracy for signals with spectral content ranging from direct current up to some upper frequency determined by the internal dynamics of the circuit elements used to implement it, typically 10 to 100 kHz.

A related operator is a *comparator,* shown in Fig. 3.10. This is essentially a limiter with a very high (ideally infinite) gain K. In Fig. 3.10 we have shown a comparison level E_R as well, to indicate that often in this application, we want to compare two signals to each other and decide which is larger. Comparators form an essential part of all analog-digital interfaces, and we will discuss this application further in Chap. 6. For the present, we want to describe how to design one. The ideal comparator can be described by

$$v_0 = \begin{cases} L_+ & v_1 > E_R \\ 0 & v_1 = E_R \\ L_- & v_1 < E_R \end{cases} \qquad (3.8)$$

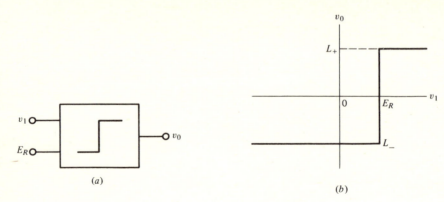

FIGURE 3.10

The comparator characteristic. (a) A typical block-diagram symbol; (b) important parameters; E_R is the comparison threshold, and L_+ and L_- are the two nominal output levels. Often one level will be near zero, e.g., when the circuit is to drive some standard digital logic gate.

Now consider Fig. 3.11. This is an operational amplifier circuit which uses a breakdown or Zener diode (we will use the latter term throughout this text). The i-v characteristic of such a diode is sketched in Fig. 3.11a, where the solid line is the ideal characteristic, and the dotted line indicates the more realistic nonideal behavior. The principal deviations from ideal will be a forward drop of perhaps 0.6 V when the diode is forward-biased, a small reverse leakage current, and often a considerable amount of rounding of the characteristics near the reverse breakdown point. If the circuit of Fig. 3.11b is constructed using this type of diode as a feedback element, we can analyze the circuit performance along the lines of Example 3.1; i.e., see Fig. 3.2 and Eq. (3.2). Figure 3.11c shows the relationship between v_0 and i_N; again

$$i_1 + i_2 + i_N = 0$$

and the summing junction voltage e_g is forced to be zero by the operational amplifier (or else the amplifier is in overload). Also (since we assume an ideal operational amplifier),

$$i_1 = \frac{v_1}{R}$$

$$i_2 = \frac{E_R}{R}$$

and

$$v_0 = f(i_N) = f\left(\frac{-v_1 - E_R}{R}\right)$$

With the Zener diode as a feedback element, then, using the ideal model of the diode,

$$f(i_N) = \begin{cases} E_z & i_N \geqslant 0 \\ 0 & i_N < 0 \end{cases}$$

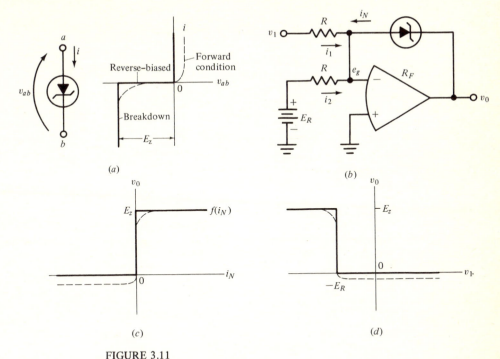

FIGURE 3.11
A simple comparator made with an operational amplifier and a Zener (break-down) diode. (*a*) An ideal model of the Zener diode; (*b*) the basic circuit; (*c*) the conductance characteristic of the diode, $f(i_N)$; (*d*) the circuit transfer characteristic, based upon $f(i_N)$.

Thus we have the overall transfer characteristic of Fig. 3.11*d*, viz.,

$$v_0 = \begin{cases} E_z & v_1 \leqslant -E_R \\ 0 & v_1 > -E_R \end{cases} \tag{3.9}$$

This is evidently a comparator circuit; except for the polarity inversion, it matches the sketch of Fig. 3.10*b*. In actuality, the output will be a few tenths of a volt negative when v_1 exceeds the threshold level $-E_R$ because of the diode forward drop. The transition between the two output levels is not discontinuous, but is made with a very steep slope, i.e., the open-loop gain of the operational amplifier.

Figure 3.12 shows some variations of the simple Zener diode comparator circuit. By reversing the diode (Fig. 3.12*a*) we get the characteristic shown. (Here we have also removed the reference-comparing voltage E_R and its associated resistance, so that the comparison is with respect to 0 V.) In Fig. 3.12*b* we employ two Zener diodes in series opposition to get the indicated bipolar characteristic. Indeed, one can even purchase so-called *double-anode Zener diodes* with both junctions incorporated in a single device.

Figure 3.13 shows a logical extension of the simple Zener diode comparator circuit to a limiter amplifier. With only the one input resistor from v_1, we have

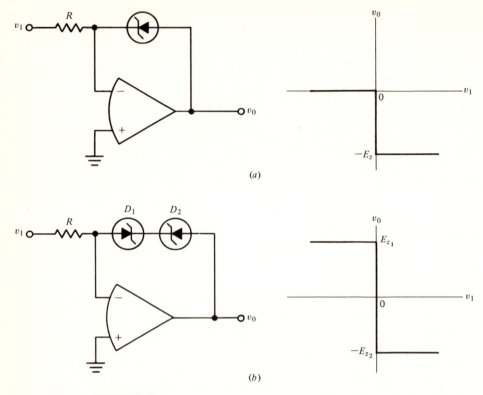

FIGURE 3.12
Variations of the Zener diode comparator circuit. (*a*) Characteristic obtained from
Fig. 3.11, but with diode polarity reversed; (*b*) characteristic obtained by two
Zener diodes in series opposition, or by a single double-anode Zener diode.

$$v_0 = -\frac{R_F v_1}{R_1} \qquad -E_{z2} < v_0 < E_{z1} \qquad (3.10)$$

However, the two Zener diodes limit the output level as indicated. This transfer
characteristic agrees with that of (3.7); of course, the gain has a negative polarity in the
amplifying region. The addition of a second input v_2, as shown in Fig. 3.13*a*, permits
forming a weighted sum of two inputs with output limiting, i.e., we have a *summer-
inverter limiter* circuit. In general, then, the two Zener diodes act as a *level clamp* or
feedback limiter to limit output excursions, but otherwise do not appreciably affect
circuit operation.

  In this circuit, of course, the limiting level is controlled by the Zener diode
breakdown voltage, and thus it is difficult to adjust it precisely. Limiting is fairly *hard*,
i.e., the slope of the transfer characteristic in the limiting regions is low. Some Zener
diodes, however, will exhibit considerable rounding near the beginning of breakdown, and
thus the limiting does not begin sharply.

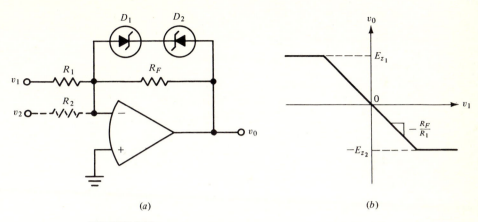

FIGURE 3.13
A simple limiter amplifier circuit made with an operational amplifier and two Zener diodes. (*a*) The basic circuit (note that summing can also be performed by adding an auxiliary input v_2); (*b*) the nominal transfer characteristic.

Figure 3.14*a* shows a different type of limiter amplifier circuit which uses ordinary *p-n* junction diodes. Its analysis is facilitated by noting that three combinations of diode states are possible:

1 D_1 and D_2 both OFF (Fig. 3.14*c*). Here obviously the diode networks have no effect on the summing junction current, and

$$v_0 = -\frac{R_F}{R_1} v_1 \qquad (3.11)$$

2 D_1 ON, D_2 OFF (Fig. 3.14*d*). If we make v_1 sufficiently positive, eventually node e_1 reaches 0 V, and can get no more negative since D_1 then turns ON. Thus we have a *breakpoint* when

$$v_0 = -\frac{E_1 R_B}{R_A} \qquad (3.12a)$$

which establishes a nominal limiting level. The reader should study this until he fully understands where this last relationship comes from. As v_1 becomes more positive, the effective feedback resistance around the amplifier is

$$R'_B = R_B \| R_F \qquad (3.12b)$$

From Fig. 3.14*d*, we can deduce that

$$v_0 = -R'_B \left(\frac{v_1}{R_1} + \frac{E_1}{R_A} \right) \qquad (3.12c)$$

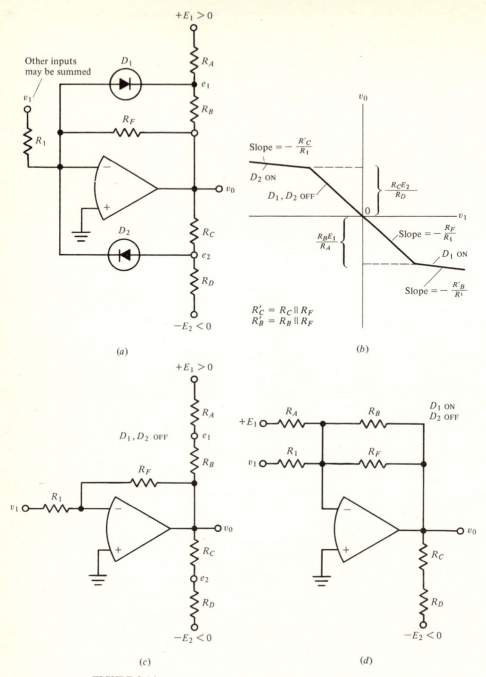

FIGURE 3.14

A general-purpose limiter made with ordinary diodes and resistor networks. (a) The basic circuit (note that dc power voltages are required); (b) the major features of the transfer characteristic; (c) equivalent circuit in the normal amplifying region, both diodes OFF; (d) equivalent circuit in the negative-limit region, D_1 ON, D_2 OFF. Note that the slopes in the limiting regions are highly dependent on the choice of resistance values; limiting levels are easily adjusted by adjusting resistance ratios.

This last equation is not itself too important; it is important to note that (3.12a) gives the breakpoint at the corner (i.e., the nominal limiting level). Also we see that the slope of (3.12c) is

$$\frac{dv_0}{dv_1} = -\frac{R'_B}{R_1} \qquad (3.12d)$$

Thus if R_B is made small compared to R_F, then the slope in the limiting region may made fairly small, and

$$R'_B \cong R_B \qquad R_F \gg R_B$$

3 D_2 ON, D_1 OFF. The analysis of this case is similar to the last; we find that there is another corner or breakpoint at

$$v_0 = \frac{+E_2 R_C}{R_D} \qquad (3.13a)$$

which establishes another nominal limiting level. The slope in the limiting region (upper left of Fig. 3.14b) is again found by defining

$$R_C' = R_C \| R_F \qquad (3.13b)$$

and then

$$\frac{dv_0}{dv_1} = -\frac{R'_C}{R_1} \qquad (3.13c)$$

Now we have a circuit that limits reasonably well if R_B and R_C are small compared to R_F; the circuit is, however, what is normally called a *soft limiter*, i.e., the limiting action is not sharp, there is some rounding at the corners, and the slope in the limiting regions is not as low as might be desired. The circuit does have the advantage that it is relatively easy to adjust the limiting levels by adjusting the resistance ratios and/or the supply voltages, as indicated in (3.12a) and (3.13a). Of course the *removal of the feedback resistor* R_F makes the circuit into a *comparator*; note that the limiting levels are still *found by (3.12a) and (3.13a)*, and the slopes in the limiting region are given by (3.12c) and (3.13c) but with R'_B and R'_C replaced by R_B and R_C, respectively.

EXAMPLE 3.5 Suppose we want a limiter amplifier to have an inverting gain of 20, to limit at ±5 V, and to have slopes in the limiting region of less than 0.1 V/V. Assume ±15 V dc is available. The largest resistor is to be 100 kΩ. Figure 3.15a shows a possible design. Hopefully it is clear that R_F is the largest resistor; thus $R_F = 100$ kΩ, and

$$R_1 = \frac{R_F}{20} = \frac{10^5}{20} = 5 \text{ k}\Omega$$

We establish the ratio of 3 to 1 for the limit-setting resistor from (3.12a) and (3.13a), that is,

$$-5 = \frac{-E_1 R_B}{R_A} = \frac{-15 R_B}{R_A}$$

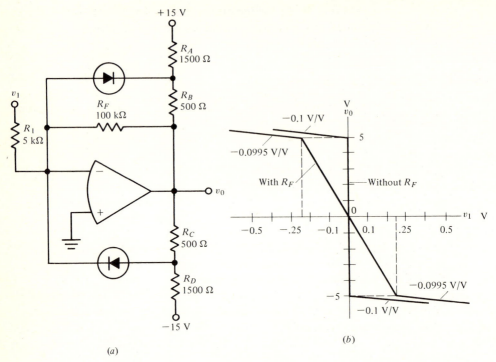

(a)

(b)

FIGURE 3.15
Refer to Example 3.5. (a) A proposed design to meet the ·specifications; (b) the resulting nominal characteristic, with and without R_F.

or

$$\frac{R_B}{R_A} = \frac{1}{3}$$

and similarly

$$\frac{R_C}{R_D} = \frac{1}{3}$$

Note that it is the +15-V dc supply which is used in establishing the negative limiting level, and vice versa. We now must find suitable values for the limiting resistors to obtain the desired slope. If we note that

$$R'_B < R_B$$

and

$$R'_C < R_C$$

we can simplify our job by setting

$$\frac{R_B}{R_1} = \frac{R_C}{R_1} = 0.1$$

or
$$R_B = R_C = 500 \ \Omega$$

and thus

$$R_A = R_D = 1500 \ \Omega$$

Actually, the slope will be less than 0.1 V/V in the limiting region.

Figure 3.15b shows the transfer characteristic of the final design (based on an ideal diode model). Included are more precise estimates of the slopes in the limiting region:

$$R_B' = R_C' = 10^5 \ \Omega \parallel 500 \ \Omega$$
$$= 497.5 \ \Omega$$

and thus the slopes are

$$-\frac{497.5}{5000} = -0.0995 \ \text{V/V}$$

Also, the limiting levels will be slightly in error, due to the forward characteristic of the diodes, and there will be some rounding at the corners.

Figure 3.15b also shows the transfer characteristic which results if R_F is removed. Note that the nominal limiting levels are unchanged, and the slopes in the limiting regions are exactly -0.1 V/V, but now the transition between the two levels is abrupt (limited only by the open-loop gain of the operational amplifier), and we have a zero-threshold comparator.   ////

Zero Limiter or Half-Wave-Rectifier Circuits

In Fig. 3.16, we see that an ordinary diode may be used to make a crude zero limiter or half-wave-rectifier operator. Ideally, we want $v_0 = -v_1$ in only one quadrant and zero in the other. As indicated by the dotted line, the forward drop of the diode will allow the output to be a few tenths of a volt in the limiting region, instead of zero. The *precision limiter circuit* of Fig. 3.17a embeds the *limiting diode* D_1 in a *feedback loop* to suppress its nonideal characteristics. D_1 now merely serves to sense the polarity of the operational amplifier output voltage. Note also that the circuit output v_0 is not taken directly from the operational amplifier. This circuit is worthy of some study. When the input is negative, the diode D_1 is ON, since v_A tends to be positive. We can look upon the forward drop of D_1 as an offset in the operational amplifier output stage; the operational amplifier (assuming it is not in overload) will do whatever is required to keep $e_g = 0$. Since v_A is positive and e_g is 0, D_2 is OFF. Thus for negative-valued input, the feedback resistor and the input resistor establish summing junction currents at e_g such that

$$v_0 = \frac{-R_F}{R_1} v_1 \qquad v_1 < 0 \qquad (3.14a)$$

If, however, v_1 becomes positive, then v_A becomes negative; v_0 cannot follow along, since D_1 can only conduct in a direction which makes i_0 positive. Thus the feedback resistor is left essentially unconnected to the amplifier output but still connected to the virtual ground at e_g, and

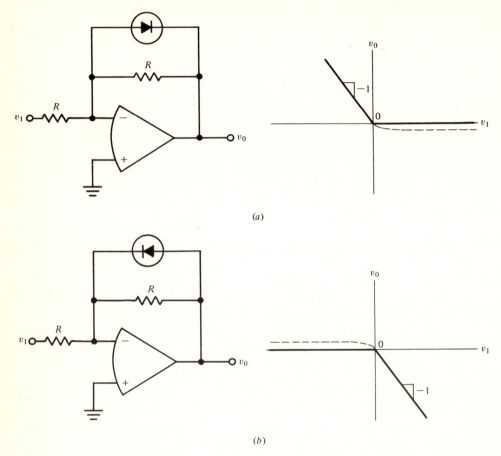

FIGURE 3.16
Crude zero limiters made with only one diode; diode forward drop will not permit precise zero limiting. Circuits (*a*) and (*b*) show the two polarities available.

$$v_0 = 0 \qquad v_1 > 0 \qquad (3.14b)$$

Some form of catching network is needed to keep the amplifier from going into overload when $v_1 > 0$; in Fig. 3.17*a*, silicon diode D_2 serves to catch the output and clamp v_A to a small negative value, viz., the forward drop across D_2, which is typically 0.6 V (see dotted line in Fig. 3.17*b*).

This circuit is quite accurate! The corner of the transfer characteristic and, at low frequencies, the ideal limiter characteristic will be implemented with an accuracy of better than 0.1 V. At higher frequencies, diode capacitance will cause some errors in the neighborhood of the corner; these errors can be minimized by using fast diodes for both D_1 and D_2, and small resistance values. Reversing both diodes will cause the circuit to limit for negative inputs and amplify for positive ones.

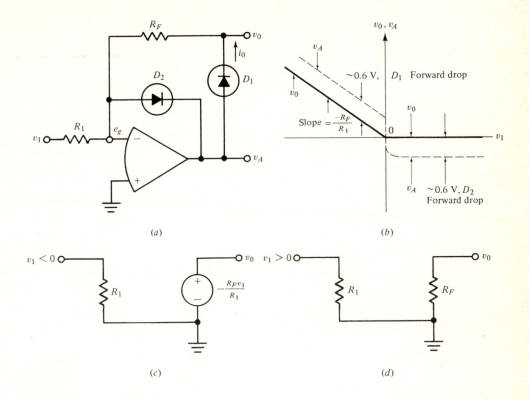

(a)

(b)

(c)

(d)

FIGURE 3.17
A precision zero limiter, made by embedding the diode D_1 in the feedback loop; D_2 merely catches the operational amplifier to prevent overload. (a) The basic circuit; (b) the resulting precise limiting characteristic; note the difference between v_0 and v_A (the dotted line is v_A); (c) the overall equivalent active circuit in the nonlimiting mode; note the low output impedance; (d) the equivalent circuit in the limiting mode; note the high output impedance.

Note that the output impedance of the circuit of Fig. 3.17a varies with circuit conditions. When v_0 is positive (D_1 ON), the output impedance will be very low because of the negative feedback provided by the amplifier. When $v_1 > 0$, v_0 will have an open-circuit voltage very close to zero, but will have the feedback resistance R_F as a Thévenin-equivalent output resistance. Thus, if any external load attached to the circuit has an internal offset voltage that tends to pull v_0 away from ground, v_0 may not stay equal to zero. Figure 3.17c and d shows the approximate equivalent circuits for this limiter in both modes of operation. It is thus often desirable to buffer the output of this circuit with an additional amplifier stage, so that external loads will not generate unwanted behavior. In the next section of this chapter, we will see how the circuit of Fig. 3.17a can be augmented to make an absolute-value or full-wave-rectifier operator.

EXAMPLE 3.6 Suppose we want to build an ac voltmeter which will provide a voltage equal to the rms value of sinusoidal input signals. We will assume that the input is a pure sinusoid with zero average value, that is,

$$v_1 = A \sin \omega t$$

Assume that v_1 can be as large as 10 V rms; thus the peak value is

$$A \leqslant \sqrt{2} \times 10 = 14.14 \text{ V} \quad \text{peak}$$

Let us use the circuit of Fig. 3.18 for this job. The output amplifier B provides a low-impedance output, but more importantly, it is used to filter the output so that only a steady-state or dc voltage appears at v_0. The filtering time constant is $R_F C_F$, which must be chosen so that

$$R_F C_F \gg \frac{1}{\omega}$$

in order to get good filtering. Suppose we have ± 10-V amplifiers to work with. We will have to be sure that neither overloads.

Figure 3.18c shows a sketch of v_2, which is the output of the half-wave rectifier. In order to keep amplifier A from overloading, we must ensure that

$$\frac{A R_2}{R_1} \leqslant 10 - \text{diode drop} \quad \text{V}$$

In this case, then, we pick

$$R_2 = \frac{R_1}{2}$$

e.g., we let $R_1 = 10 \text{ k}\Omega$ and $R_2 = 5 \text{ k}\Omega$. The average value of v_2 is

$$v_{2,\text{av}} = \frac{A R_2}{\pi R_1} = \frac{A}{2\pi}$$

Now the rms value of v_1 is

$$v_{1,\text{rms}} = \frac{A}{\sqrt{2}} = 0.707 A$$

Thus we pick the dc gain of the second stage (i.e., that dc gain from v_2 to v_0) so that

$$\frac{v_{0,\text{av}}}{v_{2,\text{av}}} = \frac{A/\sqrt{2}}{A/2\pi} = \pi\sqrt{2} = 4.44$$

or

$$R_3 = \frac{R_F}{4.44} = 0.225 R_F$$

Suppose we know that the input sinusoids have a minimum frequency of 10 Hz; then

$$\omega \geqslant 20\pi = 62.8 \text{ rad/s}$$

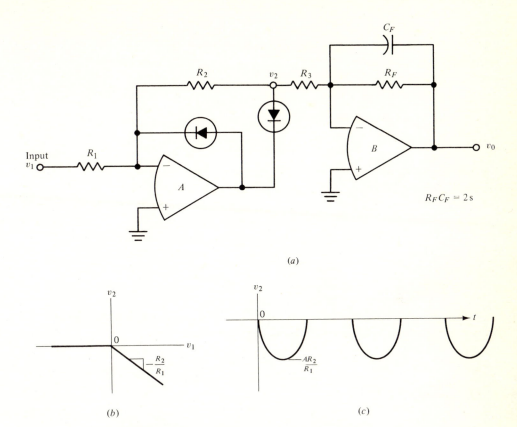

(a)

(b)

(c)

FIGURE 3.18
A basic ac voltmeter circuit. (a) Two-amplifier half-wave rectifier with filter
capacitor C_F; (b) v_2 versus v_1; (c) v_2 versus time. Output amplifier performs
averaging.

and we want

$$R_F C_F \gg \frac{1}{62.8} = 0.0159 \text{ s}$$

Let us say

$$R_F C_F = 2 \text{ s}$$

In order to keep C_F of reasonable size, we will pick

$$R_F = 1 \text{ M}\Omega \qquad C_F = 2 \text{ } \mu\text{F} \qquad R_3 = 225 \text{ k}\Omega$$

Note that we have not said anything about possible dc offset problems in the amplifiers,
which might dictate a different resistance level, with smaller resistors and a larger
capacitor. The circuit will hopefully do the specified job for all input signals above 10 Hz,
up to say 10 kHz with a fairly fast amplifier A (note that amplifier B can be slow). $\qquad$ ////

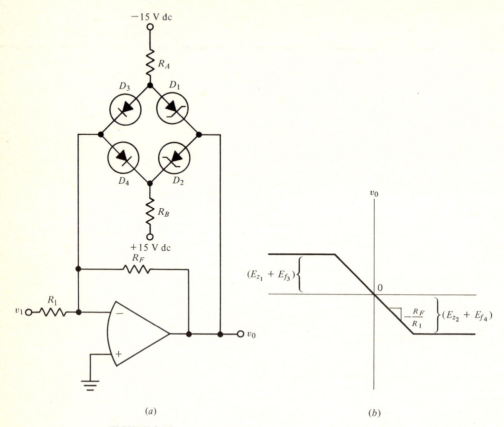

FIGURE 3.19
An improved Zener diode limiter (or comparator) circuit. (a) The current provided by the outer bridge legs via R_A and R_B provides for sharper corners and faster switching speeds; (b) approximate transfer characteristic.

Improved Limiters

Although the zero limiter of Fig. 3.17 is quite precise, the other general limiter circuits we have discussed (e.g., Figs. 3.12 to 3.14) often do not perform as well as desired. Figure 3.19 shows an improved Zener diode limiter comparator circuit which makes a sharper and faster transition between states. The use of low-leakage diodes for D_3 and D_4 reduces the effect of leakage currents on the circuit in the nonlimiting mode. The slope in the limiting region will be

$$\frac{dv_0}{dv_1} = \frac{-(R_d + R_z)}{R_1}$$

where R_d = diode forward resistance
R_z = Zener diode breakdown resistance

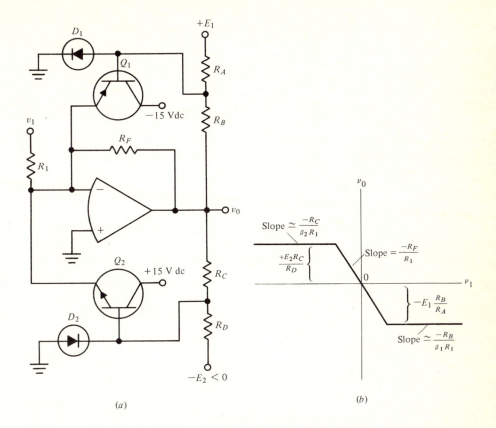

FIGURE 3.20
A transistor hard limiter, based upon the circuit of Fig. 3.14, but using the emitter-base junctions of two bipolar transistors as the limit-sensing elements. (*a*) Compare the basic circuit to Fig. 3.14; (*b*) the resulting transfer characteristic has much lower limiting slopes, which have been reduced by the factor β of the transistors. Diodes D_1 and D_2 are often needed to prevent reverse emitter-base breakdown of the transistors.

The limiting levels cannot be adjusted without changing the Zener diode breakdown voltages; also note that now the limiting levels are $+(E_{z1} + E_{f3})$ and $-(E_{z2} + E_{f4})$, where E_{z1} and E_{z2} are the Zener breakdown voltages, and E_{f3} and E_{f4} are the forward drops of D_3 and D_4.

The biasing resistors R_A and R_B are chosen to establish a few milliamperes of current through the Zener diodes, keeping them always in the breakdown mode.

Figure 3.20 shows a transistor hard limiter. The circuit operates very much like the limiter of Fig. 3.14, but the slopes in the limiting region are reduced by the forward current gain of the transistors. As in Fig. 3.14, the limiting levels are given by

$$\text{Positive limit} = \frac{+E_2 R_C}{R_D} \qquad (3.15a)$$

$$\text{Negative limit} = \frac{-E_1 R_B}{R_A} \qquad (3.15b)$$

that is, they are the same as those of the limiter of Fig. 3.14. Also the slope in the nonlimiting region is

$$\text{Nonlimiting slope} = \frac{-R_F}{R_1} \qquad (3.15c)$$

Of course, a comparator may be made by removing R_F.

In the circuit of Fig. 3.20, however, the limiting slopes are greatly reduced. In this circuit, the limiting is initiated by the emitter-base diodes of Q_1 and Q_2, as one or the other goes into the active mode. In the active mode, the emitter current is $(\beta + 1)$ times the base current provided by the feedback network. The net effect of the forward current gain of the transistors is to reduce the limiting slopes by the factor β. Thus the limiting slopes become

$$\text{Slope in positive limit} \cong \frac{-R_C}{R_1 \beta_2} \qquad (3.15d)$$

$$\text{Slope in negative limit} \cong \frac{-R_B}{R_1 \beta_1} \qquad (3.15e)$$

Since the transistor betas can be 50 to 200, the slopes are relatively small, and we have so-called *hard* limiting. Note the diodes D_1 and D_2. They are usually needed to prevent emitter-base reverse breakdown when the transistors are cut off, since in that state the bases may move several volts away from ground while the emitters are held at ground potential. Also note that Q_1 and Q_2 should be a reasonably well-matched complementary-symmetry pair.

EXAMPLE 3.7   If the circuit of Fig. 3.20 is used instead of that of Fig. 3.15, i.e., we use transistors instead of diodes but leave everything else the same, then the slopes in the limiting regions would be less. For example, if the transistor betas were 50 or more, the slopes would be about 2 mV/V.   ////

The Precision Bridge Limiter

The transistor limiter just discussed (Fig. 3.20) will serve for most accurate limiting applications. However, for high-accuracy limiting (e.g., better than 50-mV accuracy) the precision bridge limiter circuit of Fig. 3.21 is probably the best, particularly at low frequencies. This is another interesting example of how a nonlinear network (R_A, R_B, and D_1 through D_4) can be embedded in a feedback loop, to make its properties more ideal. The limiting levels are accurately established by the four-diode bridge, which is

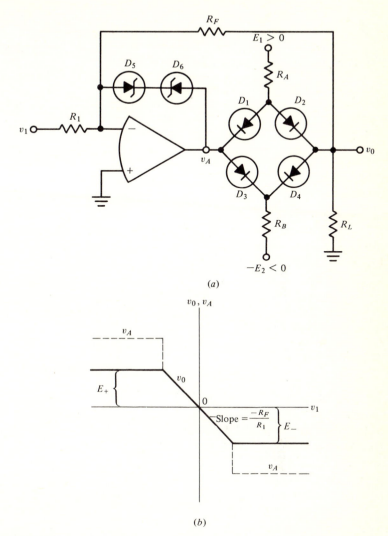

FIGURE 3.21
A precision limiter designed around a four-diode bridge embedded in a feedback loop. (*a*) In the basic circuit Zener diodes D_5 and D_6 merely serve to prevent amplifier overload; (*b*) the nominal transfer characteristic is dependent upon the values of both R_L and R_F; see Eq. (3.16). Static accuracy of 50 mV is possible.

similar to the one described in Example 3.3 and Fig. 3.8. The Zener diode catching network is merely used to keep the amplifier from going into overload when the bridge limits, since during limiting, the output of the amplifier is effectively disconnected from the feedback loop. The only tricky aspect of analyzing this circuit is the calculation of limiting levels (which, incidentally, will be fairly exact). Note that R_L and R_F together

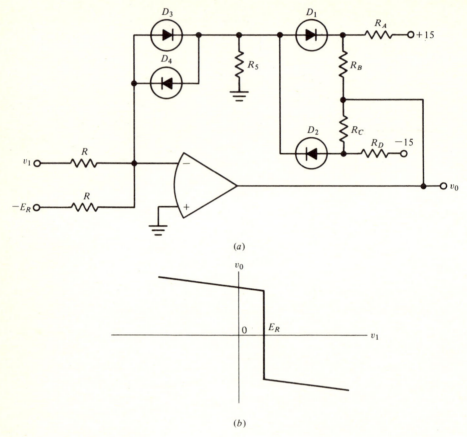

FIGURE 3.22
The use of extra diodes (D_3 and D_4) to suppress leakage currents and establish a precise threshold level. (*a*) The basic circuit, which operates in a manner similar to Fig. 3.14; (*b*) a sketch of the transfer characteristic; the comparison threshold will not be particularly sensitive to leakage current in the diodes.

act as the output load of the bridge. Thus the limiting levels are found to be (refer to Fig. 3.21)

$$E_+ = \frac{R_0 E_1}{R_0 + R_A} \qquad (3.16a)$$

and

$$E_- = \frac{-R_0 E_2}{R_0 + R_B} \qquad (3.16b)$$

where

$$R_0 = R_F \| R_L \qquad (3.16c)$$

Thus the load being supplied by the circuit affects the limiting levels, and a buffer amplifier may be advisable. One might, for example, follow the circuit by an inverting amplifier, with R_L as the input resistor.

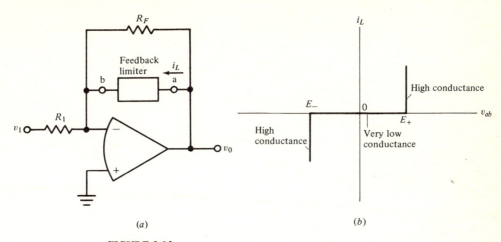

(a) (b)

FIGURE 3.23

The general feedback limiter concept. (*a*) Basic circuit structure; (*b*) required conductance characteristic of the feedback limiter.

Suppression of Leakage Currents and Stray Capacitances

In Fig. 3.22, we show a comparator made by adding leakage-suppression diodes to a clamp or limiting circuit. The network D_1, D_2, and R_A through R_D functions as a basic limiter, similar to Fig. 3.14. Diodes D_3 and D_4, along with R_5, form a tee network in the feedback loop which raises the short-circuit feedback resistance in the nonlimiting mode. This approach might be advisable in building a precision comparator, where the threshold level is to be as accurate as possible. It would suppress leakage currents from diodes D_1 and D_2, which tend to create an offset error in the circuit in a manner similar to operational amplifier offset currents. Other limiters, such as a Zener diode network, may also be used with this approach. Equations (3.11) through (3.13) still apply to the design of such a circuit. The use of relatively low resistance values would also tend to improve circuit speed and reduce the effects of offset currents and parasitic capacitances.

General Use of Feedback Limiters or Clamps

The reader has probably noticed by now that the limiter and comparator circuits thus far discussed have a number of common properties. Figure 3.23 shows the general structure of these circuits; with R_F absent we have a comparator, with it present we have a limiter. The limiter circuits in Figs. 3.12, 3.13, 3.14, and 3.20 all follow this same form, and one can usually apply one of these so-called *feedback limiters* between the output and inverting input terminals of an operational amplifier to prevent overload. If the limiting levels are set at the nominal output voltage extremes of the amplifier (±10 V), then the feedback limiter serves as an overload protector but has little, if any, effect on the amplifier circuit so long as the output level is not exceeded. Feedback limiters are also often called *clamps* or *bound* circuits; most of these feature the short-circuit conductance characteristic of Fig. 3.23*b*. Hard limiters, such as Fig. 3.20, feature higher slopes in the conductance curve.

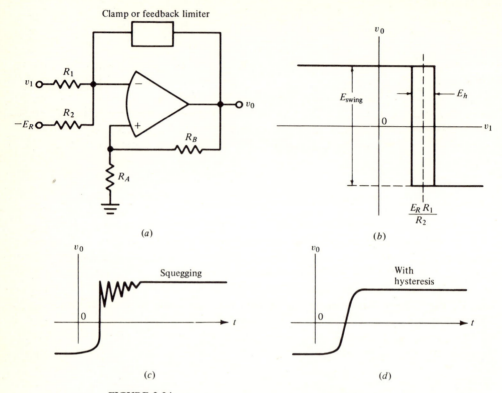

FIGURE 3.24
Improving a comparator noise immunity with hysteresis. (*a*) Resistors R_A and R_B are added to a conventional comparator circuit to provide positive feedback and thus hysteresis; (*b*) sketch of resultant transfer characteristic; E_h is the hysteresis window; (*c*) noisy transition or "squegging" without hysteresis; (*d*) monotone transition waveform with hysteresis.

Note that the precision limiters of Fig. 3.17 and 3.20 use crude limiters to prevent amplifier overload, but actually do the precise limiting by disconnecting the amplifier from the output node in the limiting regions.

Improving Comparator Noise Immunity with Hysteresis

Figure 3.24 demonstrates the use of positive feedback to add hysteresis. This provides a latching characteristic which gives the comparator some noise immunity. Without this, a fast high-gain amplifier will often have a tendency to oscillate a bit during the transition between limiting states. We might call this a sort of "squegging" (an archaic term used by electronic designers to describe parasitic oscillations in radio transmitters). The exact behavior of the circuit depends upon the type of clamp used (see Prob. 3.11 for common examples). If the amount of positive feedback is small, then the amount of hysteresis E_h is given approximately by

$$E_h = \frac{R_A E_{\text{swing}}}{R_A + R_B} \qquad (3.17)$$

Typically we make E_h 10 to 20 mV. The elimination of squegging prevents spurious triggering of logic circuits driven by the comparator. Note that the addition of the hysteresis makes the circuit behave like a bistable or flip-flop circuit; i.e., once the input is returned to zero or removed, it will latch in one state or another, depending on its past history. We have then a form of binary memory circuit, although for that purpose we would want to widen the hysteresis (see Sec. 3.3 for more on the topic of bistables).

3.3 ADDITIONAL PIECEWISE-LINEAR FUNCTIONS

Figure 3.25a shows a general piecewise-linear function $Y = F(X)$. Often we want to use such a function to approximate some smoothly varying function, such as $\sin X$; in other situations we actually do want sharp corners, e.g., in the absolute-value and dead-space amplifier functions of Fig. 3.25b and c. Whatever the original impetus for implementing the function, we want to generate $F(X)$ so that it varies linearly between *breakpoints* (Y_i and X_i, Fig. 3.25a). Sharpness at the corners may not really be desired, but it often occurs anyway.

A variety of circuit designs may be employed; some have evolved for a special nonlinearity, such as the absolute-value circuit of Fig. 3.26. General-purpose adjustable nonlinear function generation circuits may be used to approximate a broad class of nonlinear operators; some are applied in analog computers and specialized instrumentation systems. We will describe a collection of well-known piecewise-linear circuits in some detail.

Hopefully the examples we have selected here will demonstrate some techniques that the reader can apply to a broad variety of situations, and will provide a basis for developing new circuits as the need arises.

The Absolute-Value Operator

Figure 3.26 shows how the zero limiter or half-wave-rectifier circuit of the previous section (Fig. 3.17) can be augmented to implement the absolute-value, or full-wave-rectifier, operation. In Fig. 3.26a,

$$v_0 = K|v_1| \qquad K > 0 \qquad (3.19)$$

The circuit operation can be understood by looking at the block diagram of Fig. 3.26b; here we see that by summing the output of a zero limiter circuit with the input itself, along with appropriate polarity inversions, we get the desired result. In Fig. 3.26c,

$$v_2 = \begin{cases} \dfrac{-R_4 v_1}{R_3} & v_1 \geqslant 0 \\ 0 & v_1 < 0 \end{cases} \qquad (3.20)$$

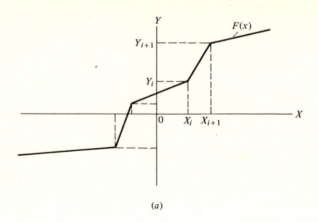

(a)

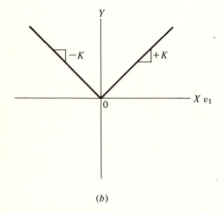

(b)

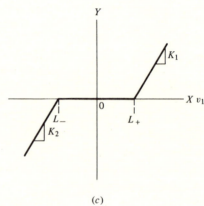

(c)

FIGURE 3.25
Piecewise-linear functions. (a) A general function $F(X)$; (b) the absolute-value operator; (c) the dead-space amplifier.

also,

$$v_0 = -R_F\left(\frac{v_1}{R_1} + \frac{v_2}{R_2}\right) \quad (3.21)$$

Now if, for example, we set

$$R_3 = R_4 = R$$

then

$$v_2 = \begin{cases} -v_1 & v_1 \geqslant 0 \\ 0 & v_1 < 0 \end{cases} \quad (3.22)$$

With

$$R_2 = \frac{R_1}{2}$$

we have

$$v_0 = \frac{R_F}{R_1}|v_1| \quad (3.23a)$$

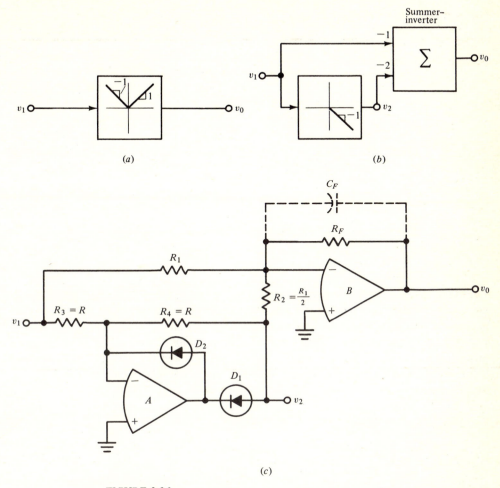

FIGURE 3.26
The absolute-value operator. (*a*) Overall operator symbol; (*b*) equivalent block diagram based upon the zero limiter of Fig. 3.17; (*c*) circuit details for a two-amplifier implementation.

which matches (3.19) with

$$K = \frac{R_F}{R_1} \quad (3.23b)$$

Note that if we reverse diodes D_1 and D_2 we get a negative constant K, and v_0 is proportional to minus the absolute value of v_1. Frequently the circuit is constructed with

$$R_1 = R_3 = R_4 = R_F = R \quad (3.24a)$$

and

$$R_2 = \frac{R}{2} \quad (3.24b)$$

in which case

$$v_0 = |v_1| \qquad (3.24c)$$

or, with the diodes reversed,

$$v_0 = -|v_1| \qquad (3.24d)$$

Measurement of Average Absolute Value

In Fig. 3.26c, if we add a capacitor C_F around the output amplifier feedback resistance R_F, then we may use the circuit to estimate the average absolute value of v_1; that is,

$$v_0 \cong |v_1|_{av} \cong \frac{1}{T} \int_0^T |v_1|\, dt \qquad (3.25a)$$

Note that this will be an approximate average, with an averaging time of the order of magnitude of

$$T \cong R_F C_F \qquad (3.25b)$$

Thus the circuit is useful for measuring such things as average absolute error (note that it does not measure true rms value for general waveforms; we will show how to do that later). The circuit is thus a precise full-wave rectifier, and the addition of C_F adds a first-order RC filter to the rectifier output. The amount of "ripple" in v_0 will, as we might expect, depend upon the relative values of $R_F C_F$ and the period of the waveform being rectified. (See Prob. 3.13 for an application of this circuit to sinusoidal ac measurements.)

The Dead-Space Operator

Figure 3.27a shows the transfer characteristic of a dead-space operator; such a transfer characteristic may be implemented in a variety of ways. The circuit of Fig. 3.27b (designed by G. A. Korn) is one of the more clever approaches. It uses a four-diode bridge to control the feedback conductance. This is a fairly difficult circuit to understand. Note that the left side of the bridge is connected to node e_g, which is always near ground potential; thus (using ideal diode models), currents I_1 and I_2 are essentially constant:

$$I_1 \cong \frac{E_1}{R_1} \qquad (3.26a)$$

$$I_2 \cong \frac{E_2}{R_2} \qquad (3.26b)$$

All four diodes will be ON whenever

$$-I_1 < I_3 < I_2 \qquad (3.27)$$

but

$$I_3 = \frac{V_1}{R_3} \qquad (3.28)$$

thus we can identify three valid modes of operation.

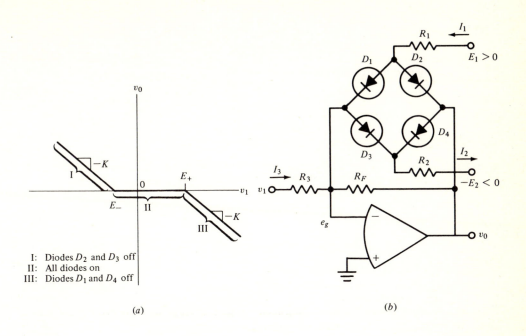

I: Diodes D_2 and D_3 off
II: All diodes on
III: Diodes D_1 and D_4 off

(a)

(b)

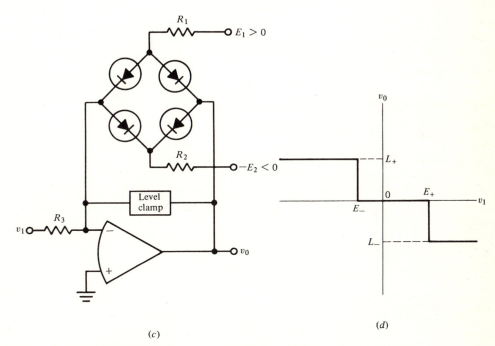

(c)

(d)

FIGURE 3.27
Dead-space operator. (a) General transfer characteristic; (b) four-diode bridge implementation; (c) modification to provide a dead-space comparator; (d) transfer characteristic of a dead-space comparator.

Region I (diodes D_2 and D_3 OFF) This region occurs when v_1 becomes sufficiently negative so that the upper bridge leg cannot supply I_3 through D_1 alone; the extra current must come via R_F, and hence v_0 becomes nonzero. The boundary of region I determines one breakpoint,

$$\frac{-E_-}{R_3} = \frac{E_1}{R_1}$$

or

$$E_- = \frac{-E_1 R_3}{R_1} \qquad (3.29a)$$

Region II (all diodes ON) This region occurs when v_1 is small enough so that (3.27) is satisfied. The feedback impedance is now very low, and $v_0 \cong 0$.

Region III (diodes D_1 and D_4 OFF) This is similar in analysis to region I; the positive breakpoint is found to be

$$E_+ = \frac{E_2 R_3}{R_2} \qquad (3.29b)$$

In Fig. 3.27c, the feedback resistor is replaced by a level clamp to obtain a *dead-space comparator* with the characteristic shown in Fig. 3.27d. The dead-space limits, E_+ and E_-, are still found from (3.29), and the limits L_+ and L_- are established by the level clamp, which may be made up of Zener diodes or any of the other feedback limiters described in Sec. 3.2.

EXAMPLE 3.8 Suppose we want the dead-space characteristic of Fig. 3.28a, with ±15 V dc available for supplying the diode bridge. Let us arbitrarily pick $R_F = 10 \text{ k}\Omega$. Then to get the desired nonzero slope of -2, we must pick

$$R_3 = \frac{R_F}{2}$$

$$= 5 \text{ k}\Omega$$

Now (3.29) predicts

$$R_1 = \frac{-15 \times R_3}{-2} = 37.5 \text{ k}\Omega$$

and

$$R_2 = 25 \text{ k}\Omega \qquad \textit{////}$$

General-Purpose Diode Function Generation

For the generation of general piecewise-linear functions (see Fig. 3.25a) we can often combine the circuits discussed above (e.g., limiters, dead-space operators, etc.) to form more complex functions. Often a clever designer can implement a complex function with only a few amplifiers. We want to now discuss some systematic approaches to the design of a general piecewise-linear function. Figure 3.29 shows an example of a general structure that is often useful. Here we sum the outputs of a set of breakpoint limiters or line-segment-generating circuits, each of which contributes to the slope of the function on

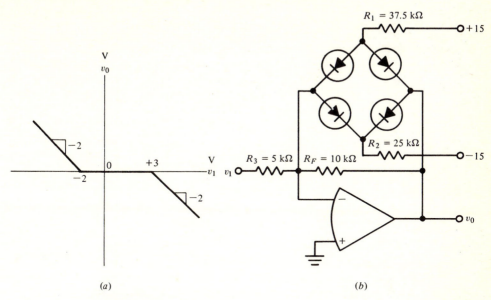

FIGURE 3.28
Refer to Example 3.8. (*a*) Desired transfer characteristic; (*b*) circuit design using a four-diode bridge.

one side of a breakpoint. The circuits of Fig. 3.30 depict one alternative for this type of approach, and we will discuss this first. The circuit of Fig. 3.30*a* by itself provides the two-segment function shown in Fig. 3.30*b*. Let us ignore the compensation diode D_C for a moment (replace it with a short circuit) and assume diode D_1 is ideal. We now have what is often called a *series limiter circuit* which establishes the two modes of operation:

1 Diode D_1 ON, D_2 OFF

$$v_0 = -R_F\left(\frac{v_1}{R_1} + \frac{E}{R_2}\right) \qquad (3.30a)$$

2 Diode D_1 OFF, D_2 ON

$$v_0 = 0 \qquad (3.30b)$$

The breakpoint occurs when $v_0 = 0$ in both parts of (3.30), and thus the breakpoint is approximately (see Fig. 3.30*b*)

$$E_x \cong \frac{-ER_1}{R_2} \qquad (3.31)$$

Typically E is a negative dc supply voltage (for example, -15 or -10 V dc), and E_x is a positive breakpoint.

The slope of the nonzero segment is found from (3.30*a*):

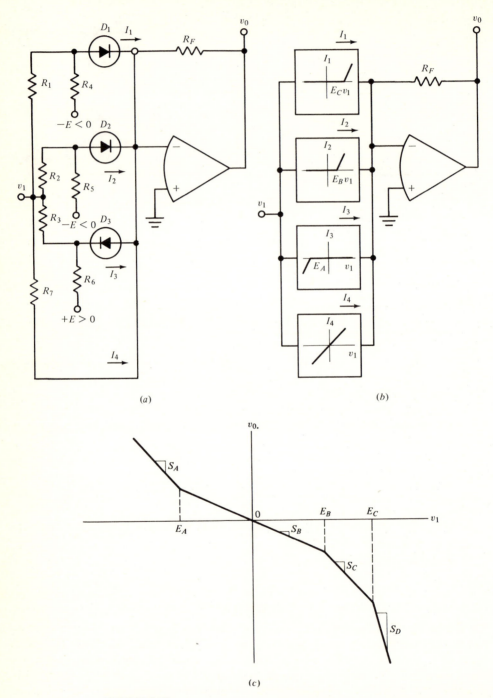

(a)

(b)

(c)

FIGURE 3.29
Parallel series limiter networks. (a) General circuit; (b) partial block diagram; (c) typical transfer characteristic.

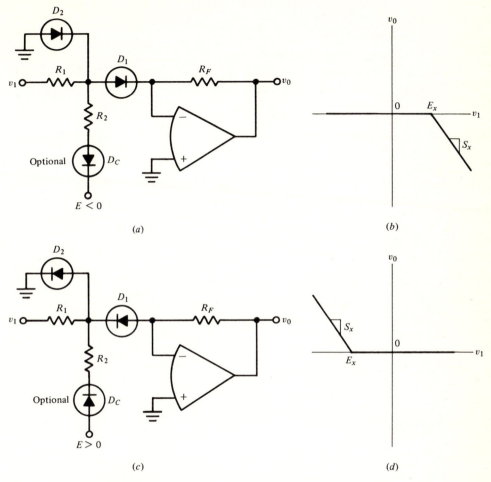

FIGURE 3.30
Series limiter circuit. (*a*) General circuit for positive breakpoint; (*b*) transfer characteristic of Fig. 3.30*a*; (*c*) general circuit for negative breakpoint; (*d*) transfer characteristic of Fig. 3.30*c*.

$$S_X = \frac{-R_F}{R_1} \qquad (3.32)$$

Thus, even with fixed values of R_F and E, we can adjust the breakpoint and the slope by picking the proper values for R_1 and R_2. The diode D_C is usually similar to D_1, and is included to provide temperature compensation for the drift in the forward drop of D_1 with temperature. Typically D_C is always conducting. The slope equation, (3.32), is fairly accurate. The breakpoint equation, (3.31), is only approximate because of the diode forward drop. If V_D is the typical drop of D_1 and D_C (assumed to be positive), then (3.31) can be easily modified to get an *improved estimate* of the breakpoint:

$$E_x \cong \frac{-ER_1}{R_2} + V_D \qquad (3.33)$$

(see Prob. 3.19). We still can estimate the slope from (3.32). Note that the use of the *catching diode* D_2 is *optional;* it is recommended for limiter circuits intended to work at high frequencies. It reduces the feedthrough when D_1 is OFF.

EXAMPLE 3.9 Suppose we want the characteristic of Fig. 3.30b to have the parameters

$$E_x = 3 \qquad S_x = -3$$

and assume that $R_F = 10 \text{ k}\Omega$, and that -15 V dc is available to supply the series limiter network of Fig. 3.31. Clearly that means that R_1 should be 3.33 kΩ. If we neglect the diode forward drop, then (3.31) predicts that R_2 should be

$$R_2 = \frac{-(-15)R_1}{+3} = 5R_1$$

$$= 16.7 \text{ k}\Omega$$

If we want a more accurate design, then the forward drop in D_1 and D_C must be included. Let us assume that both diodes have a typical forward drop of 0.6 V. Then (3.33) may be used to obtain

$$3 = \frac{-(-15)R_1}{R_2} + 0.6$$

or

$$R_2 = \frac{15R_1}{2.4} = 6.25R_1$$

$$= 20.8 \text{ k}\Omega$$

Note that (3.31) yields a rather inaccurate design, at least in terms of the breakpoint. The value for R_2 obtained from (3.33) is probably more accurate. In practice, one usually adjusts the value of R_2 empirically. ////

The breakpoint series limiter of Fig. 3.30c and d is the same as the one just discussed, except the diodes are reversed, so that we get the characteristic of Fig. 3.30d. The breakpoint for the circuit of Fig. 3.30c is found by

$$E_x = \frac{-ER_1}{R_2} - V_D \qquad (3.34)$$

and E is usually positive.

Referring to Fig. 3.29, let us see how the series limiter circuits of Fig. 3.30 may be combined to implement a piecewise-linear function. The circuit of Fig. 3.29a has three limiter networks plus a direct input resistor R_7. Each of these four paths to the summing junction has a particular short-circuit transfer conductance, as sketched in Fig. 3.29b. The

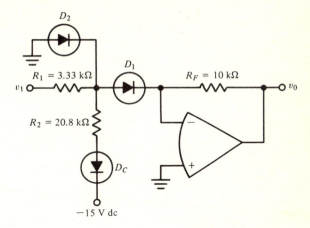

FIGURE 3.31
Circuit for Example 3.9.

upper three paths create three breakpoints E_C, E_B, and E_A, with three associated components of the output slope. The total transfer characteristic is proportional to the *total short-circuit transfer characteristic*, or *transfer conductance*, formed between v_1 and the summing junction. Figure 3.29c shows the resulting overall transfer characteristic. (Here we have assumed that the resistor values are such that $E_C > E_B$.) We can describe the characteristic by the breakpoints and the slopes associated with each line segment. Thus in this circuit network $R_3 - R_6 - D_3$ provides

$$E_A = \frac{-ER_3}{R_6} \qquad (3.35a)$$

For $v_1 < E_A$, D_3 is ON, D_1 and D_2 are OFF, and

$$S_A = -R_F\left(\frac{1}{R_3} + \frac{1}{R_7}\right) = S_B - \frac{R_F}{R_3} \qquad (3.35b)$$

that is, the slope is established by both R_3 and R_7.

In the interval $E_A < v_1 < E_B$ the slope is determined only by R_7 (i.e., all diodes are OFF); hence

$$S_B = -\frac{R_F}{R_7} \qquad (3.36a)$$

The next breakpoint, E_B, occurs when D_2 comes ON, and we have

$$E_B = \frac{+ER_2}{R_5} \qquad (3.36b)$$

and, in the interval $E_B < v_1 < E_C$,

$$S_C = -R_F\left(\frac{1}{R_2} + \frac{1}{R_7}\right) = S_B - \frac{R_F}{R_2} \qquad (3.37a)$$

The left-hand breakpoint is

$$E_C = \frac{+ER_1}{R_4} \qquad (3.37b)$$

and for $E_C < v_1$, both D_2 and D_1 are ON, and

$$S_D = -R_F\left(\frac{1}{R_7} + \frac{1}{R_2} + \frac{1}{R_1}\right) = S_C - \frac{R_F}{R_1} \qquad (3.38)$$

Let us now discuss the *shunt limiter circuit,* as depicted in Fig. 3.32. The basic circuit is shown first in Fig. 3.32a. Diode D_1 clamps node v_C, so that

$$v_C \leqslant E_C$$

This establishes a breakpoint at $v_1 = E_X$, $v_0 = E_Y$. These coordinates may be found as follows:

1 When D_1 is ON, $v_C = E_C$, and

$$v_0 = E_Y = -\frac{R_F}{R_2}E_C \qquad (3.39)$$

2 When D_1 is OFF, then

$$S_X = -\frac{R_F}{R_1 + R_2} \qquad (3.40)$$

and thus since

$$E_Y = S_X E_X$$

we have

$$E_X = \frac{E_Y}{S_X} = \frac{E_C(R_1 + R_2)}{R_2} \qquad (3.41)$$

Due to the forward drop of D_1, the above estimates will be somewhat in error for the circuit of Fig. 3.32a. A well-known device is to compensate for the forward drop by the diode network of Fig. 3.32c, or better yet, by the transistor circuit of Fig. 3.32d. Both of these approaches also provide good temperature compensation and will permit the clamping point to be accurately established at E_C, so that (3.39) and (3.41) will predict E_Y and E_X with good accuracy. The transistor emitter-follower action in the circuit of Fig. 3.32d also serves to isolate the limiter network from the source of clamping voltage. Typically, then, E_C may be derived from a medium-impedance source, such as the potentiometer shown.

The characteristic provided by the shunt limiter (Fig. 3.32b) also may be used for piecewise-linear function generation, again by summing up the effect of several such networks. Of course, by reversing the diode and power supply polarities (and using a *p-n-p* transistor in Fig. 3.32d), we can achieve limiting in the upper left quadrant.

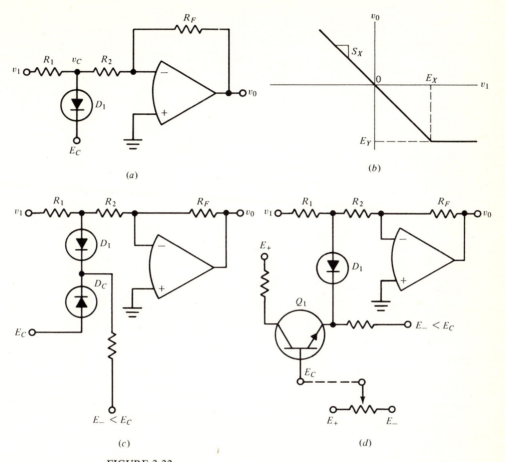

FIGURE 3.32
Shunt limiter circuit. (*a*) General circuit for positive breakpoint; (*b*) transfer characteristic; (*c*) using diode D_C to compensate for offset in D_1; (*d*) using a transistor emitter follower Q_1 to compensate for offset in D_1 and to unload node E_C.

EXAMPLE 3.10 Refer to Fig. 3.33; we want the transfer characteristic of Fig. 3.33*b*. Assume that the breakpoints are set by the two potentiometers; the emitter-follower action of the transistors will prevent the limiter networks from appreciably loading the potentiometers, and the diodes D_1 and D_2 will be clamped to reasonably stiff sources. Figure 3.33*b* shows the approximate transfer characteristic. Note that the slopes are again established by one or more networks in combination. For $2E_2 < v_1 < 2E_1$ both diodes are OFF, and the output slope is the sum of the two gains provided by the two input paths to the summing junction. If we make $R_1 = R_2$ and $R_3 = R_4$, we see that the breakpoints, from (3.41), are merely $2E_1$ and $2E_2$. ////

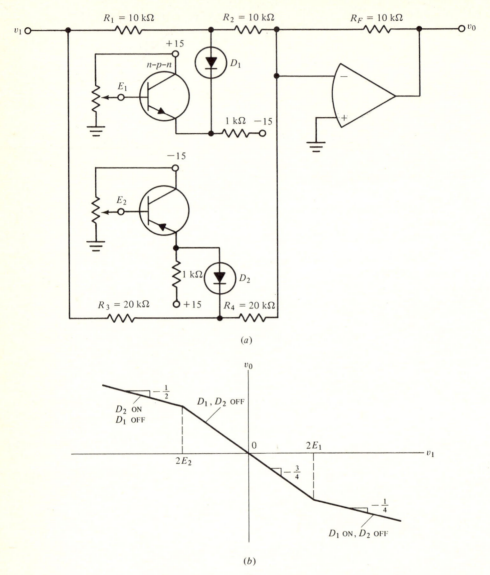

(a)

(b)

FIGURE 3.33
Refer to Example 3.10. (a) Complete circuit; (b) transfer characteristic.

Obviously, the above series and shunt limiter circuits can be combined in a variety of ways to implement a general piecewise-linear function. Commercial *diode function generator* units employ this type of circuitry to provide a general-purpose nonlinear operator with 10 to 20 line segments and adjustable slopes and/or breakpoints. Shunt limiters (such as Fig. 3.32) seem to be more commonly employed. (An interesting variation, Ref. 3, p. 257, employs the precision zero limiter of Fig. 3.17 as a basic building block for a multisegment function generator.)

Thus far, we have primarily discussed monotone functions, i.e., those with slopes of one polarity. However, the full-wave-rectifier circuit of Fig. 3.26 is definitely not monotone. Note that it was implemented by using both X and $-X$ to form $F(X)$. This general approach is used in general-purpose function generators to provide nonmonotone functions (see Ref. 3, p. 256).

EXAMPLE 3.11 Consider the two-amplifier circuit of Fig. 3.34, which uses shunt limiters to provide two breakpoints. The analysis can be organized by noting that

$$v_0 = R_2\left(-\frac{v_1}{R_4} - \frac{v_A}{R_3} + \frac{v_B}{R_1}\right)$$

$$= -\frac{1}{2}v_1 - v_A + 2v_B$$

Thus, by disabling the various signal paths, the slope of the function can be made either positive or negative. A brief inspection will lead to the conclusion that there are three regions of operation:

Region I: $v_1 < -6, D_1$ ON, D_2 OFF
Here we have

$$v_A = -3 \qquad v_B = \frac{v_1}{2}$$

and thus, from (3.42),

$$v_0 = 3 + \frac{v_1}{2}$$

Region II: $-6 < v_1 < +4, D_1$ and D_2 OFF

$$v_A = v_B = \frac{v_1}{2}$$

and thus

$$v_0 = -\frac{v_1}{2} - \frac{v_1}{2} + v_1 = 0$$

Region III: $+4 < v_1, D_1$ OFF, D_2 ON
In this region

$$v_A = \frac{v_1}{2} \qquad v_B = +2$$

and thus

$$v_0 = -\frac{v_1}{2} - \frac{v_1}{2} + 4$$

$$= -v_1 + 4$$

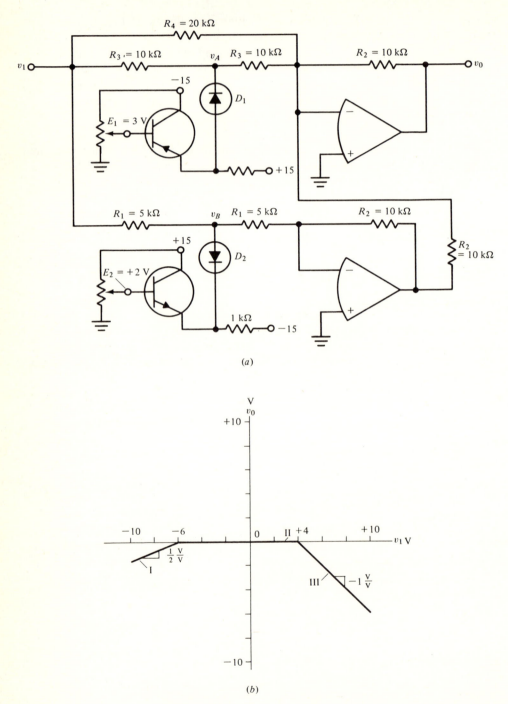

FIGURE 3.34
Refer to Example 3.11. (*a*) Complete circuit; (*b*) transfer characteristic.

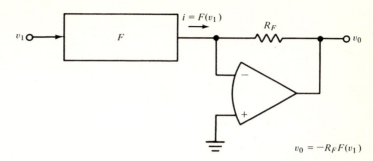

FIGURE 3.35
A general transfer conductance F in an amplifier input path, $v_0 = -R_F F(v_1)$.

Figure 3.34b shows a sketch of the resulting nonmonotone function $v_0(v_1)$. Note that if we changed the clamping voltages, the breakpoints would change, but not the individual slopes in the regions. ////

Implicit Function Generation

The previous discussion described ways to build a network with a desired short-circuit conductance, approximating some nonlinearity by piecewise-linear functions. Figure 3.35 illustrates the general approach, where the network F has a short-circuit conductance described by

$$i = F(v_1) \qquad (3.42a)$$

In Fig. 3.35, we have then

$$\frac{v_0}{R_F} + F(v_1) = 0$$

or

$$v_0 = -R_F F(v_1) \qquad (3.42b)$$

and the network F could be implemented by, e.g., the diode limiter circuits just described. Now suppose we embed the network F in an operational amplifier feedback loop, as in Fig. 3.36. This situation is basically an extension of the feedback limiter concepts discussed in Sec. 3.2. Indeed, one could view the limiter circuit of Fig. 3.14 as being made up of a pair of series limiter circuits similar to those in Fig. 3.30. In the configuration of Fig. 3.36,

$$F(v_0) + \frac{v_1}{R_1} = 0$$

or

$$v_0 = F^{-1}\left(-\frac{v_1}{R_1}\right) \qquad (3.43)$$

where F^{-1} denotes the *inverse* function to F, that is

$$F^{-1}[F(x)] = x$$

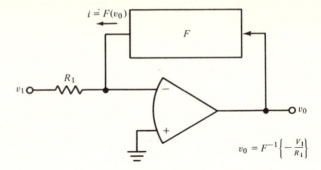

FIGURE 3.36
A general transfer conductance F in an amplifier feedback path, $v_0 = F^{-1}(-v_1/R_1)$.

Hence a network designed to implement an approximation to the sine function in a circuit like Fig. 3.35 could be used to implement arcsin in Fig. 3.36. We will use this technique of *implicit function generation* to advantage in Sec. 3.5, where we get $\sqrt{X}$ from X^2.

Amplitude Selectors and Peak Detectors

Figure 3.37 shows two ways to implement the maximum-selection operator

$$v_3 = \max(v_1, v_2)$$

Note that the circuit of Fig. 3.37a is a simple diode AND gate. For analog signals v_1 and v_2 the output of such a circuit will be approximately equal to the maximum of the two, but there will be an error in the operation resulting from the diode forward drop. In Fig. 3.37b, however, we use two precision limiters (see Fig. 3.17) in a *precision selector* circuit that yields the desired maximum-selection operation with high accuracy. Note, unfortunately, that the inputs must now be $-v_1$ and $-v_2$ (and both inputs must be negative in polarity). The 1-$M\Omega$ resistor is used as a "keep-alive" resistor to ensure that one diode is always ON. Diodes D_3 and D_4 act as clamps to prevent overload. Note that the circuit always has a low output impedance. Additional diodes (up to perhaps five) can be added to the circuits of Fig. 3.37 to obtain

$$v_3 = \max(v_1, v_2, \ldots)$$

Note also that if all diode and power-supply polarities are reversed, then we obtain

$$v_3 = \min(v_1, v_2)$$

Figure 3.38a shows an (inverting) *peak detector* circuit (Ref. 4, p. 138). A typical waveform is illustrated in Fig. 3.38b. Switch S is used to reset the circuit output to zero before the history of v_1 is observed. Note that $-v_1$ must be provided and must be one polarity. Capacitor C is chosen as a compromise between slewing rate and holding ability;

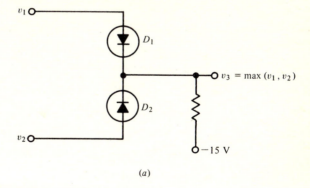

(a)

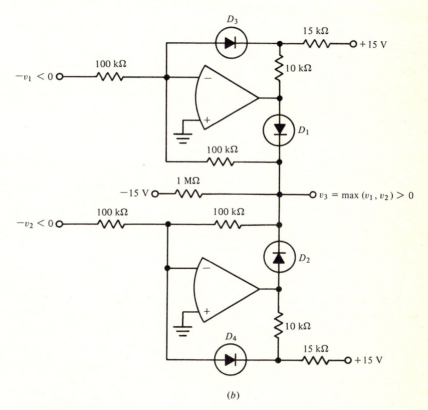

(b)

FIGURE 3.37

Selection circuits. (a) A simple diode maximum selector; (b) precision level selection using operational amplifiers. Reversal of diodes and power polarities provides min (v_1, v_2).

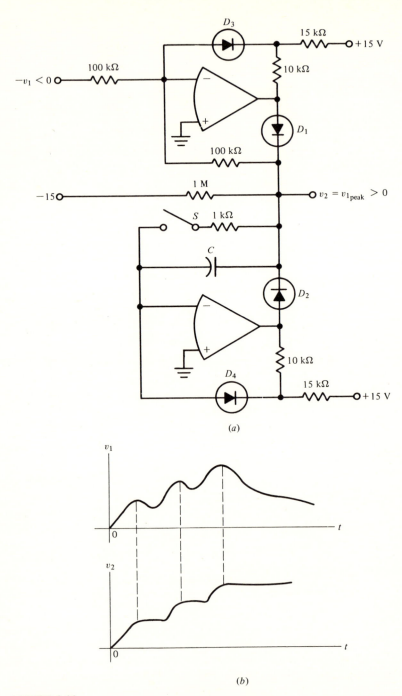

(a)

(b)

FIGURE 3.38
Two-amplifier inverting peak detector circuit. (a) The circuit combines two precision limiters with a capacitor to retain the peak value; (b) typical waveforms.

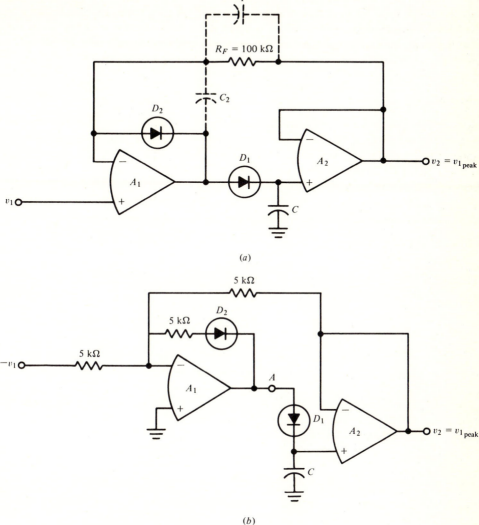

FIGURE 3.39
Two more peak detector circuits. (*a*) A noninverting circuit; C_1 and C_2 may be required for stability; (*b*) a high-speed inverting circuit; slewing rate may be improved by adding current boosting at point A.

0.01 μF is typical for signals in the audio-frequency range. Figure 3.39 shows two other useful circuits. Figure 3.39*a* (see Ref. 3, p. 356) shows a noninverting circuit; capacitor C is charged through D_1 by A_1. Amplifier A_2 serves as a buffer between C and the output when D_1 is OFF. D_2 prevents A_1 from overloading. Capacitors C_1 and C_2 may be needed to improve stability and transient response. A_1 must have good common-mode rejection and be stable with C as a load. A_2 should have a high input impedance (e.g., an FET or varactor input stage). The circuit of Fig. 3.39b (Ref. 6) is an inverting circuit which has been used in high-speed applications; it is more stable than the circuit of Fig. 3.39a. The circuit slewing rate may be improved by adding a current-boosting emitter follower at point A.

Bistables

Figures 3.40 and 3.41 show two implementations of a bistable or flip-flop (see also Ref. 3, p. 363, and Ref. 4, p. 139). The two-amplifier circuit of Fig. 3.40a provides the noninverting transfer characteristic shown in Fig. 3.40b. The tops of the characteristic will be fairly flat, even with a simple pair of diode clamps on the output, because of the low-amplitude drive from A_1. If D_1 and D_2 are replaced by transistors (as in Fig. 3.20), then the tops will be very flat. Normally R_5 is made large compared to R_2 and R_3. The dimensions of the hysteresis loop are determined by the following equations:

$$L_+ = \frac{15R_2}{R_1} \tag{3.44a}$$

$$L_- = \frac{-15R_3}{R_4} \tag{3.44b}$$

$$S_+ = \frac{R_6|L_-|}{R_7} = \frac{15R_6R_3}{R_7R_4} \tag{3.44c}$$

and

$$S_- = \frac{R_6L_+}{R_7} = \frac{-15R_6R_2}{R_7R_1} \tag{3.44d}$$

EXAMPLE 3.11 The circuit of Fig. 3.40 can be made to implement a 10- by 10-V square hysteresis loop by making

$$R_2 = R_3 = 10 \text{ k}\Omega$$
$$R_1 = R_4 = 15 \text{ k}\Omega \quad \text{(adjust to get exact limiting)}$$
$$R_6 = R_7 \quad \text{(typically 10 k}\Omega\text{)}$$
$$R_5 = 100 \text{ k}\Omega \quad \text{(not critical)}$$

With these values, and ±15-V supplies,

$$L_+ = +10 \quad L_- = -10$$
$$S_+ = +10 \quad S_- = -10 \qquad \text{////}$$

The one-amplifier circuit of Fig. 3.41 is actually an extension of the technique of using positive feedback to add hysteresis to a comparator (Fig. 3.24). With simple diode clamps, the output levels will vary some with the input. By adding better limiting, e.g., by using a transistor limiter (see Fig. 3.20), one can obtain an almost square characteristic (with a polarity inversion).

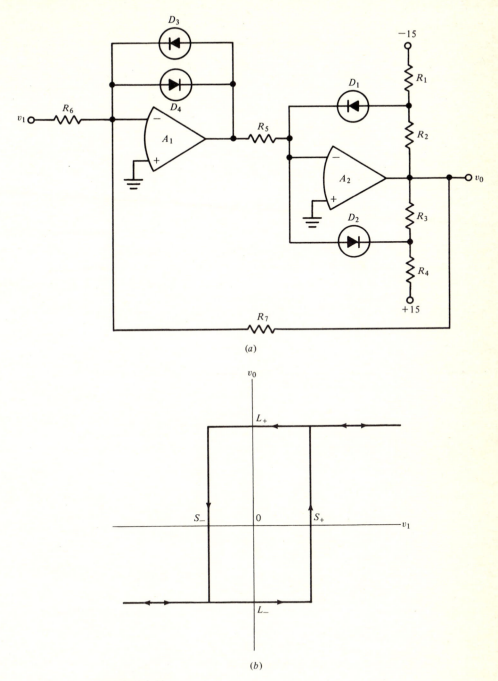

(a)

(b)

FIGURE 3.40

A two-amplifier bistable circuit. (a) Output levels are set by soft limiters on A_2, and diodes D_3 and D_4 limit the output of A_1 and improve output level regulation and circuit speed; (b) typical output characteristic; note that it is noninverting.

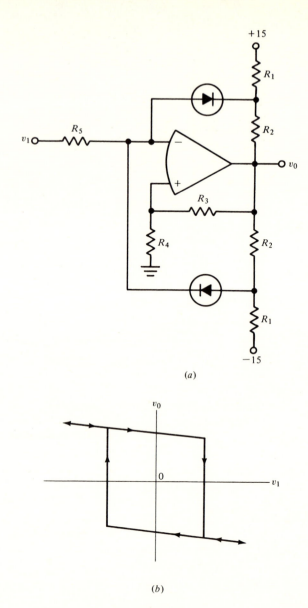

(a)

(b)

FIGURE 3.41

A simple one-amplifier bistable. (a) Positive feedback via R_3 and R_4 adds hysteresis; (b) output characteristic is not well limited and is inverting.

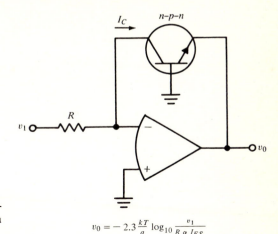

FIGURE 3.42
A basic log operator circuit; the transistor is used in the transdiode connection for improved dynamic range.

$$v_0 = -2.3 \frac{kT}{q} \log_{10} \frac{v_1}{R \, \alpha \, I_{ES}}$$

We shall see in Sec. 3.6 that these bistable circuits are used in the design of precision waveform generators (square and triangle waves).

3.4 LOGARITHMIC OPERATORS

In the past six or seven years, a number of specialized log and antilog function generation circuits have been developed based upon the exponential behavior of a *p-n* junction, or more specifically, the exponential behavior of the emitter-base junction of a transistor (Fig. 3.42). It is not our intention here to provide all the information required to design these circuits so that they perform well. The interested reader can learn more about this from other sources (see Ref. 3, sec. 7.4; Ref. 7, sec. 5.3; and Ref. 8). We will attempt to describe the basic manner in which these circuits work; further, we will show a number of ways in which logarithmic operators may be used to implement a variety of nonlinear functions.

In Fig. 3.42, we see a basic log operator circuit. The transistor is connected in the so-called *transdiode* configuration. In this connection, the Ebers-Moll model of a bipolar transistor predicts (Ref. 7, pp. 87ff.):

$$I_C = \alpha I_{ES} (e^{-q v_{EB}/kT} - 1) \qquad (3.45)$$

where v_{EB} = emitter-base voltage (v_0 in Fig. 3.42)

T = temperature, K

k = Boltzmann's constant (1.38×10^{-23} J/K)

q = electron charge (1.6×10^{-19} C)

I_{ES} = reverse saturation current of the emitter-base junction, with the collector connected to the base

α = forward grounded-base current gain of the transistor

NOTE: $kT/q \cong 26$ mV at $27°$C.

Whenever $v_{EB} < 100$ mV or so, (3.45) becomes

$$I_C \cong \alpha_F I_{ES} e^{-q v_{EB}/kT} \qquad (3.46)$$

Thus in Fig. 3.42, we can show that since $v_{EB} = v_0$,

$$v_0 \cong \frac{-kT}{q} \ln \frac{v_1}{R \alpha I_{ES}}$$

or

$$v_0 \cong \frac{-2.3kT}{q} \log_{10} \frac{v_1}{R \alpha I_{ES}} \qquad (3.47)$$

Thus we obtain an output voltage proportional to the logarithm of the input voltage. This equation is obviously very temperature-sensitive. Moreover, as the input signal varies over a wide range, the small-signal or incremental feedback resistance provided by the transistor will change, possibly by several orders of magnitude, that is,

$$r_{ac} \cong \frac{kT}{qI_C} = \frac{0.026}{I_C} \Omega \qquad (3.48)$$

This means that for small currents (low input signals), the feedback impedance is possibly quite high, and thus any stray capacitance associated with the circuit may interact to yield poor frequency response for small signals. On the other hand, for large signals, the feedback impedance is quite low, and this may lead to stability problems. Of course, the ideal exponential relationship will not hold exactly; in properly selected transistors, however, reasonably accurate "logging" can be performed over 6 to 8 decades of input signal level.

The circuit of Fig. 3.43 shows a more practical version of the logarithmic operator. It may be shown (see Prob. 3.26) that

$$v_0 \cong -\frac{R_1 + R_2}{R_2} \frac{kT}{q} \ln \frac{i_1}{i_2} \qquad (3.49a)$$

or

$$v_0 = -\frac{R_1 + R_2}{R_2} \frac{kT}{q} \ln \frac{v_1 R_3}{v_2 R_5} \qquad v_1, v_2 > 0 \qquad (3.49b)$$

Thus we have a *log ratio* operator. Such a circuit might be packaged as a single module, in which case the resistance values would be selected for some nominal range of operation. The capacitances C_1 and C_2 are usually required for stability, and they limit the speed of operation. R_4 permits amplifier A_2 to operate over its normal output range.

This circuit has a transfer characteristic

$$v_0 = -K_1 \ln \frac{K_2 v_1}{v_2} = \frac{-K_1}{0.433} \log_{10} \frac{K_2 v_1}{v_2} \qquad v_1, v_2 > 0 \qquad (3.50)$$

with

$$K_1 = \frac{R_1 + R_2}{R_2} \frac{kT}{q} \qquad (3.51a)$$

and

$$K_2 = \frac{R_3}{R_5} \qquad (3.51b)$$

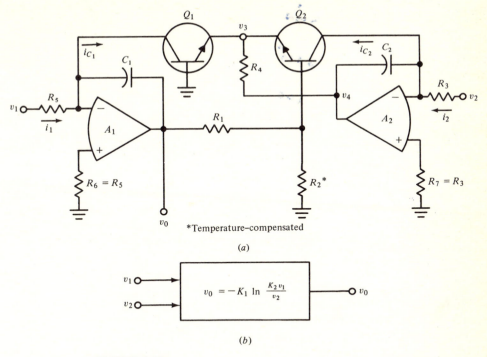

FIGURE 3.43
A practical log operator circuit. (*a*) This two-amplifier arrangement permits finding the log ratio; (*b*) a block-diagram symbol.

Often one input, for example v_2, is held at a reference potential, and the circuit merely generates an output proportional to the log of the input, i.e., it is a *log operator* or *log amplifier.* Appropriate scale factors are chosen. Such a circuit, with careful design and selection of critical components, can be made accurate to perhaps 1 percent over an input dynamic range of perhaps 6 decades. Note that R_2 is usually made from some form of temperature-sensitive element, e.g., a thermistor, and is designed to provide temperature compensation for changes in temperature, in order to hold K_1 in (3.51a) reasonably constant. Due to the capacitative compensation, such circuits have rather slow settling times, particularly when the input signals are small.

EXAMPLE 3.12 Evaluate K_1 and K_2 if, in Fig. 3.43,

$$R_1 = 15.7 \text{ k}\Omega \qquad R_3 = 1.5 \text{ M}\Omega$$

$$R_2 = \quad 1 \text{ k}\Omega \qquad R_5 = \quad 10 \text{ k}\Omega$$

$$\frac{kT}{q} = 0.026 \text{ V} \qquad V_2 = +15 \text{ V dc}$$

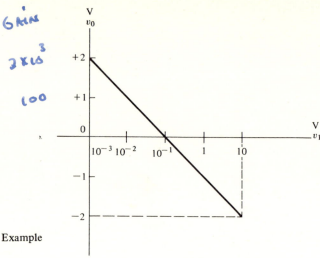

FIGURE 3.44
The transfer characteristic of Example 3.12, showing 4-decade range.

From (3.51) we have that

$$K_1 = 0.433 \qquad K_2 = 150$$

Thus (3.50) becomes

$$v_0 = -0.433 \ln 10v_1$$
$$= -\log_{10} 10v_1 \qquad (3.52)$$

Figure 3.44 shows the plot of (3.52) over 4 decades of input values (see also Prob. 3.27).

////

Antilog or Exponentiation Operators

The circuit of Fig. 3.45 does the inverse operation to that of Fig. 3.43. Here we can show that

$$v_0 = \frac{v_2 R_6}{R_3} \exp\left(\frac{-q}{kT} \frac{R_2}{R_1 + R_2} v_1\right) \qquad v_1 > 0 \qquad (3.53)$$

In this circuit v_2 is usually a fixed reference voltage, but it could be a time-varying signal. The general form of the operation is thus

$$v_0 = K_1 v_2 e^{-K_2 v_1} \qquad v_1 > 0 \qquad (3.54a)$$

with

$$K_1 = \frac{v_2 R_6}{R_3} \qquad (3.54b)$$

and

$$K_2 = \frac{q}{kT} \frac{R_2}{R_1 + R_2} \qquad (3.54c)$$

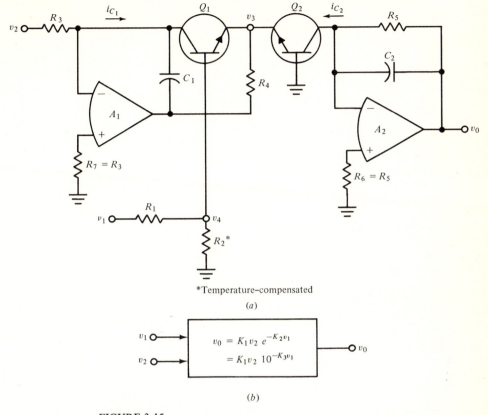

FIGURE 3.45
A practical exponentiation or antilog operator. (a) In this two-amplifier circuit v_2 is often a fixed reference, but it may be time-varying; (b) block-diagram symbol.

Again at room temperature (27°C) this last equation becomes

$$K_2 = \frac{R_2}{0.026(R_1 + R_2)} \qquad (3.54d)$$

Alternatively, we can express (3.53) in the form

$$v_0 = v_2 K_1 \times 10^{-K_3 v_1} \qquad v_1 > 0 \qquad (3.55a)$$

where K_1 is given by (3.54b) and

$$K_3 = \frac{q}{2.3kT} \frac{R_2}{R_1 + R_2} \qquad (3.55b)$$

$$= \frac{R_2}{0.0598(R_1 + R_2)} \qquad \text{at } 27°C \qquad (3.55c)$$

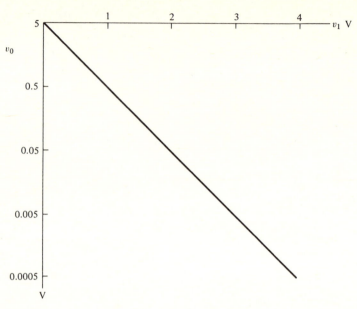

FIGURE 3.46
The transfer characteristic of Example 3.13; note the log scale for the output v_0.

EXAMPLE 3.13 In Fig. 3.45, if

$$v_2 = +15 \text{ V dc}$$
$$R_3 = R_7 = 30 \text{ k}\Omega$$
$$R_4 = 2 \text{ k}\Omega \quad \text{(not critical)}$$
$$R_1 = 15.7 \text{ k}\Omega \quad R_2 = 1 \text{ k}\Omega$$
$$R_5 = R_6 = 10 \text{ k}\Omega$$

then we have, from (3.55),

$$v_0 = 5 \times 10^{-v_1} \quad v_1 > 0 \qquad (3.56)$$

Fig. 3.46 shows a sketch of (3.56).   ////

Special Nonlinear Operations

Of course, the log and antilog circuits described above can be combined to implement a wide variety of special nonlinear operations. Consider the block diagrams of Fig. 3.47. Here the blocks labeled "Log" represent circuits similar to Fig. 3.43, and the blocks labeled "Antilog" represent circuits similar to Fig. 3.45. The other elements are simple summing and coefficient operations.

Thus we see in Fig. 3.47a how a log and an antilog circuit can be combined to yield a general *exponentiation operator*

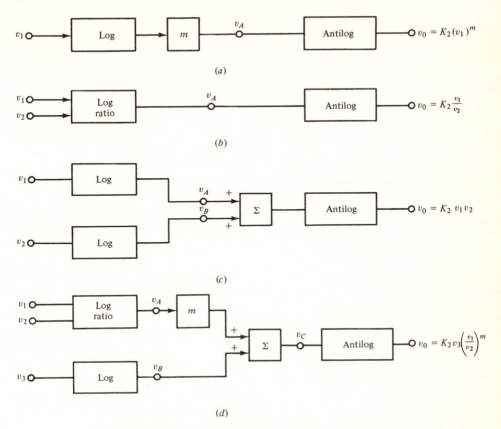

FIGURE 3.47
Combining log and antilog operators. (*a*) One log and one antilog operator do general exponentiation (raising to a power); (*b*) one log ratio and one antilog do one-quadrant division; (*c*) similarly, two log operators and an antilog do one-quadrant multiplication; (*d*) a general multipurpose nonlinear operator; see Table 3.1 for typical capabilities.

$$v_0 = K_2 v_1^{\ m} \qquad (3.57a)$$

In principle, m can be either positive or negative; typically $0.2 < m < 5$. Note also that v_1 is of one polarity, $v_1 > 0$.

In Fig. 3.47*b*, we combine a log ratio circuit with an antilog circuit to obtain a *division operator* or *divider*

$$v_0 = K_2 \frac{v_1}{v_2} \qquad (3.57b)$$

Again the polarity of the inputs is restricted, typically $v_1 > 0$ and $v_2 > 0$.

Similarly, Fig. 3.47*c* shows how the output of two log operators may be summed, and then an antilog circuit used to form a *multiplication operator* or *multiplier*

$$v_0 = K_2 v_1 v_2 \qquad (3.58)$$

Obviously more inputs can be added; again $v_1 > 0$ and $v_2 > 0$ (a one-quadrant multiplier).

As a final example, consider Fig. 3.47d. This shows the block diagram of a general *multipurpose nonlinear module* that implements

$$v_0 = K_2 v_3 \left(\frac{v_1}{v_2}\right)^m \qquad (3.59)$$

Such a unit can be used in a variety of applications (Refs. 9 and 10). Of course, with $m = 1$ and $v_3 =$ constant, we have a divider; with $m = 1$ and $v_2 =$ constant, we have a multiplier. Moreover, since m is adjustable, this provides a great deal of flexibility.

Table 3.1 shows some of the salient specifications of the Analog Devices Model 433 Programmable Multifunction Module, which is similar in structure to Fig. 3.47d. It implements the general function

$$v_0 = \frac{10}{9} v_y \left(\frac{v_z}{v_x}\right)^m \qquad 0.2 \geqslant m \geqslant 5.0$$

using log and antilog techniques.

The exponent m is set by adding two external resistors, or in the case of $m = 1$, by making a simple connection. Note the specifications, which indicate an accuracy of about 0.5 percent of full-scale or better at 25°C. The principal limitations on this unit are that the inputs must be of one polarity (hence a one-quadrant multiplier or divider may be implemented), and that the speed is somewhat restricted when the inputs are small in magnitude (400-Hz bandwidth for inputs of the order of 0.01 V). This is, however, characteristic of such units based upon logarithmic operations.

Table 3.1 CHARACTERISTICS OF THE ANALOG DEVICES MODEL 433 PROGRAMMABLE MULTIFUNCTION MODULE*

General expression	$v_0 = +10/9 \, v_y (v_z/v_x)^m$
Typical signal range for spec. error	$0.01 \text{ V} \leqslant v_y, v_z \leqslant 10 \text{ V}$ $0.1 < v_x < 10 \text{ V}$
Range of m (adjust with external resistors)	$0.2 \leqslant m \leqslant 5$
Accuracy as multiplier or divider (referred to output)	Typical = 5 mV ± 0.3 percent of output at 25°C Drift ±1 mV/°C Maximum ±50 mV
DC offset	±10 mV, ±1 mV/°C
Small-signal bandwidth v_y, v_z	100 kHz, all inputs $\cong$ 10 V 400 Hz, all inputs $\cong$ 0.01 V

*Analog Devices, Inc., Norwood, Mass.

Reference 10 illustrates how this type of operator can be used to implement polynomial approximations to transcendental functions. For example, noting that

$$\sin \theta \cong \theta - \frac{\theta^{2.827}}{2\pi}$$

(which is accurate to 0.25 percent, $0 \geqslant \theta \geqslant \pi/2$ rad), we can implement the sine function with one such module and an external summing amplifier (see also Prob. 3.31).

3.5 MULTIPLIERS, DIVIDERS, AND APPLICATIONS

We often need to perform the following analog signal operations:

Multiplication:

$$v_m = K_m v_1 v_2 \qquad (3.60)$$

Division:

$$v_d = \frac{K_d v_1}{v_2} \qquad (3.61)$$

We have already seen how the logarithmic operators of the last section may be used to implement these operations. In this section, we will describe additional important methods of doing these operations, and how functional modules that can do multiplication and division may be used in a variety of nonlinear circuit applications.

Scale Factors

We normally assume that general-purpose analog signal operators should be able to accommodate analog signals with a ±10-V range; similarly, their output range is typically ±10 V. Thus *amplitude scaling* considerations must be used in selecting the gain factors K_m and K_d above, and in using the operators for designing signal-processing systems. The proper choice for the multiplier gain factor is fairly obvious: If the analog signal range is ±E, then the multiplier should implement

$$v_m = \frac{v_1 v_2}{E} \qquad (3.62)$$

that is, $K_m = 1/E$. This assures that

$$|v_0| \leqslant E \qquad (3.63)$$

Thus for a ±10-V analog system, we would normally design the multiplier so that

$$v_m = 0.1 v_1 v_2 \qquad (3.64)$$

A truly general-purpose multiplier should feature *four-quadrant* operation, i.e., no limitations should be placed on the polarity of either input. Thus for a ±10-V analog system, (3.63) would be satisfied, and $|v_0| \leqslant 10$ with the inputs permitted to range into four quadrants: $-10 \text{ V} \leqslant v_1, v_2 \leqslant 10 \text{ V}$. Note that the *logarithmic multiplier* described in Sec. 3.4 *does not* meet this requirement.

In the case of a divider, the choice of the gain factor K_d is less obvious; it is dependent on the anticipated range of values of the inputs, in particular on the *minimum magnitude* of the *divisor* signal v_2. Typically, for a 10-V analog system, we would pick $K_d = 10$ and implement

$$v_d = 10 \frac{v_1}{v_2} \qquad (3.65)$$

Obviously, this choice is based on the assumption that the ratio v_1/v_2 never exceeds 1. Moreover, we usually have to restrict the polarity and range of the divisor v_2. Typically, for a 10-V system, we would see specifications such as

$$v_d = 10 \frac{v_1}{v_2} \qquad \left| \frac{v_1}{v_2} \right| \leqslant 10 \qquad \text{and} \qquad 1 \text{ V} \leqslant v_2 \leqslant 10 \text{ V} \qquad (3.66)$$

These specifications assure that overload will never occur, and they are typical of the allowable ranges within which divider accuracy will be guaranteed. The above equation describes a *two-quadrant* divider, in which the numerator input may take on either polarity, but the denominator input is restricted to one polarity.

Multiplier and divider performance *specifications* usually fall into two categories: *static accuracy* and *frequency response*. Manufacturers' specifications are usually given in terms that make the multiplier look as good as possible.

Static accuracy is usually given in terms of percent of the nominal output range. Thus, for a 10-V multiplier, a 1 percent static accuracy means that the output will not differ from the ideal by more than 0.1 V, so long as the inputs are within their normal limits. *Note that for small outputs,* this could be a relatively *large true percentage error.* Suppose, for example, that $v_1 = v_2 = 1 \text{ V}$ in (3.64). Then a nominal 1 percent static accuracy specification could mean that the output error might be as much as 0.1 V, which is as large as the desired value; i.e., we might have a 100 percent error relative to the proper value!

Dynamic accuracy or frequency response is also usually specified in an optimistic way. Often one will see a specification of -3-dB bandwidth. This is an almost useless measure of multiplier speed, since errors due to phase shift might be quite high long before the -3-dB frequency was reached.

Consider Fig. 3.48; here the desired waveform is

$$v_d = A \sin \omega t \qquad (3.67a)$$

but the actual waveform might be different, due to phase shift, even though the amplitude was correct, for example, if

$$v_a = A \sin (\omega t - \theta) \qquad (3.67b)$$

then the error would be

$$\delta = A \left[\sin \omega t - \sin (\omega t - \theta) \right] \qquad (3.68)$$

For small θ, $\sin \theta \cong \theta$ and $\cos \theta \cong 1$; thus, using (3.68) and some well-known identities,

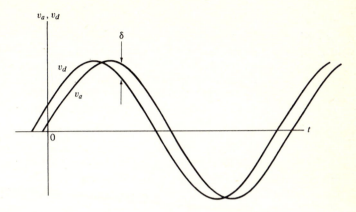

FIGURE 3.48
The effect of time delay or phase shift between two sinusoidal waveforms; v_d is a desired signal, and v_a is the actual (delayed) signal. The absolute difference δ is the true total error.

$$\delta = A\,[\sin \omega t - \sin \omega t \cos(-\theta) - \cos \omega t \sin(-\theta)]$$
$$\cong A\theta \cos \omega t \qquad (3.69)$$

That is, the peak error is

$$\delta_{\text{peak}} \cong A\theta \qquad (3.70)$$

Thus the maximum error is 100θ percent, where θ is in radians, or about 1.8 percent per *degree* of phase error. Suppose now we take a multiplier (or divider) and hold one input constant, while the other is varied sinusoidally. A typical first-order frequency response will have at least 45° phase lag at the −3-dB frequency f_0 Hz. At frequencies well below f_0, we can often estimate the phase shift as

$$\theta \cong -f/f_0 \text{ rad } f \ll f_0 \qquad (3.71)$$

Thus we will have typically 1 percent error at 0.01 f_0 Hz! This well-known result (see Ref. 4, Sec. 3.2) applies to linear operators in general, and shows how important knowledge of phase shift in amplifiers and filters may be! Thus one should not expect high accuracy from a multiplier *unless input frequencies are quite small compared to the −3-dB frequency.*

Instead of merely giving the −3-dB frequency, a cooperative manufacturer will give additional information such as slew rate, settling time, and frequencies below which the total error due to phase and/or amplitude response can be assumed to be less than some guaranteed amount. Table 3.2 lists typical figures for various popular types of multipliers.

Major Types of Multipliers

Analog multiplication may be implemented in a variety of ways. We have already described one-quadrant multiplication in the previous section, using logarithmic techniques. Frequently we want full four-quadrant multiplication (division, as we shall soon see, is usually implemented implicitly, by putting a multiplier in an operational amplifier feedback path). Currently, most four-quadrant multipliers are made using one of three major techniques:

    1 Quarter-square (diode function generation) methods
    2 Pulse-width-height modulation time-division methods
    3 Current ratioing (variable transconductance) methods

We will describe these approaches in summary form, in an attempt to permit the reader to draw broad comparisons. Other methods have been used (see, e.g., Ref. 4, Chap. 6); however, the above represent the most commonly used approaches. Table 3.2 compares the performance of typical units.

Quarter-square multipliers Refer to Fig. 3.49, which sketches a block diagram of the quarter-square multiplication technique. This general approach has been very popular for the past 10 to 15 years, but has perhaps become obsolescent. Diode function generation networks (see Sec. 3.3) are used to approximate the squaring operation by piecewise-linear approximation, often using 10 or more segments. The system is usually built with discrete components, except possibly for the operational amplifiers. Clever designs have evolved which use only three operational amplifiers, with the rest of the operations performed with special function generation networks. The approach relies upon the basic relationship

$$\frac{(X + Y)^2}{4} - \frac{(X - Y)^2}{4} = XY \qquad (3.72)$$

hence the jargon quarter-square. In Fig. 3.49 we see that sum and difference operations are performed on the two inputs. Then nonlinear networks are used to form the square of the absolute value of each. A final summing operation combines the two squares to implement (3.60) by means of (3.72). Reasonable static accuracy can be achieved (0.1 percent at one temperature, and perhaps 0.25 percent over a reasonable range of temperatures). Speed may be fairly good, with a 1 percent bandwidth of 10 to 20 kHz.

Table 3.2 COMPARISON OF ANALOG MULTIPLIERS*

Type	Static error, %	−3-dB bandwidth	1% bandwidth	Slew rate, V/s	Accuracy stability, %/°C	Cost, $
Quarter-square or piecewise linear	0.25–0.5	1–2 MHz	30 kHz	10^7	0.05	300–600
Pulse-width-height modulation	0.1	100 kHz	700 Hz	10^6	0.02	125–150
Current ratioing or transconductance	0.5–2	1–10 MHz	50 kHz	10^8	0.05	5–20

*Figures are typical of current (1973) units, and not representative of any one manufacturer's product.

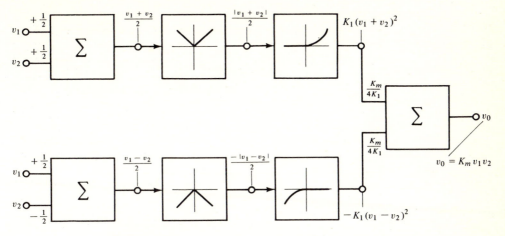

FIGURE 3.49
Block diagram of a quarter-square multiplier; the absolute-value and squaring operations are often implemented by a single diode function generation network. Two such networks and three amplifiers form a complete four-quadrant multiplier.

Cost of fabrication is relatively high. A large number of manufacturers have marketed multipliers based on this principle for at least 10 years.

Pulse-width-height modulation multipliers In Fig. 3.50, we see a block diagram of this type of multiplier. Often the output summing and filtering are combined, and the combining of the triangle wave with v_2 is performed at the comparator input. We have split these operations up for clarity. In Fig. 3.50b, we see typical waveforms within the unit. The triangle wave v_T is usually generated by using a precision bistable in a loop with an integrator (see the next section for more detail). The comparator output establishes a duty cycle for the switching operation such that

$$T_1 = \frac{v_2 + E}{2E} T \qquad (3.73a)$$

and

$$T_2 = \frac{E - v_2}{2E} T \qquad (3.73b)$$

where T is the period of the triangle waveform. The action of the switch causes the waveform at v_4 to appear as shown, with the peak-to-peak amplitude of v_4 equal to $2v_1$, and the area under v_4 proportional to v_2. Thus v_4 has an average value

$$v_{4,\mathrm{av}} = K v_1 v_2$$

The output low-pass filter removes the short-term variations and delivers the average value to the multiplier output. This type of multiplier normally has superior static accuracy (0.1 percent over a reasonable temperature range). The method has severe dynamic

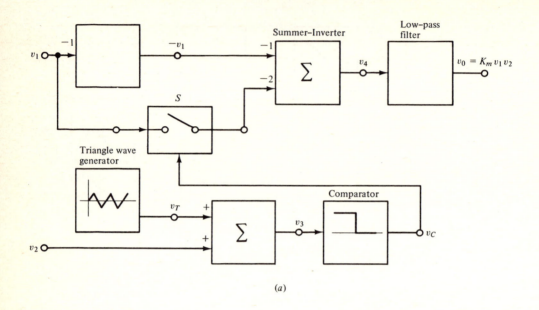

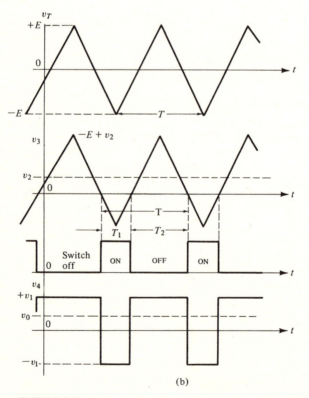

FIGURE 3.50
Pulse-width-height modulation. (*a*) Block diagram; (*b*) typical internal waveforms.
Note that v_0 is proportional to the average value of v_1.

limitations, however, due to the phase shift inherent in the output filter. Triangle wave frequencies in the megahertz range still yield a multiplier with a limited bandwidth; the typical -3-dB bandwidth may be under 1 kHz. Nevertheless, this approach is often employed where high static accuracy is called for.

Current ratioing or variable transconductance multipliers Following the announcement of the basic method by B. Gilbert in 1968 (Ref. 12), this type of multiplier has become increasingly more popular. Currently the entire unit is fabricated on a single monolithic IC chip. Precise matching of device characteristics is required for this type of multiplier, and IC fabrication is almost mandatory. This type of multiplier usually has moderate static accuracy (1 to 2 percent) but quite good speed (-3-dB bandwidths of over 10 MHz are readily achieved). Moreover, large-quantity production costs are far below those of comparable units based upon other design approaches, such as those described above. Figure 3.51 shows the essential details of this type of multiplier.

The operation of this type of circuit is dependent upon precise matching of device characteristics. Figure 3.51a shows the basic *gain cell*, which is the heart of the multiplier circuit. Let us assume that D_1, D_2, Q_1, and Q_2 are all matched, and that the current gain of the transistor is very high; then we can write

$$i_7 \cong e^{\alpha v_{D2}} \qquad (3.74a)$$

$$i_8 \cong e^{\alpha v_{D1}} \qquad (3.74b)$$

$$i_3 \cong e^{\alpha v_{BE1}} \qquad (3.74c)$$

$$i_4 \cong e^{\alpha v_{BE2}} \qquad (3.74d)$$

where $\alpha = q/kT$

Kirchhoff's voltage law gives

$$v_{D1} + v_{BE1} = v_{BE2} + v_{D2} \qquad (3.75)$$

or

$$v_{D2} - v_{D1} = v_{BE1} - v_{BE2} \qquad (3.76)$$

Thus

$$\frac{i_7}{i_8} = e^{\alpha(v_{D2} - v_{D1})} \qquad (3.77a)$$

and

$$\frac{i_3}{i_4} = e^{\alpha(v_{BE2} - v_{BE1})} \qquad (3.77b)$$

or

$$\frac{i_3}{i_4} = \frac{i_7}{i_8} \qquad (3.78)$$

Thus these two current ratios are held equal. Similarly, in Fig. 3.51b,

$$\frac{i_3}{i_4} = \frac{i_6}{i_5} = \frac{i_7}{i_8} \qquad (3.79)$$

Also, in Fig. 3.51b,

$$i_1 = i_3 + i_4 \qquad\qquad i_2 = i_5 + i_6$$

$$i_9 = i_3 + i_5 \qquad\qquad i_{10} = i_4 + i_6$$

$$i_1 + i_2 = i_A \qquad\qquad i_7 + i_8 = i_B$$

$$v_1 = R_1(i_8 - i_7) \qquad\qquad v_2 = R_1(i_1 - i_2)$$

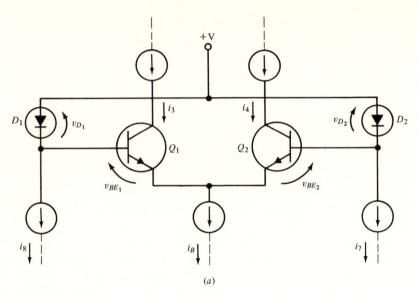

(a)

FIGURE 3.51
The current ratioing or transconductance type of multiplier. (a) Basic gain cell;
(b) a complete four-quadrant multiplier circuit. The entire circuit may be
implemented on one IC chip.

or

$$i_8 - i_7 = \frac{v_1}{R_1} \qquad\qquad i_1 - i_2 = \frac{v_2}{R_1} \qquad (3.80)$$

$$v_A - v_B = R_2(i_{10} - i_9)$$

Also,

$$v_0 = K(v_A - v_B)$$

or

$$v_0 = KR_2(i_{10} - i_9) \qquad (3.81)$$

From the above, one can obtain

$$\frac{i_4}{i_5} = \frac{i_5}{i_6} = \frac{i_8}{i_7} = \frac{i_B + v_1/R_1}{i_B - v_1/R_1} = F_1 \qquad (3.82)$$

and

$$\frac{i_1}{i_2} = \frac{i_A + v_2/R_1}{i_A - v_2/R_1} = F_2 \qquad (3.83)$$

Also,

$$i_A = i_3 + i_4 + i_5 + i_6 = i_9 + i_{10} = i_1 + i_2$$

After some more algebra, one can obtain

$$i_3 = \frac{i_A F_2}{(1 + F_1)(1 + F_2)} \qquad i_4 = \frac{i_A F_1 F_2}{(1 + F_1)(1 + F_2)}$$

$$i_5 = \frac{i_A F_1}{(1 + F_1)(1 + F_2)} \qquad i_6 = \frac{i_A}{(1 + F_1)(1 + F_2)}$$

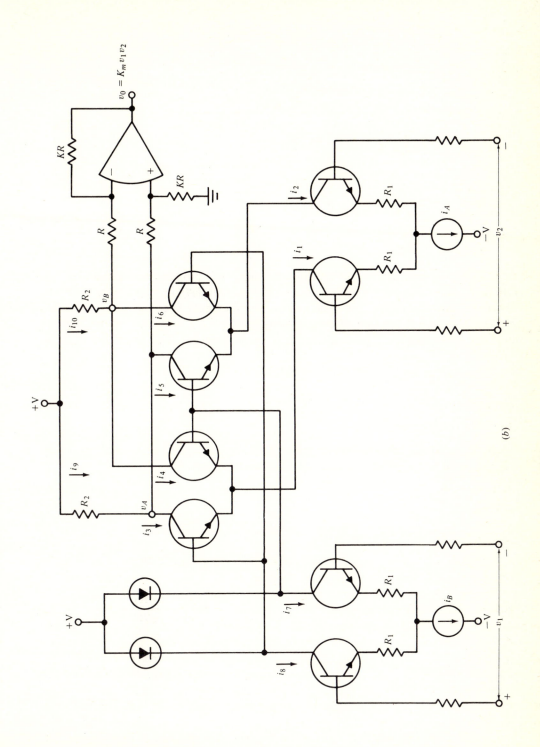

(b)

Thus
$$i_{10} - i_9 = i_4 + i_6 - i_3 + i_5 = \frac{i_A(1 + F_1 F_2 - F_1 - F_2)}{1 + F_1 + F_2 + F_1 F_2}$$

From (3.82) and (3.83) we can show that

$$i_{10} - i_9 = \frac{(v_1/R_1)(v_2/R_1)}{i_B}$$

or from (3.81)

$$v_0 = \frac{KR_2(v_1/R_1)(v_2/R_1)}{i_B} \qquad (3.84)$$

Thus we have a multiplier, as described in (3.60), with

$$K_m = \frac{KR_2}{i_B R_1^{\ 2}} \qquad (3.85)$$

This somewhat amazing result indicates that, with properly matched devices, we can implement multiplication with a circuit that has potentially very high speed, since the devices may be made physically small and in close proximity to each other. As indicated in Table 3.2, such is the case. Moreover, many good-quality multipliers of this type are currently being marketed for under $50 in small quantities, and much less in larger quantities.

Let us now see how multipliers may be used to implement a variety of useful nonlinear operations.

Multiplier Applications

In the following discussion we will assume that we are dealing with ±10-V analog systems, and that we have available four-quadrant multipliers that implement

$$v_0 = \frac{v_1 v_2}{10} \qquad (3.86)$$

Figure 3.52a illustrates our block-diagram symbol for a multiplier. Note that we are assuming *no polarity inversion* of the product in (3.86); in many cases one will find that commercial multipliers may actually have a polarity inversion in their transfer characteristic, or they may permit the user to choose the polarity of the generated product. In any event, it should be possible to use either type of multiplier in any of the applications described here simply by using appropriate inverters.

Any of the three types of multipliers discussed above in this section may be used in the applications described, although, of course, some will have superior accuracy and/or speed advantages. Nevertheless, they will all tend to behave in about the same way. Typical commercial multipliers will have an output driving capability and impedance similar to an operational amplifier, i.e., they will deliver a ±10-V signal and an output current of the order of 5 to 10 mA. Input impedances are typically greater than 10 kΩ.

Implicit division   Often analog division is performed by using a circuit similar to Fig. 3.52b; indeed, many manufacturers internally connect the multiplier in the feedback loop

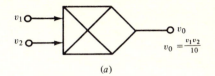

(a)

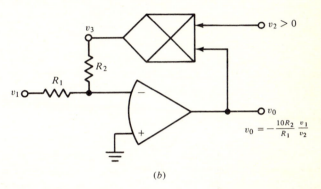

(b)

FIGURE 3.52
Multiplication and implicit division. (a) Typical multiplier block-diagram symbol;
(b) the use of a multiplier in a feedback loop permits implicit division.

of an operational amplifier so that the unit may be set up by the user as either a
multiplier or a divider. The circuit of Fig. 3.52b may be analyzed as follows:

The feedback path around the amplifier hopefully prevents amplifier overload; if
that is true, then, assuming an ideal operational amplifier,

$$\frac{v_3}{R_2} + \frac{v_1}{R_1} = 0 \qquad (3.87)$$

However, the multiplier forces

$$v_3 = \frac{v_0 v_2}{10} \qquad (3.88)$$

Combining these last two equations, we can obtain

$$v_0 = \frac{-10 R_2 v_1}{R_1 v_2} \qquad v_2 > 0 \qquad (3.89)$$

Thus we have implemented division. Note that the overall scale factor may be adjusted
using the resistance ratio. If $R_1 = R_2$, then (3.89) becomes similar to (3.65) except for
the polarity inversion.

Note also that it is necessary to restrict the polarity of the divisor; here we must
assure that v_2 is positive in order to get the proper feedback polarity. Thus we have a
two-quadrant inverting divider. If a polarity inversion is introduced into the feedback
path, then the polarity of the division operation changes, and the input v_2 must be
strictly negative (see Prob. 3.33).

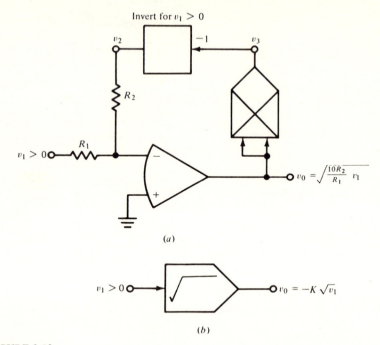

(a)

(b)

FIGURE 3.53
Implementing the square-root operation. (a) A multiplier (or squaring operator) in a feedback loop implements the operation; (b) a typical block-diagram symbol. Note that only one polarity of the input may be accommodated; removal of the inverter between v_3 and v_2 permits the use of negative inputs.

Again, it is necessary to restrict the *minimum* value of v_2 so that the output does not overload; e.g., in (3.89) we must ensure that

$$\frac{R_2 v_1}{R_1 v_2} \leqslant 1 \qquad (3.90)$$

for a 10-V analog system.

Square-rooting In Fig. 3.53 we see how a multiplier may be used to perform the square-root operation. Here again we have to be careful about the polarity of the input (as we should expect, we cannot take the square root of a negative number). If the circuit input v_1 is positive, we will find that we need to invert the multiplier output in Fig. 3.53, that is,

$$v_2 = \frac{-v_0{}^2}{10} \qquad (3.91)$$

Again, since feedback forces

$$v_2 = \frac{-R_2}{R_1} v_1 \qquad (3.92)$$

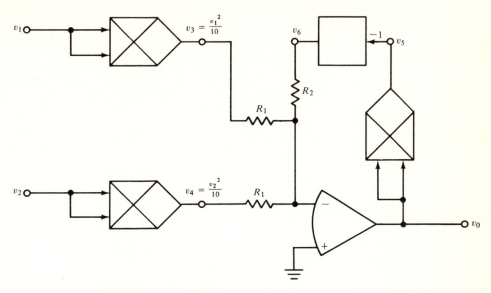

FIGURE 3.54
A two-input rms operator; additional inputs may be added, provided proper amplitude scaling is maintained.

we can show that

$$v_0 = -K\sqrt{v_1} \qquad v_1 > 0 \qquad (3.93a)$$

where

$$K = \sqrt{\frac{10R_2}{R_1}} \qquad (3.93b)$$

The minus sign in (3.93a) is required for stability (see Prob. 3.32). Note that if v_1 is strictly negative, then we can implement

$$v_0 = K\sqrt{-v_1}$$

by removing the inverter following the multiplier in Fig. 3.53a; K is still given by (3.93b). Figure 3.53b shows a typical block-diagram symbol for a square-rooter. Usually we set $R_1 = R_2$ and $K = \sqrt{10}$, and so (3.93a) becomes

$$v_0 = -3.16\sqrt{v_1} \qquad (3.94)$$

that is, 10 V in yields in 10 V out.

Rms operations In Fig. 3.54, we have a circuit that will form the rms combination of two inputs. The basic operating equations are

$$v_3 = \frac{v_1^2}{10} \qquad (3.95a)$$

$$v_4 = \frac{v_2{}^2}{10} \qquad (3.95b)$$

$$v_6 = -\frac{v_0{}^2}{10} \qquad (3.95c)$$

$$v_6 = -\frac{R_2}{R_1}(v_3 + v_4) \qquad (3.95d)$$

From these equations, we obtain

$$v_0{}^2 = \frac{R_2}{R_1}(v_1{}^2 + v_2{}^2)$$

or

$$v_0 = -\sqrt{\frac{R_2}{R_1}} \sqrt{v_1{}^2 + v_2{}^2} \qquad (3.96)$$

Note that the two inputs may have either polarity. Also, (3.96) must have the minus sign to ensure stability.

EXAMPLE 3.14   In Fig. 3.54, if we know that both inputs may range over ±10 V, then we must pick R_2/R_1 so that the overall transfer characteristic is properly amplitude-scaled, that $v_0 < 10$ V. In this instance

$$v_1{}^2 + v_2{}^2 \leqslant 200$$

and in (3.96) we require

$$\frac{R_2}{R_1} \leqslant 0.5$$

Thus, picking $R_1 = 50$ kΩ and $R_2 = 100$ kΩ,

$$v_0 = -0.707\sqrt{v_1{}^2 + v_2{}^2}$$

which is properly scaled.   ////

Note that proper amplitude scaling is important, in order to obtain full static accuracy. If the multipliers are used in such a way that their inputs are never more than a small fraction of the allowable 10-V range, then the percent static accuracy (which is only guaranteed in terms of full-scale output magnitude) may be very poor! The problem is the same as the one encountered in scaling analog computers.

EXAMPLE 3.15   In Fig. 3.55 we see a design of a system to measure true rms values. Here we want to average out short-term fluctuations of the input. The circuit shown has a 10-s averaging time constant. The input voltage should be limited to an rms value that is perhaps 0.2 to 0.33 times the allowable peak input value, so that the input squarer will

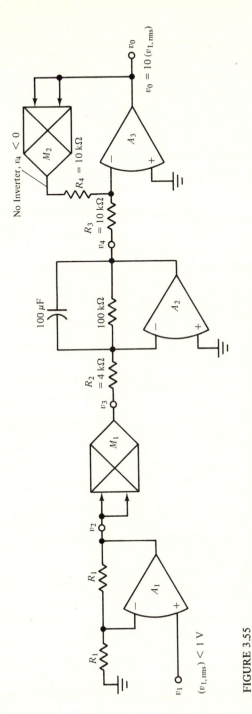

FIGURE 3.55

A true rms voltmeter. The input buffer amplifier is optional. Care must be taken to avoid internal overloads due to momentary excursions of the signals.

normally not overload due to temporary excursions of the input. The circuit of Fig. 3.55 is designed to accommodate an input with an rms value of less than 1 V, and to provide a true rms reading regardless of signal waveform. Of course, the bandwidth of the input must be consistent with the speed of the input multiplier. Also we want a gain of 10, that is, $v_0 = 10$ V when $v_{1,\text{rms}} = 1.0$ V; this assures that the output square-rooting operation is done fairly accurately.

Note that we have selected the various subsystem gains in an attempt to properly scale. For example, $v_2 = 2v_1$, and v_2 will have a value of about 2 V rms and a peak value of no more than 6 to 10 V in most cases. The input amplifier A_1 provides a high input impedance to the signal being observed and amplifies it slightly. Also

$$v_3 = \frac{v_2{}^2}{10}$$

$$= 0.4v_1{}^2$$

The low-pass filter associated with A_2 averages the square of v_1 with a 10-s time constant, and thus v_4 is a slowly varying quantity with

$$v_4 = \frac{-10^5}{R_2} v_{3,\text{av}} = \frac{-0.4 \times 10^5}{R_2} v_{1,\text{av}}^2$$

We choose R_2 so that

$$v_4 = -10 \text{ V} \quad \text{when} \quad v_1 = +1 \text{ V rms}$$

Thus

$$-10 = \frac{-10^5}{R_2} 0.4$$

or

$$R_2 = 4 \text{ k}\Omega$$

The output amplifier should have the proper gain so that $v_0 = 10$ V when $v_{1,\text{rms}} = 1.0$ V. Note that since v_4 is negative, we *do not* want an inverter following multiplier M_2.

In this case

$$v_0 = \sqrt{\frac{10R_4}{R_3}} \sqrt{-v_4}$$

and we want

$$v_0 = +10 \text{ V} \quad \text{when} \quad v_4 = -10 \text{ V}$$

Thus we set

$$R_4 = R_3 = 10 \text{ k}\Omega$$

Since, on the average, v_3 is about 0.4 V, it seems that M_1 is considerably underscaled. Note, however, that if v_1 momentarily became 3 V, then v_3 would accordingly be 3.6 V, etc. We have left considerable allowance for temporary excursions of v_1 well above the rms value, which we should do if we have no prior knowledge of the input waveform. If, for example, we knew that the input to an rms measuring circuit was a random signal with a gaussian or normal amplitude distribution, then typically the input would seldom be more than 3 times the rms value. Such information could be used to better scale the input gain.   ////

3.6 WAVEFORM GENERATION

This topic could well occupy a full chapter (see, e.g., Chap. 10 of Ref. 3). Our main goal here is to present a few representative techniques for generating commonly needed waveforms (sine square and triangle waves, and controlled pulses). Many circuits have been developed to do these jobs. The ones we have chosen to describe hopefully represent basic design approaches, and are relatively easy to build with a minimum of special debugging.

Sine-Wave Oscillators

The generation of sinusoidal waveforms with precisely controlled amplitude and low harmonic distortion is an important requirement in many communication systems. For radio frequencies (say above 1 MHz), one usually uses high-Q resonant circuits (LC or quartz-crystal resonators) in some form of simple transistor oscillator circuit. We are more interested in the generation of sinusoids in the frequency range of, say, 0.001 Hz to 1 MHz. Here operational amplifier circuits are generally employed. Below about 10 Hz, we recommend using the method, described below, of generating a triangle waveform and then using a sine function generator or bandpass filter to get the final waveform. Above 10 Hz, circuits similar to the block diagram of Fig. 3.56 are generally satisfactory. Figure 3.57 shows details of typical circuit implementations.

This is the so-called analog-computer approach, also called the quadrature oscillator. Both configurations in Fig. 3.56 implement the differential equation

$$\frac{d^2v_0}{dt^2} + \omega_0{}^2 v_0 = 0 \qquad (3.97)$$

The solution to this equation is

$$v_0 = A \sin(\omega_0 t + \theta) \qquad (3.98)$$

where θ depends upon the initial conditions.

In actuality, some form of regeneration must be added to ensure that the oscillation will build up. In some cases, the generation will occur due to the inherent phase lag in analog integrators (Ref. 4, Sec. 3.24). It is a good idea, however, to provide for some regeneration. Thus in Figs. 3.56a and 3.57a, we are actually setting up the equation

$$\frac{d^2v_0}{dt^2} - \epsilon\omega_0 \frac{dv_0}{dt} + \omega_0{}^2 v_0 = 0 \qquad (3.99)$$

which will build up slowly for small positive ϵ, with a frequency of oscillation that is still close to ω_0. Usually ϵ is adjusted empirically for proper performance. In Figs. 3.56b and 3.57b, we show an alternative way to implement (3.97), using one inverting and one

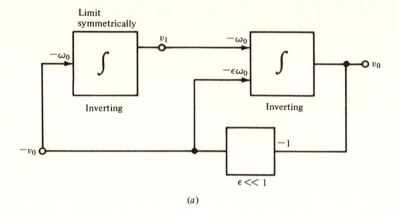

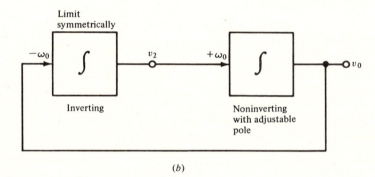

FIGURE 3.56
Analog-computer or sine-loop type of sinusoidal oscillator. (a) Two inverting integrators and a unity gain inverter solve the harmonic oscillator differential equation; (b) one inverting and one noninverting integrator also may be used.

noninverting integrator. To obtain regeneration, we usually can adjust the location of the pole of the noninverting integrator (i.e., move into the left half of the s plane). In Fig. 3.57 this is done by adjusting R_3. Note that the noninverting integrator is the one previously discussed in Chap. 1 (Fig. 1.22, Prob. 1.16). In both parts of Fig. 3.57, the individual integrator gains establish

$$\omega_0 = \frac{1}{R_1 C_1} \quad \text{rad/s} \quad (3.100)$$

Both integrator outputs will have approximately the same amplitude, but will be 90° out of phase.

Another requirement is to limit the amplitude of the oscillation, hopefully without adding distortion. One fairly simple limiting method is to provide feedback limiting

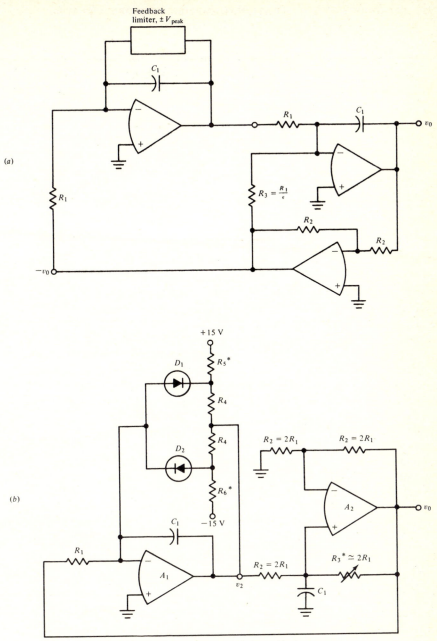

(a)

(b)

*Trim R_5 and R_6 for desired output level
and symmetrical limiting. Trim R_3 for slight
regeneration.

FIGURE 3.57
Sinusoidal quadrature oscillator, based upon Fig. 3.56; the main integrator gains
should be equal, and regeneration is adjusted by R_3. (a) Three-amplifier circuit
based on Fig. 3.56a; R_3 is usually very large; (b) two-amplifier circuit based upon
Fig. 3.56b. Details of soft limiter are shown; R_3 is adjusted to control
regeneration, and resistors R_5 and R_6 are adjusted to control amplitude and to
limit symmetrically.

around the integrator *not* being used to provide the main output. Any distortion due to clipping in this integrator output will then be suppressed by the output integrator. The limiter should be fairly soft and adjustable. The diode feedback limiter illustrated in Fig. 3.57*b* is often used.

EXAMPLE 3.16 Suppose we want to use the circuit of Fig. 3.57*b* to design a 1-kHz oscillator with a ±10-V amplitude. Using $C_1 = 0.001$ μF, we have from (3.100)

$$R_1 = \frac{1}{\omega_0 C_1} = \frac{1}{2\pi \times 10^3 \times 10^{-9}}$$

$$= 159 \text{ k}\Omega$$

Thus $$R_2 = 318 \text{ k}\Omega$$

and R_3 is made adjustable in the neighborhood of 318 kΩ to establish the desired amount of regeneration. If the limiter on A_1 is set to clip at ±10 V, v_0 will also have a ±10-V amplitude. With the ±15-V supplies, typical values are $R_5 = 1.7$ kΩ and $R_4 = 1$ kΩ. The value of the two resistors labeled R_4 should be trimmed to get the desired amplitude, with symmetrical limiting. ////

Note that the circuit of Fig. 3.57*a* requires an extra operational amplifier, but it can be set up on an analog computer, which normally does not have noninverting amplifier terminals available. Improved automatic gain control (AGC) circuits for regulating the amplitude of the waveform may be employed (see, e.g., Ref. 3, p. 390).

Frequency-modulated sine-wave oscillators Figure 3.58 shows a modification to the system of Fig. 3.56 which permits generation of FM waveforms. Normally two multipliers should be used to modulate the loop gain and keep the magnitude of the integrator outputs properly scaled. The instantaneous angular frequency of oscillation will be

$$\omega_0(t) = \frac{K[v_m(t) + A]}{10} \qquad \text{rad/s} \qquad (3.101)$$

for a 10-V system. Limiting and regeneration will be added in a manner similar to Example 3.16.

Other sine-wave oscillators The operational amplifier may be used as the active element in a variety of oscillator circuits. Figure 3.59 shows two common approaches:

1 The Wien bridge oscillator (Fig. 3.59*a*)
2 The phase-shift oscillator (Fig. 3.59*b*)

These types are more fully discussed in Ref. 3, Chap. 10. We present them here mainly for completeness. Although they employ only one amplifier, their design takes some care if well-controlled amplitude with low distortion is to be achieved. Moreover, they do not provide the two-quadrature outputs (sine and cosine waves) available from the three-amplifier circuits discussed above.

The Wien bridge circuit oscillates at approximately

$$\omega_0 \cong \frac{1}{RC} \qquad \text{rad/s} \qquad (3.102a)$$

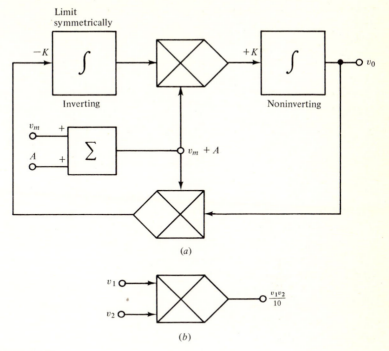

FIGURE 3.58
Generation of frequency-modulated (FM) waveforms. (*a*) General block diagram; (*b*) explanation of multiplier symbol.

with

$$R_2 \cong 2R_1 \quad (3.102b)$$

The amplitude of oscillation may be regulated by using some form of nonlinear resistor for R_1. For example, if R_1 is a thermistor or simple incandescent lamp, R_1 will increase with the power dissipated in R_1, and this will tend to regulate the amplitude.

The phase-shift oscillator of Fig. 3.59*b* oscillates at approximately

$$\omega_0 \cong \frac{1}{\sqrt{3}\,RC} \quad (3.103)$$

and the feedback resistance R_1 should be slightly greater than $12R$.

Precision Square- and Triangle-Wave Generation

Figure 3.60 shows the block diagram of a system for generating square and triangle waves. It uses a precision bistable circuit (see, e.g., Fig. 3.40) with output level $L_+ > 0$ and $L_- < 0$, and switching points $S_+ > 0$ and $S_- < 0$. The inverting integrator output is a triangle wave, as shown in Fig. 3.60*b*, with peak values determined by S_+ and S_-.

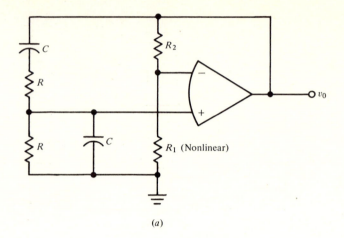

(a)

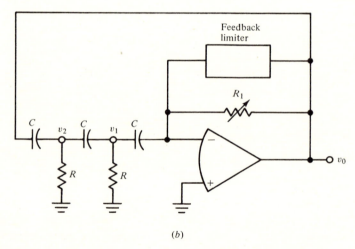

(b)

FIGURE 3.59
One-amplifier sinusoidal oscillators. (a) Wien bridge type; (b) phase-shift type.

Similarly, the square wave has peak values determined by L_+ and L_-. We have shown a symmetrical case. The period of the waveform is

$$T = \frac{4S}{LK} \qquad s \qquad (3.104)$$

where $L = L_+ = |L_-|$
$\quad S = S_+ = |S_-|$

Figure 3.60c shows a typical situation, where the bistable has $L = S = 10$ V, and $K = 1/RC$; in this case

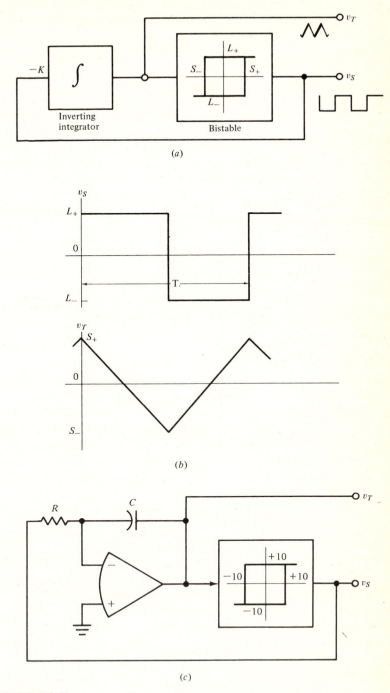

FIGURE 3.60
Generation of square and triangle waves. (*a*) General block diagram; (*b*) typical
waveforms; (*c*) typical circuit parameters.

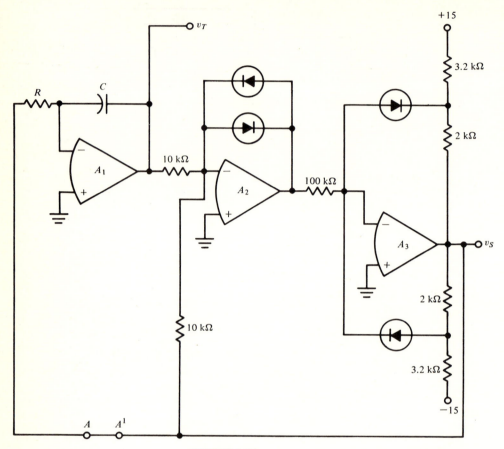

FIGURE 3.61
A complete circuit for generating square and triangle waveforms.

$$T = 4RC \quad \text{s}$$

and we get ±10-V square and triangle waves.

Figure 3.61 shows a more complete circuit. The frequency of oscillation is

$$f = \frac{1}{4RC} \quad \text{Hz} \quad (3.105)$$

The bistable is based upon Fig. 3.40.

In Fig. 3.62, we see some interesting modifications that may be made to the basic system of Fig. 3.60. First, we have shown the addition of a sine function generator. This could be implemented, for example, by the piecewise-linear function generation techniques of Sec. 3.3. Usually five to seven segments are used to cover the two quadrants

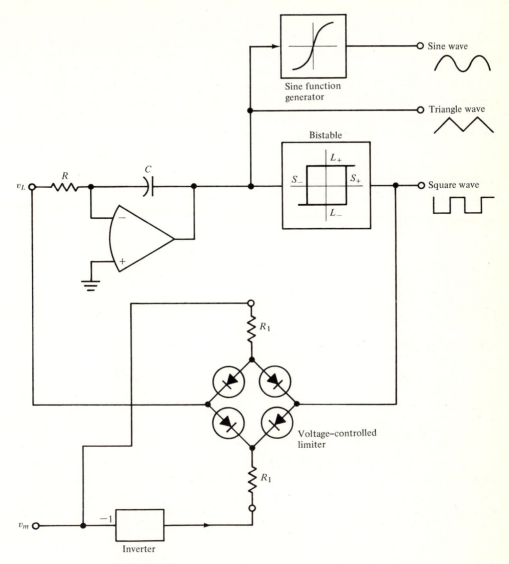

FIGURE 3.62
Modifications to the basic square- and triangle-wave generator to obtain sine waves as well, and to frequency-modulate via v_m and the four-diode bridge limiter.

of the function. The reader may well comment that this seems to be a difficult way to generate sine waves! Nevertheless, this approach is generally accepted as *one of the best ways* of generating stable *low-frequency* (e.g., less than 10 Hz) *sine waves.* Commercial low-frequency function generators for generating sine, square, and triangle waves over a wide frequency range are usually designed with these techniques.

Also we have added a four-diode bridge limiter in the oscillator feedback path of Fig. 3.62. This limiter is driven by a source of modulating signal v_m, so that signal v_L is a square wave of amplitude $\pm v_m$. Essentially we are modulating the loop gain of the circuit, and the result is a *VCO* (voltage-controlled oscillator) or *FM oscillator*. An analog multiplier can be used instead of the bridge and inverter, but it is not needed here. The amplitudes of the triangle wave will still be controlled by the bistable switch points S_+ and S_-, and the amplitude of the square-wave output will still be controlled by the bistable output levels L_+ and L_-.

EXAMPLE 3.17 Suppose we use the basic circuit of Fig. 3.61, but add a four-diode bridge at point A-A'; from (3.104) with $S = 10, K = 1/RC$, and

$$L = v_m \frac{R}{R + R_1}$$

then

$$T = \frac{4RC \times 10(R + R_1)}{v_m R}$$

or

$$T = \frac{4C(R + R_1) \times 10}{v_m}$$

Thus the frequency of oscillation is

$$f = \frac{v_m}{40(R + R_1)C}$$

We see that the circuit has very linear frequency modulation, and thus makes a very nice VCO.   ////

This general technique for generating square, triangle, and sine waves has become so well established that one can now buy commercial integrated circuits that provide a source of square, triangle, or sine waves which may be amplitude- or frequency-modulated. For example, the EXAR XR-205 monolithic waveform generator can generate 3-V peak-to-peak waveforms with frequencies up to 4 MHz.

Monostable (Precision Pulse) Circuits

Figure 3.63 shows a three-amplifier circuit for generating pulses of precise amplitude and duration. Essential features are shown in Fig. 3.63a, and the associated waveforms in Fig. 3.63b. The bistable circuit is similar to that of Fig. 3.61, except a trigger pulse v_T is introduced into the input of the bistable at the summing junction of A_2. Normally in the stable state the output v_0 is negative, and amplifier A_1 is driven so that D_2 is ON; v_2 is approximately $+0.6$ V. Thus D_1 is OFF, and R_2 keeps C_1 discharged so that v_1 is zero. A positive trigger pulse drives the output of the bistable around to -10; now v_2 comes down to zero, D_1 turns ON, and A_1 starts to integrate v_0 downward. This continues until v_1 reaches -10 V (point A), at which time the bistable changes state again (halfway

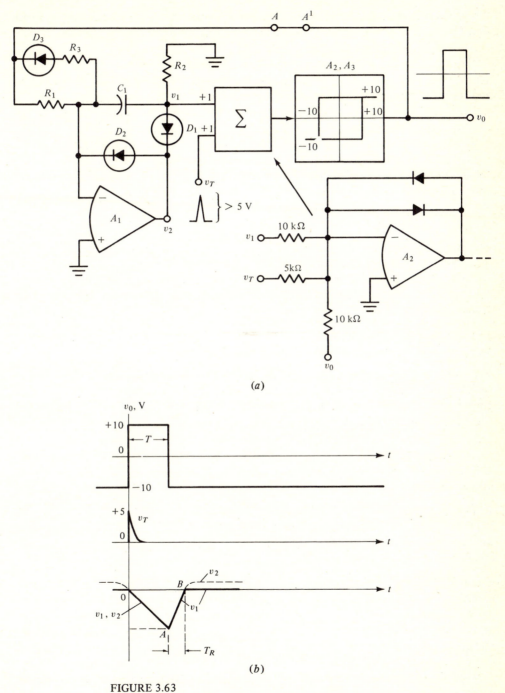

FIGURE 3.63
Three-amplifier monostable pulse generator. (*a*) Partial circuit diagram; note that positive trigger pulse must exceed 5 V; (*b*) typical waveforms.

through the pulse duration). Now v_1 comes back up until it reaches point B, at which time D_1 turns OFF and the cycle is completed. During the pulse interval, we have

$$\frac{dv_1}{dt} = \frac{-10}{R_1 C_1} \qquad (3.106)$$

and the pulse duration is then

$$T = R_1 C_1 \qquad (3.107)$$

with considerable accuracy, perhaps one- or two-tenths of a percent. The time it takes for the circuit to reset (T_R in Fig. 3.63b) can be made much less than T by using D_3 and R_3 to increase the integrator gain during this interval. Now we have

$$T_R \cong R'C_1 \qquad (3.108a)$$

where

$$R' = R_1 \| R_3 < R_1 \qquad (3.108b)$$

and thus $T_R < T$.

Note that the pulse duration T can be adjusted by changing the positive level returned from the bistable to the integrator. Thus if a voltage-controlled limiter (similar to that in Fig. 3.62) is added to point $A\text{-}A'$, we could make this circuit into a precision *voltage-time* converter.

One-Amplifier Astable and Monostable Circuits

Figure 3.64a shows a typical one-amplifier square-wave generator or astable (free-running multivibrator, Ref. 3, Sec. 1.1.1, and Ref. 7, Sec. 7.2.1). To understand its operation, consider the simplified circuit in Fig. 3.64b. The waveforms for v_1 and v_0 are shown in Fig. 3.64c. The resistance ratio

$$\alpha = \frac{R_3}{R_3 + R_2} \qquad (3.109)$$

establishes

$$v_2 = \alpha v_0 \qquad (3.110)$$

In Fig. 3.64b, we assume that the amplifier output saturates at $\pm E$ V, symmetrically, and with reasonably short overload recovery time. (Many amplifiers fulfill this requirement satisfactorily at frequencies below 100 kHz.) The output will switch states when

$$v_1 = \pm \alpha E$$

With this in mind, we can show that the period of oscillation is

$$T = 2R_1 C \ln \frac{1 + \alpha}{1 - \alpha} \qquad (3.111)$$

(see Prob. 3.45). If α is chosen, for example, to be 0.473, then

$$T = 2R_1 C \qquad \alpha = 0.473$$

In Fig. 3.64a we have added a few components. The double-anode breakdown diode D_1 will establish the output levels, and R_5 will limit the overload output current

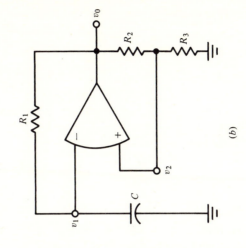

(b)

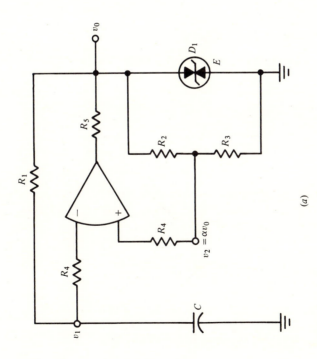

(a)

FIGURE 3.64

A one-amplifier stable pulse generator (less precise timing). (a) Circuit diagram;
(b) simplified diagram; (c) typical waveforms; (d) use of transistor current-source
charging to yield a better triangle wave at v_1; (e) an asymmetrical version; (f)
waveforms of asymmetrical circuit.

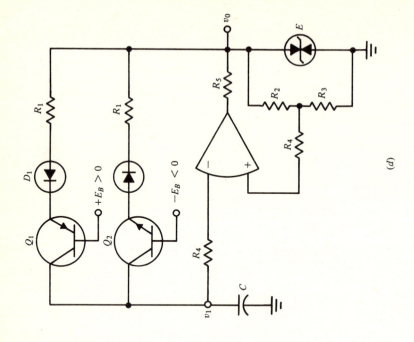

(d)

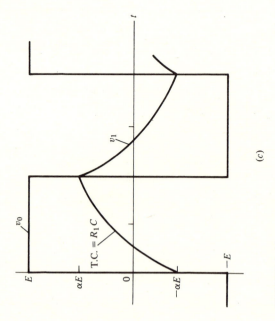

(c)

FIGURE 3.64 (continued)

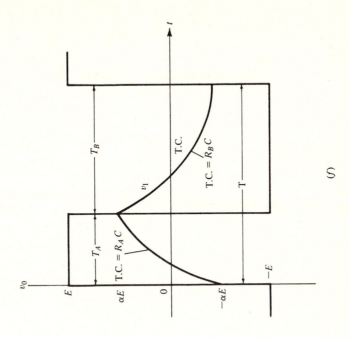

(f)

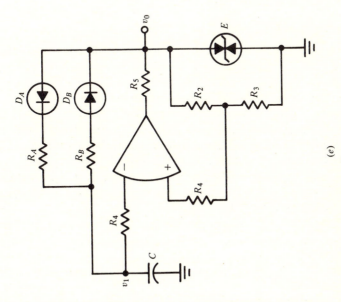

(e)

FIGURE 3.64 (continued)

233

from the amplifier. The resistors labeled R_4 are usually made about 100 kΩ, and maintain a high input impedance across the amplifier input at all times. Resistor R_5 and diode D_1 may be omitted if the amplifier overload behavior is satisfactory.

In Fig. 3.64d, we have replaced R_1 by a bipolar constant current source. Assuming that $E_B < E$, the capacitor charging current will be (approximately)

$$|i_{charge}| = \frac{E - E_B}{R_1}$$

and the waveform v_1 will be a triangle wave, rather than the exponential transfer waveform of Fig. 3.64c. Diodes D_1 and D_2 prevent emitter-base breakdown of the transistors.

In Fig. 3.64e, we have replaced R_1 with the two charging resistors R_A and R_B and their associated diodes D_A and D_B. We now have an *asymmetrical multivibrator*, with an output waveform depicted in Fig. 3.64f; here we have assumed $R_A < R_B$, and we can show that, with α defined by (3.109), the pulse lengths are given by

$$T_A = R_A C \ln \frac{1 + \alpha}{1 - \alpha} \quad (3.112a)$$

and

$$T_B = R_B C \ln \frac{1 + \alpha}{1 - \alpha} \quad (3.112b)$$

and of course, the total period is

$$T = T_A + T_B \quad (3.112c)$$

which agrees with (3.111) if $R_A = R_B$.

While the one-amplifier astable circuits of Fig. 3.64 will not generate waveforms quite as precisely as the three-amplifier units of Figs. 3.60 to 3.62, they often will serve quite nicely. If precise amplitude and/or frequency over a wide frequency range is required, the multiamplifier approach may be preferred. Figure 3.65 shows a *one-amplifier monostable* circuit. Its behavior is quite similar to the asymmetrical astable circuit of Fig. 3.64e, except now we use diode D_1 to prevent the inverting terminal of the amplifier from going very far negative. The associated waveforms are shown in Fig. 3.65b. The amplifier output is held at $-E$, with v_1 at $-V_F$, where V_F is the diode forward drop, about 0.6 V. A *positive trigger of magnitude greater* than αE at v_T causes the output to switch to $+E$, and v_1 begins charging as shown, with time constant $R_1 C$. When v_1 reaches $+E$, the output switches to $-E$, and v_1 returns to the original stable state. The use of D_3 and R_7 permits the return to be relatively fast. Network C_T and R_6 couple the trigger into the circuit, and the element values are best determined empirically, depending on amplifier speed and trigger rise time.

The pulse duration may be estimated by

$$T \cong R_1 C \ln \frac{E + V_F}{E(1 - \alpha)} \quad (3.113)$$

where α is given by (3.109).

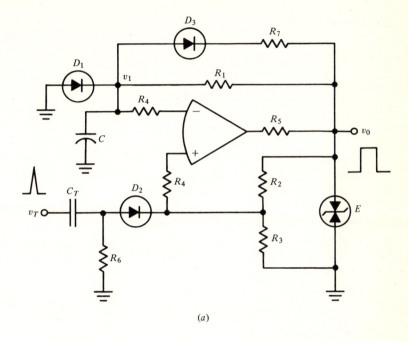

(a)

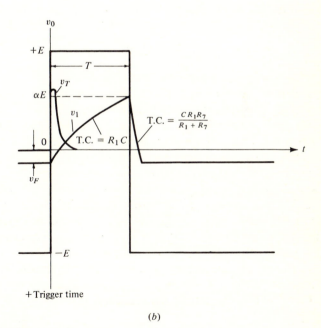

(b)

FIGURE 3.65
A one-amplifier monostable circuit. (a) Circuit diagram; (b) typical pulse waveforms; note that trigger must be greater than βE.

REFERENCES AND BIBLIOGRAPHY

1 *Electronics:* A biweekly journal for electronics engineers; in particular, see the Circuit Design features, McGraw-Hill, Inc.

2 *Electronic Design:* A biweekly journal for electronics engineers; in particular, see the articles in the Technology section. Hayden Publ. Co.

3 TOBEY, G. E., J. G. GRAEME, and L. P. HUELSMAN: "Operational Amplifiers," McGraw-Hill, New York, 1971.

4 KORN, G. A., and T. M. KORN: "Electronic Analog and Hybrid Computers," 2d ed., McGraw-Hill, New York, 1972.

5 SEARLE, C. L., A. R. BOOTHROYD, ET AL.: "Elementary Circuit Properties of Transistors," Wiley, New York, 1964.

6 CALLAHAN, M. J., R. L. PECK, and J. A. REAGAN: A Lidar Output Energy Digital Detector, *Abstracts, 4th Conference on Laser Radar Studies of the Atmosphere, University of Arizona,* June 27, 1972, pp. 14–16.

7 CLAYTON, G. B.: "Operational Amplifiers," Butterworth, London, 1971.

8 DOBKIN, R. G.: Logarithmic Converters, *IEEE Spectrum,* November 1969, pp. 69–72.

9 *Analog Dialogue,* vol. 6, nos. 2 and 3, Analog Devices, Inc.

10 SHEINGOLD, D. H.: Approximate Analog Functions with a Low-cost Multiplier/Divider, *Electronic Design News,* Feb. 5, 1973, pp. 50–52.

11 CATE, T.: Modern Techniques of Analog Multiplication, *The Electronic Engineer,* April 1970, pp. 75–79.

12 GILBERT, B.: A Precise Four-quadrant Multiplier with Subnanosecond Response, *IEEE Journal of Solid-State Circuits,* December 1968, pp. 365–373.

13 SCHWARTZ, M.: "Information, Transmission, Modulation, and Noise," 2d ed., McGraw-Hill, New York, 1970.

PROBLEMS

3.1 (Sec. 3.1) Refer to Fig. P3.1*a*; the nonlinear network that is placed in the amplifier feedback path is sketched in Fig. P3.1*b*. Assume an ideal operational amplifier.

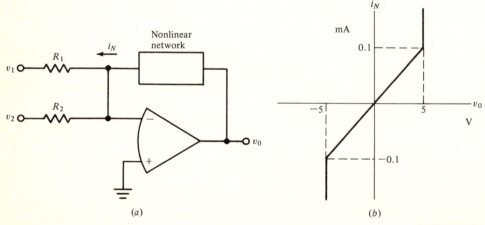

(a)   *(b)*

FIGURE P3.1

(a) If $R_1 = R_2 = 10$ kΩ and $v_2 = 0$ V dc, sketch v_0 versus v_1, $|v_1| \leqslant 10$ V.

(b) Repeat part a if $v_2 = +5$ V dc.

(c) Repeat part a if $v_2 = +5$ V dc and $R_2 = 20$ kΩ.

3.2 *(Sec. 3.1)* Refer to Fig. P3.2. Find v_0 versus v_1 for this more general case of Example 3.2.

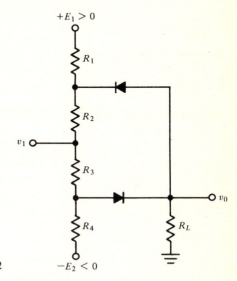

FIGURE P3.2

3.3 *(Sec. 3.1)* Refer to Fig. P3.3; assume an ideal operational amplifier and diodes.

(a) Sketch v_0 versus v_1.

(b) Repeat part a, but with diodes D_3 and D_4 and their associated resistor and power source removed; i.e., with a half-bridge.

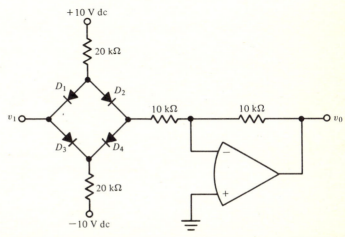

FIGURE P3.3

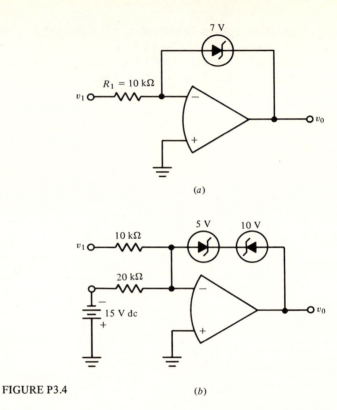

FIGURE P3.4

3.4 *(Sec. 3.2)* Sketch the transfer characteristic, v_0 versus v_1, for the circuits of Fig. P3.4. The Zener diode voltages are labeled. Cover the range $-10 \text{ V} \leqslant v_1 \leqslant +10 \text{ V}$.

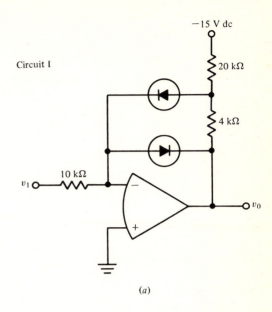

Circuit I

(a)

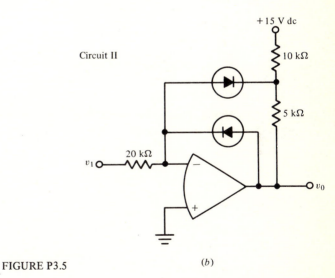

Circuit II

FIGURE P3.5

(b)

3.5 *(Sec. 3.2)*

(*a*) Sketch the transfer characteristic, v_0 versus v_1, for the circuits of Fig. P3.5. Assume ideal diodes and operational amplifier. Cover the range -10 $V \leqslant v_1 \leqslant +10$ V.

(*b*) Repeat the above, but add a 10-kΩ resistor between the operational amplifier output and the inverting terminal.

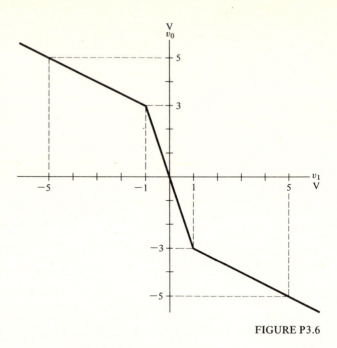

FIGURE P3.6

3.6 *(Sec. 3.2)* Using the circuit of Fig. 3.14, design a limiter amplifier to match the transfer characteristic of Fig. P3.6; use an ideal diode model.

3.7 *(Sec. 3.2)* Use the basic circuit of Fig. 3.17 to design circuits for the following transfer functions:

$$(a)\ v_0 = \begin{cases} -2v_1 & v_1 < 0 \\ 0 & v_1 > 0 \end{cases}$$

$$(b)\ v_0 = \begin{cases} -4v_1 & v_1 > 0 \\ 0 & v_1 < 0 \end{cases}$$

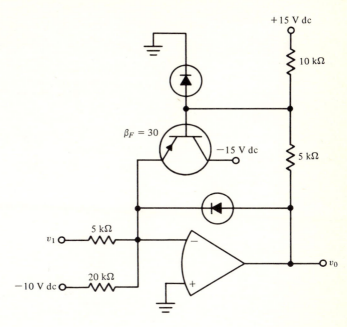

FIGURE P3.8

3.8 *(Sec. 3.2)* Sketch v_0 versus v_1 for the circuit of Fig. P3.8. Try to make accurate estimates of slopes, etc. Cover the range $-10 \text{ V} \leqslant v_1 \leqslant +10 \text{ V}$.

3.9 *(Sec. 3.2)* Refer to Fig. 3.21; assume ideal diodes and the following parameters:

$$E_1 = +15 \text{ V dc} \qquad -E_2 = -15 \text{ V dc}$$
$$R_A = R_B = 10 \text{ k}\Omega \qquad R_L = 5 \text{ k}\Omega$$
$$R_F = 20 \text{ k}\Omega \qquad R_1 = 5 \text{ k}\Omega$$

Both Zener diodes break down at a reverse voltage of 9 V.

(*a*) Sketch v_0 versus v_1, $-10 \text{ V} \leqslant v_1 \leqslant +10 \text{ V}$.

(*b*) Repeat part *a* if D_3, D_4, and R_B are removed. Does this answer suggest another way to build a precision half-wave rectifier?

3.10 *(Sec. 3.2)* Figure P3.10 shows two different comparator circuits; sketch v_0 versus v_1 for each circuit, and compare. Assume the Zener diodes have an incremental resistance of 50 Ω in the breakdown region, and that the other diodes have an ac resistance of 25 Ω and a forward drop of 0.6 V.

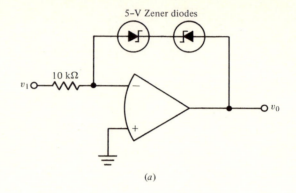

(a)

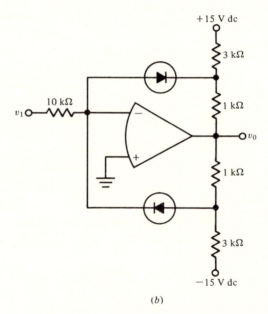

(b)

FIGURE P3.10

3.11 (Sec. 3.2) Figure P3.11 shows two different level clamps used to form a comparator, with positive feedback for hysteresis. Use the following symbols:

$$K_1 = \frac{R_2}{R_1 + R_2}$$

$$K_2 = \frac{R_4}{R_3 + R_4}$$

(a) Compare the two circuits, that is, derive expressions for both circuits for estimating

Circuit I

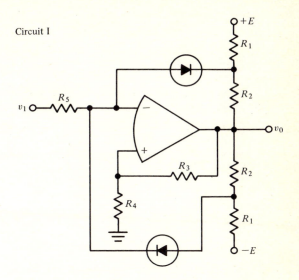

Circuit II

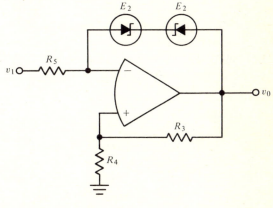

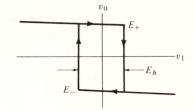

FIGURE P3.11

$$E_+ = \text{positive output level}$$

$$E_- = \text{negative output level}$$

$$E_h = \text{hysteresis width}$$

in terms of the rest of the symbols; verify (3.17) for both circuits. Assume both Zener diodes in circuit II break down at E_z V. Note: The sketch of v_0 versus v_1 is idealized; circuit I would not limit perfectly, but that need not be considered in part a.

(b) Sketch v_0 versus v_1 accurately (both circuits) if

$$E = 15 \text{ V dc} \qquad E_z = 5 \text{ V}$$
$$R_1 = 3 \text{ k}\Omega \qquad R_3 = 100 \text{ k}\Omega$$
$$R_2 = 1 \text{ k}\Omega \qquad R_4 = 1 \text{ k}\Omega$$
$$R_5 = 10 \text{ k}\Omega$$

3.12 *(Sec. 3.2)* Redesign the circuit of Example 3.6 so that the output will be one-tenth of the rms value of the input. Assume the input is a sinusoid with zero average value and a peak value of up to 120 V. Assume ±10-V amplifiers, and be sure to avoid overloads.

3.13 *(Sec. 3.3)* Redesign the circuit for Example 3.6 to make use of the full-wave-rectifier circuit of Fig. 3.26c; comment on the advantages of this approach.

3.14 *(Sec. 3.3)* Design a circuit based on Fig. 3.26c which will implement accurately the transfer characteristic

$$v_0 = \begin{cases} -2v_1 & v_1 < 0 \\ 3v_1 & v_1 > 0 \end{cases}$$

3.15 *(Sec. 3.3)* Design a circuit to accurately implement the transfer characteristic of Fig. P3.15; D can be adjusted by a potentiometer between 0 and 10.0 V. Assume ±10-V dc power is available; use no more than two operational amplifiers. The output impedance need not be low.

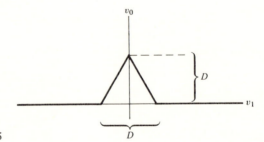

FIGURE P3.15

3.16 *(Sec. 3.3)* Adapt the absolute-value circuit of Fig. 3.26 to provide the *window comparator* characteristic of Fig. P3.16a. Figure P3.16b shows a block diagram; complete the design of the entire circuit, using two operational amplifiers,

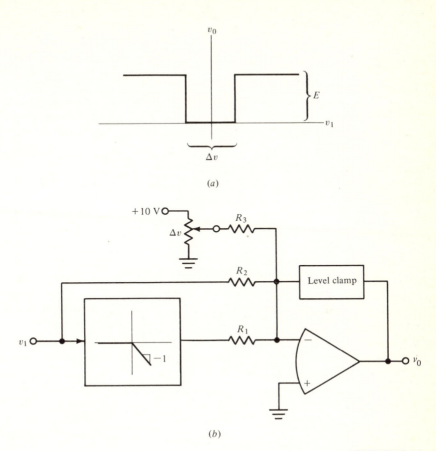

FIGURE P3.16

resistances in the range of 10 to 100 kΩ, switching diodes, and a Zener diode for the level clamp. Assume ΔV is set by the potentiometer, and $0 < \Delta V \leqslant 5$ V. The window comparator is very useful for precise measurement of limits on a signal or for signaling an overload or out-of-tolerance condition. Assume ±15 V dc is available.

3.17 *(Sec. 3.3)* Design a dead-space circuit to implement the characteristic of Fig. P3.17. Assume ±15 V dc is available.

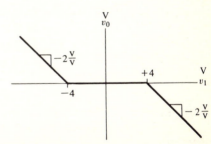

FIGURE P3.17

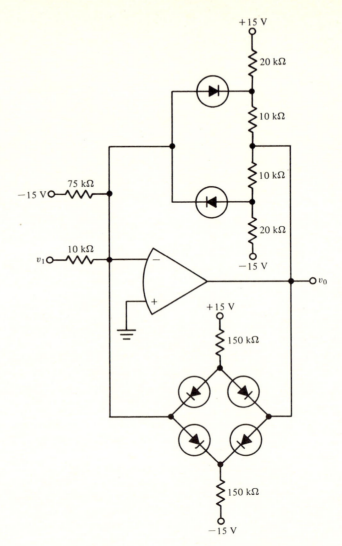

FIGURE P3.18

3.18 *(Sec. 3.3)* Sketch the transfer characteristic of the circuit in Fig. P3.18, -10 V $\leqslant v_1 \leqslant +10$ V. Calculate the slopes of the characteristic carefully.

3.19 *(Sec. 3.3)* Refer to Fig. P3.19.

    (a) Use the series limiter circuit of Fig. 3.29 to implement the nonlinearity of Fig. P3.19; assume a 10-kΩ feedback resistor. Design the circuit, using the ideal diode model.

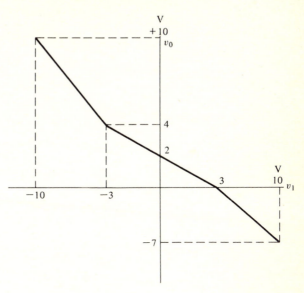

FIGURE P3.19

(b) Repeat the design, now including compensation and catching diodes (see Fig. 3.30) and assuming 0.6-V diode forward drops [use Eq. (3.33)].

3.20 *(Sec. 3.3)* Refer to Fig. P3.20. Here we want to design a series limiter circuit to approximate

$$v_0 = \frac{-v_1^2}{10} \qquad 0 < v_1 < +10$$

using three straight-line segments, as shown.

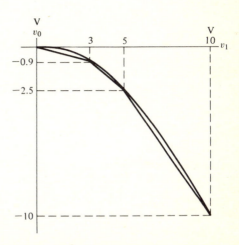

FIGURE P3.20

(a) Design the circuit using series limiters (Fig. 3.30). Assume 0.6-V diode forward drops.

(b) Assuming that the straight-line approximation of Fig. P3.20 is accurately achieved, estimate the maximum error in volts in generating the desired squaring operation.

3.21 *(Sec. 3.3)* Refer to Fig. P3.21. Assume you have ±15 V dc available. Design a circuit based on the shunt limiter of Fig. 3.32*d* to implement the desired characteristic. Use a 10-kΩ feedback resistor.

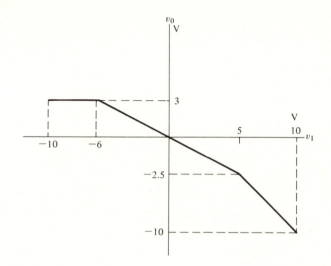

FIGURE P3.21

3.22 *(Sec. 3.3)* Refer to Fig. P3.22.

(a) Sketch the transfer characteristic, v_0 versus v_1, if the network of Fig. P3.22*a* is used in an amplifier input path, as in Fig. P3.22*b*.

(b) Again sketch v_0 versus v_1 if the network is used in a feedback path, as in Fig. P3.22*c*.

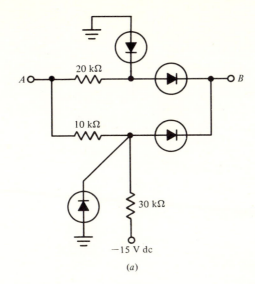

(a)

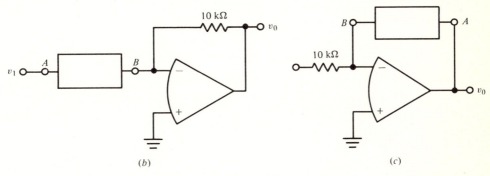

(b)

(c)

FIGURE P3.22

3.23 *(Sec. 3.3)* A diode function generation network (similar to that of Fig. 3.29) is designed to implement an approximation to the sine function. When it is used in the circuit of Fig. P3.23a,

$$v_0 = -10 \sin \frac{v_1}{K_2} \qquad |v_1| \leqslant 10 \text{ V}$$

and it is designed so that $v_0 = -10$ V when $v_1 = +10$ V, that is, ± 10 V at the input corresponds to $\pm 90°$.

(a) Find K_2.

(b) Find the *short-circuit transfer conductance* of the diode network, i.e., in Fig. P3.23a find

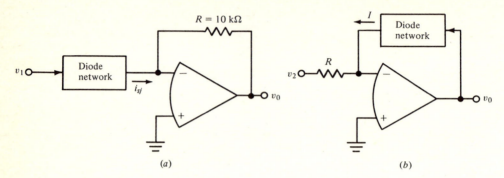

(a)   (b)

FIGURE P3.23

$$i_{SJ} = G(v_1)$$

Be sure to specify units.

(c) If the same diode network is used in the circuit of Fig. P3.23b, find v_0 as a function of v_2. Specify the allowable ranges of v_0 and v_2.

3.24 *(Sec. 3.3)* Using the circuit of Fig. 3.40, design a bistable circuit with the following specifications:

$$L_+ = 10 \text{ V} \qquad L_- = -5 \text{ V}$$

$$S_+ = +3 \text{ V} \qquad S_- = 0 \text{ V}$$

HINT: You may have to introduce an offset into the circuit by providing a second input to the summing junction of A_1.

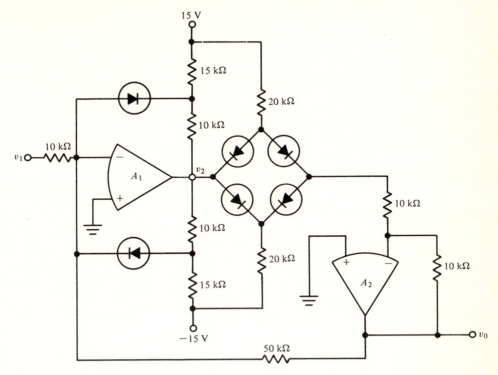

FIGURE P3.25

3.25 *(Sec. 3.3)*

(*a*) Analyze the circuit of Fig. P3.25; sketch v_0 versus v_1. What type of circuit is this?

(*b*) Sketch v_2 versus v_1.

3.26 *(Sec. 3.4)* Given the circuit of Fig. 3.43, develop Eq. (3.49). Assume very high transistor current gains and ideal operational amplifiers, and use the current and voltage labels of the figure.

3.27 *(Sec. 3.4)* Redesign the circuit of Example 3.12 so that the input-output characteristic is described by

v_1	v_0
10 V	−10 V
10^{-2} V	0 V
10^{-5} V	+10 V

You may alter R_1, R_2, R_3, and R_5, but R_1 must be less than 20 kΩ.

3.28 *(Sec. 3.4)* Given the circuit of Fig. 3.45, develop Eq. (3.53). Assume very high transistor current gains and ideal operational amplifiers, and use the current and voltage labels of the figure. Why do we make $R_3 = R_7$ and $R_5 = R_6$?

3.29 *(Sec. 3.4)* The circuit of Fig. 3.45 can also be described by

$$v_0 = K_1 A^{-K_4 v_1} \qquad v_1 > 0$$

Find K_1 and K_4 at $27°C$.

3.30 *(Sec. 3.4)* Redesign the circuit in Example 3.13 to obtain

    (a) $v_0 = 5 \times 10^{-0.1 v_1} \qquad v_1 > 0$

    (b) $v_0 = 10 \times 2^{-v_1} \qquad v_1 > 0$

3.31 *(Sec. 3.4)* Suppose we have a general-purpose module that can implement

$$v_4 = v_3 \left(\frac{v_1}{v_2}\right)^m$$

where $0.2 \leqslant m \leqslant 5.0$

$$0 < v_1, v_2, v_3, v_4 < 10 \text{ V}$$

Assume that m can be adjusted by some externally provided potentiometer to any value in the range. Also assume that the module has a 10-kΩ input impedance at each input terminal, and a very low output impedance. The terminals are numbered as shown in Fig. P3.31.

    (a) Use the module, along with an extra operational amplifier and resistors, to implement the following approximation to $\sin \theta$ (good to 0.25 percent, Ref. 10).

$$\sin \theta \cong \theta - \frac{\theta^{2.827}}{2\pi} \qquad 0 < \theta \leqslant \frac{\pi}{2}$$

Represent θ by a signal scaled so that $+10$ V corresponds to $90°$; implement this approximation so that the output is $+10$ V when the input is $+10$ V; be sure no output overloads.

    (b) Repeat the above for the cosine approximation (good to 1 percent)

$$\cos \theta \cong 1 + 0.235\theta - \frac{\theta^{1.504}}{1.445}$$

Assume ± 10 V dc is available; set it up so that the output is $+10$ V when the input is zero and a $+10$-V input represents $\pi/2$ rad.

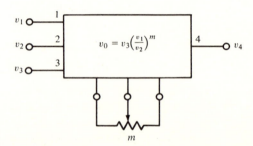

FIGURE P3.31

3.32 *(Sec. 3.5)* Show why there must be a minus sign on the right side of Eq. (3.96). HINT: Find the small-signal short-circuit transfer resistance in the feedback path around some nominal value of v_0.

3.33 *(Sec. 3.5)* In Fig. 3.52b, suppose we have an inverting multiplier, that is,

$$v_3 = \frac{-v_2 v_0}{10}$$

Repeat the development of the resulting circuit transfer equation, and comment on the signal polarity restrictions.

3.34 *(Sec. 3.5)* Redesign the circuit of Example 3.14 to implement vector magnitude resolution. That is, assume

$$v_1 \leqslant 10 \qquad v_2 \leqslant 10$$

and also

$$v_1^2 + v_2^2 \leqslant 100$$

so that the desired operation is

$$v_0 = -\sqrt{v_1^2 + v_2^2}$$

Make $R_2 = 10\ \text{k}\Omega$.

3.35 *(Sec. 3.5)* Using noninverting ± 10-V multipliers, implement

$$v_0 = -K\sqrt{v_1^2 + v_2^2 + v_3^2}$$

so that the output never exceeds 10 V if all inputs are less than 10 V.

3.36 *(Sec. 3.5)* Refer to Fig. 3.55 and Example 3.15. Suppose we know that the input signal to be measured never exceeds 4 V in magnitude, but may have an rms value as large as 3 V. Redesign the circuit to implement

$$v_0 = 3v_{1,\text{rms}}$$

Include a redesign of the input amplifier so that the input multiplier M_1 is driven over its 10-V range.

3.37 *(Sec. 3.5)* In the circuit of Example 3.15 (Fig. 3.55), suppose that the input was a 1-V rms, 100-Hz sine wave. Sketch the waveforms at points v_2, v_3, v_4, and v_0. Include an estimate of the ripple signal on v_4 and v_0.

$$(\sin \omega t)^2 = \frac{1 - \cos 2\omega t}{2}$$

3.38 *(Sec. 3.6)*

   (*a*) Given the block diagram of Fig. 3.56a, show that (3.99) holds.

   (*b*) If, in the block diagram of Fig. 3.56b, the noninverting integrator has the transfer function

$$\frac{V_0(s)}{V_2(s)} = \frac{\omega_0}{(s - \delta)}$$

   show that (3.99) holds, and find ϵ in terms of δ.

(c) Find the location of the poles of $V_0(s)$ in the Laplace transform of (3.99), and show that a positive value of ϵ leads to regeneration.

3.39 *(Sec. 3.6)* Design a sine-wave oscillator, based on Fig. 3.57b, that produces a 500-Hz, ±8-V waveform. Use 0.01-μF capacitors. Assume that ±15 V dc is available for the limiter circuit; clipping level can be approximate.

3.40 *(Sec. 3.6)* Refer to Fig. 3.59a. Derive the conditions for oscillation of the Wien bridge oscillator (3.102). The trick is to assume that the voltage on the two amplifier input terminals is the same.

3.41 *(Sec. 3.6)* Refer to Fig. 3.59b.

(a) Derive the conditions for oscillation for the phase-shift oscillator, that is,

$$\omega_0 \cong \frac{1}{\sqrt{3}\,RC}$$

and $R_1 \cong 12R$.

(b) Find $v_1(t)$ in terms of $v_0(t)$.

3.42 *(Sec. 3.6)* Refer to the FM oscillator circuit block diagram of Fig. 3.58.

(a) Show that (3.101) gives the frequency of oscillation.

(b) If $|v_m| \leqslant 1$-V peak, design an oscillator circuit, based on Example 3.16, that produces an FM waveform that varies ±100 Hz from a nominal center frequency of 1000 Hz. NOTE: be sure $(v_m + A) \leqslant 10$ V at all times.

3.43 *(Sec. 3.6)*

(a) Using the circuit of Fig. 3.61, design an oscillator that will generate ±5-V square and triangle waves, with a frequency that may be switched between 1 and 10 kHz (manual switching of one resistor).

(b) Modify the design so that the circuit generates ±10-V square waves, but ±5-V triangle waves.

3.44 *(Sec. 3.6)* Refer to Fig. 3.60.

(a) Demonstrate that (3.104) holds for the case where the bistable has symmetrical switch points and limiting levels.

(b) Suppose we have an asymmetrical bistable. Develop a general expression for the period of the resulting square wave; also find the expression for the time intervals of the positive and negative portions of the square wave. Call T_1 the duration of the positive part and T_2 the duration of the negative part. Of course, assume $L_+ > 0, L_- < 0, S_+ > 0, S_- < 0$.

3.45 *(Sec. 3.6)* Design a circuit based on Figs. 3.60 to 3.62 which will generate an FM waveform. The modulating signal (to which a constant must be added) should produce the following amount of frequency change:

Value of v_m, V	Output frequency, H_z
−1	900
0	1000
+1	1100

The output waveform should be a square wave with a ±5-V peak value.

3.46 *(Sec. 3.6)* Derive Eq. (3.111).

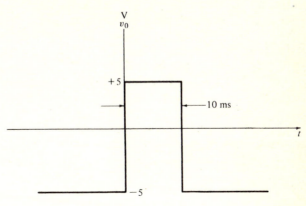

FIGURE P3.47

3.47 *(Sec. 3.6)* Refer to Fig. P3.47; suppose we want to design a monostable pulse generator to provide the waveform shown.

(a) Using the three-amplifier circuit of Fig. 3.63 (see also Fig. 3.61 for details of the bistable subsystem), design a circuit to meet the specifications of Fig. 3.47. Sketch the entire circuit, including all important waveforms. What value of α did you use?

(b) Using the one-amplifier circuit of Fig. 3.65, repeat the above.

Assume ±15 V dc is available.

3.48 *(Sec. 3.6)* Derive Eq. (3.113).

3.49 *(Sec. 3.6)* Sketch the output of the circuit of Fig. 3.65 when a positive 6-V trigger is applied with

$$R_1 = 100 \text{ k}\Omega \qquad R_7 = 10 \text{ k}\Omega$$
$$C = 0.001 \ \mu\text{F} \qquad R_2 = R_3 = 10 \text{ k}\Omega$$
$$E = 8 \text{ V} \qquad V_F = 0.6 \text{ V}$$
$$R_A = 10 \text{ k}\Omega$$

4

ACTIVE RC FILTERS

In the preceding chapter we showed how operational amplifiers could be applied to realizing many useful circuit functions such as comparators, limiters, multipliers, etc. These circuit functions are all characterized by the fact that they are *nonlinear,* i.e., that superposition does not apply. In this chapter we shall continue the presentation of operational amplifier applications. The subject here, however, will be a quite different one, namely the realization of *linear* circuit functions. Such functions are widely used whenever it is necessary to control or shape the time or frequency domain characteristics of signals. Circuits which perform such shaping are called *filters*. Originally, filter circuits were composed only of passive network components, namely, resistors, capacitors, and inductors. In more recent years, however, it has been found that if active elements such as operational amplifiers are also used, then either type of reactive element, i.e., the capacitors or the inductors, may be eliminated without loss of generality in circuit performance. Of the two types of reactive elements, inductors are by far the more logical elements to eliminate, since the typical inductor is considerably poorer than the typical capacitor with respect to linearity, dissipation, size, and weight. In addition, while capacitors may be directly included in integrated circuit realizations using either thick-film, thin-film, or monolithic techniques, inductors are usually feasible only as discrete elements which must be "wired" into the circuit. Thus they are considerably less attractive. Filter circuits which consist of resistors, capacitors, and active elements are usually referred to as *active* RC *filters.* There are many advantages and disadvantages to the use of these filters rather than their passive *RLC* counterparts. Among the advantages are small size and weight, simplicity of synthesis and tuning, isolation of stages, and large available gain. Among the disadvantages are the limitations on voltage and current levels, inability to "float" a given realization, and the parasitic effects induced by the active element, such as dc output voltage offset, input bias currents, etc. In most applications the advantages outweigh the disadvantages, and thus active *RC* filters are becoming increasingly important in a wide range of engineering applications. They are available both as stock items and as custom-manufactured ones from a number of large and small manufacturers. The production and marketing of such filters is a large business, with annual production running into the tens of millions. In this chapter we shall introduce the subject of active *RC* filters. The subject is a large one, and several books have been devoted entirely to it. Thus our treatment here must necessarily be a limited one. We shall, however, present details of many of the best-known and most useful active *RC* circuits. Suggestions for additional reading are given at the end of the chapter, and these will provide additional references for the reader who wishes to pursue the subject in more depth.

4.1 GENERAL DESIGN AND EVALUATION TECHNIQUES–SENSITIVITY

In this section we present a discussion of some general design and evaluation techniques which are applicable to many active *RC* filters. We will illustrate in detail their application to a specific filter, the second-order low-pass noninverting low-gain-amplifier one. In the following sections we will discuss the results obtained by applying these techniques to other types of filters.

A general active *RC* filter configuration is shown in Fig. 4.1. Other circuit configurations are, of course, possible, but this one has been shown to be very practical in the realization of voltage transfer functions, one of the most commonly required filtering applications. In the figure, the square box with the three numbered terminals represents the passive portion of the filter. It is comprised of resistors and capacitors. The triangle represents a VCVS (Voltage-Controlled Voltage Source) of infinite input impedance, zero output impedance, and finite gain. Such a VCVS can, of course, be effectively approximated using an operational amplifier and the circuits shown in Fig. 1.21 (for the noninverting or positive-gain case) and Fig. 1.6 (for the inverting or negative-gain case). To analyze the circuit shown in Fig. 4.1, we may begin by first considering only the passive network. The properties of this are readily defined using a set of *y* parameters similar to those used for two-port networks. Here, however, there are three voltage and current variables, so a 3 by 3 matrix of *y* parameters is required rather than the usual 2 by 2 matrix used in the two-port case. We now define the passive network by the relations

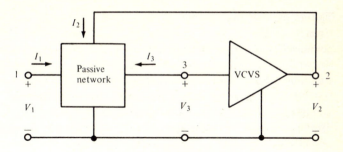

FIGURE 4.1
General active *RC* filter configuration.

$$
\begin{bmatrix} I_1(s) \\ I_2(s) \\ I_3(s) \end{bmatrix} = \begin{bmatrix} y_{11}(s) & y_{12}(s) & y_{13}(s) \\ y_{12}(s) & y_{22}(s) & y_{23}(s) \\ y_{13}(s) & y_{23}(s) & y_{33}(s) \end{bmatrix} \begin{bmatrix} V_1(s) \\ V_2(s) \\ V_3(s) \end{bmatrix}
\tag{4.1}
$$

where the quantities $V_i(s)$ and $I_i(s)$ are defined in Fig. 4.1. The square matrix in (4.1) is called the y-*parameter matrix*. Note that it is symmetrical. This is the result of the reciprocal nature of the passive network. The y parameters are readily found using the relation

$$
y_{ij}(s) = \frac{I_i(s)}{V_j(s)} \bigg|_{V_k(s) = 0} \qquad k = 1, 2, 3; k \neq j
\tag{4.2}
$$

The requirement that the voltages $V_k(s) = 0$ for all terminals except the one at which the excitation $V_j(s)$ is applied is, of course, satisfied by shorting these terminals to ground.

Now let us consider the effect that the VCVS has on the relations of (4.1). In terms of the variables defined in Fig. 4.1, the VCVS imposes the constraints

$$
V_2(s) = KV_3(s) \qquad \text{and} \qquad I_3(s) = 0
\tag{4.3}
$$

where K is called the *gain* of the VCVS. Substituting these relations in (4.1), we obtain

$$
\frac{V_2(s)}{V_1(s)} = \frac{-Ky_{13}(s)}{y_{33}(s) + Ky_{23}(s)}
\tag{4.4}
$$

as the overall voltage transfer function for the circuit of Fig. 4.1. In the following paragraphs we shall show how this relation forms a basis for finding active *RC* network configurations for a variety of specific filtering characteristics.

The first filter characteristic that we shall consider realizing, using the general active *RC* configuration shown in Fig. 4.1, is the second-order low-pass one. This has the form

$$
\frac{V_2(s)}{V_1(s)} = \frac{H}{s^2 + a_1 s + a_0}
\tag{4.5}
$$

The dc gain of this function is H/a_0, and the values of the coefficients a_1 and a_0 are usually chosen so as to achieve some specific low-pass characteristic. Several of the commoner ones are summarized in Table 4.1. Now let us consider what form the passive network must have to produce the function given in (4.5). First we note that since the function is second-order, a minimum of two capacitors are required. Second, comparing the relations given in (4.4) and (4.5), we see that the numerator of the transfer admittance $y_{13}(s)$ must be a constant.† Since any series capacitor placed between terminals 1 and 3 will produce a transmission zero at the origin and thus a term s in the numerator of $y_{13}(s)$, we conclude that no such series capacitors can occur. An active RC network configuration whose passive portion satisfies the above requirements is shown in Fig. 4.2.‡ The numerators $n_{ij}(s)$ of the y parameters $y_{ij}(s)$ for the passive portion of this network are

$$n_{11}(s) = \frac{1}{R_1}\left(s + \frac{1}{R_2 C_1}\right) \qquad n_{12}(s) = \frac{-s}{R_1} \qquad n_{13}(s) = \frac{-1}{C_1 R_1 R_2}$$

$$n_{22}(s) = s\,\frac{R_1 + R_2}{R_1 R_2} \qquad\qquad n_{23}(s) = \frac{-s}{R_2} \tag{4.6}$$

$$n_{33}(s) = C_2\left[s^2 + s\left(\frac{1}{R_1 C_1} + \frac{1}{R_2 C_1} + \frac{1}{R_2 C_2}\right) + \frac{1}{R_1 R_2 C_1 C_2}\right]$$

> †In general, except when private poles are present the denominators in all the y parameters will be the same. Thus, in (4.4), the denominators of the parameters will cancel, and the numerators may be directly equated with the polynomials in the numerator and denominator of (4.5).
>
> ‡ This and several of the other active RC circuits presented in this and the following section were originally described in R. P. Sallen and E. L. Key, A Practical Method of Designing RC Active Filters, *IRE Transactions on Circuit Theory*, vol. CT-2, no. 1, pp. 74–85, March 1955.

Table 4.1 LOW-PASS NETWORK FUNCTION CHARACTERISTICS FOR (4.5)*

No.	Name	Characteristic	a_1	a_0	Pole locations
1	Butterworth (maximally flat magnitude)	At 1 rad/s down 3.01 dB from zero frequency value	1.41421	1.00000	$-0.70711 \pm j0.70711$
2a	Chebychev (equal-ripple magnitude)	$\frac{1}{2}$-dB ripple in passband from 0–1 rad/s	1.42562	1.51620	$-0.71281 \pm j1.00404$
2b		1-dB ripple in passband from 0–1 rad/s	1.09773	1.10251	$-0.54887 \pm j0.89513$
3	Thompson (maximally flat delay, linear phase)	1-s delay, unity phase constant	3.00000	3.00000	$-1.50000 \pm j0.86603$

*A more detailed description of these characteristics may be found in any standard passive network synthesis book. See, for example, L. Weinberg, "Network Analysis and Synthesis," McGraw-Hill, New York, 1962.

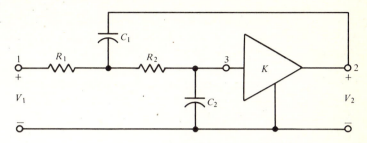

FIGURE 4.2
Second-order noninverting low-pass filter.

The denominators $d_{ij}(s)$ of the parameters are all the same:

$$d_{ij}(s) = s + \frac{R_1 + R_2}{C_1 R_1 R_2} \qquad (4.7)$$

Inserting these values in (4.4), we obtain the voltage transfer function for the network

$$\frac{V_2(s)}{V_1(s)} = \frac{K/R_1 R_2 C_1 C_2}{s^2 + s[1/R_1 C_1 + 1/R_2 C_1 + (1-K)/R_2 C_2] + 1/R_1 R_2 C_1 C_2} \qquad (4.8)$$

This obviously has the required low-pass form of (4.5).

It is interesting to note the manner in which the complex conjugate poles required by the filter characteristics defined in Table 4.1 are produced by the function given in (4.8). Passive *RC* networks, of course, can only have poles on the negative real axis. The same restriction holds for the zeros of *RC* driving-point functions such as $y_{33}(s)$. Thus the numerator of this y parameter must have the form $C_2(s - \sigma_1)(s - \sigma_2)$ where σ_1 and σ_2 are functions of the passive network elements $R_1, R_2, C_1,$ and C_2. Now consider $y_{23}(s)$. From the circuit shown in Fig. 4.2, we note that one series capacitor is present in the path from terminal 2 to terminal 3. Thus, $y_{23}(s)$ must have a zero of transmission at the origin. Verifying this from (4.6), we see that its numerator is $-s/R_2$. Inserting these results in the denominator of (4.4) and comparing this with the denominator of (4.5), we obtain

$$(s - \sigma_1)(s - \sigma_2) - \frac{sK}{R_2 C_2} = s^2 + a_1 s + a_0 \qquad (4.9)$$

In order to produce complex conjugate poles, it is readily shown that K must be positive, i.e., the VCVS must be noninverting. When $K = 0$, the poles of the low-pass network function (the zeros of $s^2 + a_1 s + a_0$) are at σ_1 and σ_2. As K increases, the poles follow a circular locus as indicated in Fig. 4.3. Obviously, any desired pole location can be obtained by choosing an appropriate value for K and, if necessary, making a frequency normalization. It is also apparent from the locus that too high a value of K will produce instability, i.e., right half-plane poles.

The values of the network elements required to realize a specific low-pass characteristic can be readily determined by assigning some of the elements fixed values

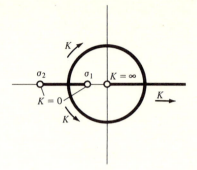

FIGURE 4.3
Locus of pole positions for the filter of
Fig. 4.2.

and solving for the values of the other ones. Since, in the network shown in Fig. 4.2, there are five element values to be selected and only three specifications to be met [the values of a_0, a_1, and H in (4.5)], obviously there are two degrees of freedom which may be used to arbitrarily specify element values or impose restrictions relating the values of different elements. In some cases, the values of more than two elements may be chosen. For example, choosing all passive elements to be unity, i.e., setting $R_1 = R_2 = C_1 = C_2 = 1$ (ohms and farads), (4.8) becomes

$$\frac{V_2(s)}{V_1(s)} = \frac{K}{s^2 + s(3 - K) + 1} \qquad (4.10)$$

Thus a Butterworth characteristic with a bandwidth or -3-dB frequency of 1 rad/s (no. 1 in Table 4.1) is obtained by setting $3 - K = 1.414$, that is, $K = 1.586$. In this case the constant H of (4.5) is also 1.586. More generally, if we set $R_1 = C_1 = 1$, then for any arbitrary value of C_2 (or R_2) we may use the design equations†

$$R_2 = \frac{1}{a_0 C_2} \qquad K = 1 + C_2 - \frac{a_1}{a_0} + \frac{1}{a_0} \qquad (4.11)$$

Some other solutions are summarized in Table 4.2.

From the entries of Table 4.2 we see that many different combinations of values for the elements of the circuit shown in Fig. 4.2 may be used to obtain a given filter characteristic. The question naturally arises as to how to select the best of these possibilities, or whether to look for still other combinations. To answer such a question, the network designer must first decide what he means by the word "best," i.e., what criterion of performance is to be optimized. If ease of fabrication is the dominant criterion, then realizations in which all elements of one type have the same value may be the best choice. Several such combinations are given in Table 4.2. Another criterion which is frequently important is the use of integer values for K, since these are usually more easily realized than irrational values. For example, for realizations in which an operational amplifier is to be used, the choice of $K = 2$ is especially attractive since the

†Some of the design equations and examples given in this and the following section are taken from Ref. 2, chap. 2.

circuit shown in Fig. 1.21 can then be constructed with equal-valued resistors. Such resistors can be more closely matched in tolerance and in aging and temperature characteristics than those of unequal value, and thus circuit stability is improved. Several solutions for integer values of *K* are given in Table 4.2.

One of the most important criteria used in evaluating active *RC* filter designs is *sensitivity*. This is a measure of the amount that some characteristic of the filter changes as a result of a change in some circuit parameter. The filter characteristic most frequently of interest is the position of the poles of the network function. The most critical circuit parameter is usually the gain of the active element. Using these quantities we now define the sensitivity $S_K^{P_0}$ of a pole at $p_0 = \sigma_0 + j\omega_0$ to the VCVS gain *K* as

$$S_K^{P_0} = \frac{\partial \sigma_0}{\partial K} \frac{K}{\sigma_0} + j \frac{\partial \omega_0}{\partial K} \frac{K}{\omega_0} \qquad (4.12)$$

Thus the real part of the sensitivity gives the ratio of the normalized (or percentage) change in the real part of the pole position to a normalized (or percentage) change in the gain *K*, while the imaginary part of the sensitivity gives the corresponding ratio between the imaginary part of the pole position and the gain.† As an example of the determination

> †This definition of sensitivity is due to W. J. Kerwin, L. P. Huelsman, and R. W. Newcomb, State Variable Synthesis for Insensitive Integrated Transfer Functions, *IEEE Journal of Solid-State Circuits,* vol. SC-2, no. 3, pp. 87–92, September 1967. For a more detailed discussion of the various types of sensitivities and the theory of sensitivity minimization, see Ref. 1, chap. 2.

Table 4.2   ELEMENT VALUES FOR LOW–PASS NETWORK OF FIG. 4.2 (ohms and farads)*

Characteristic no. (from Table 4.1)	R_1	R_2	C_1	C_2	K
1	1.00000	1.00000	1.00000	1.00000	1.58578
	1.00000	1.00000	1.41421	0.70711	1.00000
	1.00000	1.00000	0.87403	1.14412	2.00000
	0.70711	1.41421	1.00000	1.00000	2.00000
2a	0.81220	0.81204	1.00000	1.00000	1.84213
	1.00000	1.00000	1.40289	0.47013	1.00000
	1.00000	1.00000	0.77088	0.85557	2.00000
	0.70145	0.94026	1.00000	1.00000	2.00000
2b	0.95237	0.95237	1.00000	1.00000	1.95444
	1.00000	1.00000	1.82192	0.49783	1.00000
	1.00000	1.00000	0.93809	0.96688	2.00000
	0.91096	0.99567	1.00000	1.00000	2.00000
3	0.57735	0.57735	1.00000	1.00000	1.26795
	1.00000	1.00000	0.66667	0.50000	1.00000
	1.00000	1.00000	0.45743	0.72871	2.00000
	0.33333	1.00000	1.00000	1.00000	2.00000

*Some of the values given in this table and the tables of the following section have been adopted from R. S. Aikens, "Canonic Active *RC* Networks," Master of Science thesis, University of Arizona, 1972.

of $S_K{}^{P_0}$, consider the circuit given in Fig. 4.2 with all the passive elements set to unity. For this case the voltage transfer function is given in (4.10). Applying (4.12), we obtain

$$S_K{}^{P_0} = \frac{K}{K-3} + j\frac{K(3-K)}{4-(3-K)^2} \qquad (4.13)$$

As an example of the use of (4.13), if we desire to realize the network function

$$\frac{V_2(s)}{V_1(s)} = \frac{H}{s^2 + 0.1s + 1} \qquad (4.14)$$

with this circuit, then $K = 2.9$, and from (4.13) $S_K{}^{P_0} = -29 + j0.0725$. Since the imaginary part of this sensitivity is small compared to the real part, we conclude that the pole displacement produced by a change in K is very nearly parallel to the real axis. In addition, since the real part of the sensitivity is negative, and since σ_0 is negative, we conclude that an increase in K produces an increase in σ_0, i.e., the pole moves toward the $j\omega$ axis. Both these conclusions are verified by the pole locus shown in Fig. 4.3.

The sensitivity computed in the above example is so high that the network realization is on the borderline of practical value. It is possible, however, to choose different values for the passive network elements so as to reduce the sensitivity. For example, consider choosing $R_1 = C_1 = 1$, $R_2 = 10$, and $C_2 = 0.1$ (ohms and farads) in the circuit shown in Fig. 4.2. The voltage transfer function of (4.8) now becomes

$$\frac{V_2(s)}{V_1(s)} = \frac{H}{s^2 + s(2.1 - K) + 1} \qquad (4.15)$$

and the sensitivity becomes

$$S_K{}^{P_0} = \frac{K}{K-2.1} + j\frac{K(2.1-K)}{4-(2.1-K)^2} \qquad (4.16)$$

If we use this circuit to realize the network function given in (4.14), a gain of 2 is required. From (4.16), the sensitivity is $S_K{}^{P_0} = -20 + j0.05$. Thus, this new choice of element values produces two effects: (1) it lowers the sensitivity, and (2) it permits the use of an integer value of gain (which is more readily stabilized than a noninteger value). From a nonrigorous viewpoint we may say that the reduction in sensitivity has been brought about by raising the impedance of the passive network at terminal 3, thus more closely matching it to the high input impedance of the VCVS. Although the sensitivity has been lowered, it is still somewhat high for practical use. It may be shown that other combinations of passive elements bring only minor additional sensitivity reduction, and thus we conclude that the circuit configuration of Fig. 4.2 is only marginally practical for the network function of (4.14). Since the real part of the sensitivity defined in (4.12) has the quantity σ_0 in its denominator, obviously, as the poles are moved further away from the $j\omega$ axis, the sensitivity will decrease. Thus, we can conclude that, in general, low-Q network functions will have lower sensitivities than high-Q ones, for a given active RC

circuit configuration.† As an example of this, for the Butterworth network function defined as no. 1 in Table 4.1, using the first realization given in Table 4.2, we find that $S_K^{P_0} = -1.12 + j1.12$, which is a quite reasonable value of sensitivity for a practical realization. The sensitivities of most of the other low-pass functions given in Table 4.1 are correspondingly low. In a later section of this chapter we shall present some active *RC* circuit configurations which are specially designed to realize high-*Q* network functions with low sensitivities.

An alternative method of specifying the low-pass function given in (4.5) is

$$\frac{V_2(s)}{V_1(s)} = \frac{H_c \omega_c^2}{s^2 + \alpha \omega_c s + \omega_c^2} \tag{4.17}$$

The dc gain of this function is H_c, ω_c is the nominal cutoff frequency (which is not necessarily at the −3-dB point), and $\alpha = 1/Q$. For this characterization, the poles are at

$$p_0 = \sigma_0 + j\omega_0 = \frac{-\alpha \omega_c}{2} \pm j\omega_c \sqrt{1 + \left(\frac{\alpha}{2}\right)^2} \tag{4.18}$$

The sensitivities usually used with the function given in (4.17) are readily related to the pole sensitivity given in (4.12). The first of these is a *Q* sensitivity S_K^Q defined as

$$S_K^Q = \frac{\partial Q}{\partial K} \frac{K}{Q} = -\frac{\partial \alpha}{\partial K} \frac{K}{\alpha} \tag{4.19}$$

Since, from (4.18) and (4.12),

$$\text{Re } S_K^{P_0} = \frac{\partial \sigma_0}{\partial K} \frac{K}{\sigma_0} = \frac{\partial \alpha}{\partial K} \frac{K}{\alpha} + \frac{\partial \omega_c}{\partial K} \frac{K}{\omega_c} \tag{4.20}$$

for the case where ω_c is not a function of K, as is true for the low-pass network described here,

$$S_K^Q = -\text{Re } S_K^{P_0} \tag{4.21}$$

The second sensitivity function usually used with (4.17) is the cutoff frequency sensitivity $S_K^{\omega_c}$, defined as

$$S_K^{\omega_c} = \frac{\partial \omega_c}{\partial K} \frac{K}{\omega_c} \approx \text{Im } S_K^{P_0} \tag{4.22}$$

where, from (4.18), the approximate equality can be readily seen to hold for all high-*Q* cases.

---

†Note that the *Q* of a network function is a measure of the sharpness of its resonance and is determined primarily by the positions of its poles. For a network function with a denominator polynomial having the form $s^2 + a_1 s + a_0$, we may define $Q = \sqrt{a_0}/a_1$. See L. P. Huelsman, "Basic Circuit Theory with Digital Computations," pp. 469–470, Prentice-Hall, Englewood Cliffs, N.J., 1972.

An important part of the actual fabrication of any active RC filter is a final tuning to compensate for inaccuracies in the network element values, parasitic effects, etc. In the low-pass low-gain-amplifier filter, tuning is readily accomplished since the adjustment of the $Q(=1/\alpha)$ and ω_c of (4.18) may be made separately without interaction occurring. Specifically, equal percentage changes in the values of R_1 and R_2 will vary ω_c without affecting Q, and changes in K will vary Q without affecting ω_c. Some additional information on tuning procedures may be found in Sec. 4.5.

In this section we have presented our first active RC circuit configuration and showed how to evaluate its performance by determining its sensitivity to the gain of the active element. A complete analysis of a design using such a circuit should also include the determination of the passive element sensitivities $S_x{}^{P_0}$ where x is successively set to R_1, R_2, C_1, and C_2. In general, these passive element sensitivities are considerably lower than $S_K{}^{P_0}$. The actual computation of some typical cases is left to the reader as an exercise.

4.2   SECOND-ORDER LOW-GAIN-AMPLIFIER FILTERS

In the preceding section we presented a discussion of a general approach to the design and evaluation of active RC filters, using a specific low-pass network configuration as an example. In this section we will apply the same technique to realize several other network functions, restricting our interest to realizations which require a VCVS of relatively low gain, i.e., one realized by the circuits given in Figs. 1.6 and 1.21.

High-pass Noninverting Filter

The first filter function that we shall consider in this section is a second-order high-pass voltage transfer function having the form

$$\frac{V_2(s)}{V_1(s)} = \frac{Hs^2}{s^2 + a_1 s + a_0} = \frac{Hs^2}{s^2 + \alpha \omega_c s + \omega_c{}^2} \qquad (4.23)$$

Table 4.3   HIGH–PASS NETWORK FUNCTION CHARACTERISTICS FOR (4.23)

No.	Name	Characteristic	a_1	a_0	Pole locations
1	Butterworth (maximally flat magnitude)	At 1 rad/s, down 3.01 dB from infinite frequency value	1.41421	1.00000	$-0.70711 \pm j0.70711$
2a	Chebychev (equal-ripple magnitude)	$\frac{1}{2}$-dB ripple in passband from 1 rad/s to infinity	0.94026	0.65954	$-0.47013 \pm j0.66221$
2b		1-dB ripple in passband from 1 rad/s to infinity	0.99566	0.90702	$-0.49783 \pm j0.81191$

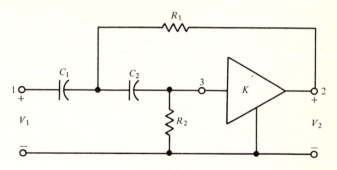

FIGURE 4.4
Second-order noninverting high-pass filter.

where the second form is useful in tuning procedures. The high-frequency gain of the function is H, and the cutoff frequency is ω_c. In general, high-pass functions may be determined directly from low-pass ones by substituting $1/s$ for the complex frequency variable s. Thus, if a network function $N(s)$ is a low-pass function, the function $N(1/s)$ will be a high-pass one. Such a substitution inverts the magnitude characteristic of the low-pass function around the frequency 1 rad/s. Thus a low-pass function having a 1-dB ripple in its passband from 0 to 1 rad/s would be transformed into a high-pass function having a 1-dB ripple in its passband from 1 to infinite rad/s, etc. This means that the low-pass characteristics given in Table 4.1 may be directly related to corresponding high-pass characteristics. The results for cases 1 and 2 are shown in Table 4.3.

To find an active *RC* configuration capable of realizing a voltage transfer function having the form of (4.23), we note from (4.4) that, for the general active *RC* configuration shown in Fig. 4.1, the numerator of the transfer admittance $y_{13}(s)$ must be some multiple of s^2, that is, it must have a second-order transmission zero at the origin, which means that there must be two independent series capacitors in the path between terminals 1 and 3. Such a condition is satisfied by the network configuration shown in Fig. 4.4. By direct analysis we find that the voltage transfer function for this circuit is

$$\frac{V_2(s)}{V_1(s)} = \frac{Ks^2}{s^2 + s[1/R_2C_1 + 1/R_2C_2 + (1/R_1C_1)(1-K)] + 1/R_1R_2C_1C_2} \quad (4.24)$$

This obviously is of the required form. The decomposition of the denominator follows a form similar to that given in (4.9). Thus, K is required to be positive, the pole locus has the form shown in Fig. 4.3, and the comments on sensitivity, as well as the numerical examples given in that section, are directly applicable.

The determination of the values of the elements of the network shown in Fig. 4.4 is similar to the process used in the preceding section. For example, setting $R_1 = R_2 = C_1 = C_2 = 1$ (ohms and farads) in (4.24) gives

$$\frac{V_2(s)}{V_1(s)} = \frac{Ks^2}{s^2 + s(3-K) + 1} \quad (4.25)$$

More generally, if we set $R_1 = C_1 = 1$, then for any arbitrary value of R_2 (or C_2) we may use the design equations

$$C_2 = \frac{1}{a_0 R_2} \qquad K = 1 + a_0 + \frac{1}{R_2} - a_1 \qquad (4.26)$$

Some other solutions are summarized in Table 4.4. It is easily shown that, in general, the network configuration of Fig. 4.4 is suitable only for low-Q realizations.

The second function given in (4.23) may be used to specify a tuning procedure for the network. The procedure is similar to the one specified for the low-pass network of Sec. 4.1, namely, equal percentage changes in R_1 and R_2 will change ω_c without affecting α, and changes in K will vary α without affecting ω_c. More details on tuning may be found in Sec. 4.5.

Bandpass Noninverting Filter

Techniques similar to those used above can be used to obtain a second-order bandpass noninverting low-gain-amplifier filter. The network function for such a filter has the form

$$\frac{V_2(s)}{V_1(s)} = \frac{Hs}{s^2 + a_1 s + a_0} = \frac{H_c \alpha \omega_c s}{s^2 + \alpha \omega_c s + \omega_c^2} \qquad (4.27)$$

where the second function is useful in tuning procedures. The gain at resonance is $H/a_1 = H_c$. The resonant frequency is $\omega_c = \sqrt{a_0}$. An active RC network capable of realizing such a function may be found by noting from (4.4) that, for the general active RC configuration shown in Fig. 4.1, the numerator of the transfer admittance $y_{13}(s)$ must be some multiple of s, that is, it must have a single transmission zero at the origin, and thus a single series capacitor is required in the path between terminals 1 and 3. Such a

Table 4.4 ELEMENT VALUES FOR HIGH–PASS NETWORK OF FIG. 4.4 (ohms and farads)

Characteristic no. (from Table 4.3)	R_1	R_2	C_1	C_2	K
1	1.00000	1.00000	1.00000	1.00000	1.58579
	0.70711	1.41421	1.00000	1.00000	1.00000
	1.00000	1.00000	1.41421	0.70711	2.00000
	1.14412	0.87403	1.00000	1.00000	2.00000
2a	1.23134	1.23134	1.00000	1.00000	1.84222
	0.71281	2.12707	1.00000	1.00000	1.00000
	1.00000	1.00000	1.42563	1.06354	2.00000
	1.29722	1.16881	1.00000	1.00000	2.00000
2b	1.05001	1.05001	1.00000	1.00000	1.95455
	0.54886	2.00872	1.00000	1.00000	1.00000
	1.00000	1.00000	1.09772	1.00436	2.00000
	1.06599	1.03426	1.00000	1.00000	2.00000

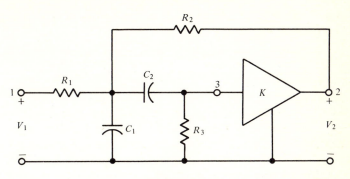

FIGURE 4.5
Second-order noninverting bandpass filter.

condition is satisfied by the configuration shown in Fig. 4.5.† By direct analysis we find that the voltage transfer function for this circuit is

$$\frac{V_2(s)}{V_1(s)} = \frac{sK/R_1C_1}{s^2 + s[1/R_1C_1 + 1/R_3C_1 + 1/R_3C_2 + (1-K)/R_2C_1] + (1/R_1 + 1/R_2)/R_3C_1C_2}$$

$$(4.28)$$

The comments on the denominator decomposition are identical with those given for the high-pass network described above. Various combinations of element values for different values of Q for a center frequency of 1 rad/s ($a_0 = 1$) are given in Table 4.5. A useful set

---

†Many other bandpass configurations are possible. This one is due to R. P. Sallen and E. L. Key, *op. cit.*, and has been shown to be of practical value.

Table 4.5 ELEMENT VALUES FOR BANDPASS NETWORK OF FIG. 4.5 with a_0 of (4.27) equal to unity (ohms and farads)

Case 1: All elements of same type equal-valued $R_i = 1.41421, C_i = 1.00000$			
$Q(= 1/a_1)$ 2	5	10	20
K 3.29289	3.71716	3.85858	3.92928

Case 2: Integer gain, $K = 2, R_1 = C_1 = C_2 = 1$			
$Q(= 1/a_1)$ 2	5	10	20
R_2 0.74031	0.63439	0.60471	0.59076
R_3 2.35078	2.57630	2.65367	2.69274

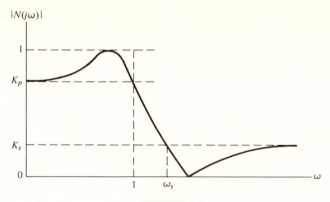

FIGURE 4.6
Elliptic filter magnitude characteristic.

of design formulas may also be derived by setting $R_1 = R_2 = C_1 = 1$. In this case, for any arbitrary value of R_3 (or C_2) we find that

$$C_2 = \frac{2}{R_3 a_0} \qquad K = 2 + \frac{a_0}{2} + \frac{1}{R_3} - a_1 \qquad (4.29)$$

As an example of the use of these equations, for a network function with a Q of 10 and a center frequency of 1 rad/s ($a_1 = 0.1$, $a_0 = 1$), to minimize K we might choose $R_3 = 10$ Ω, for which $C_2 = 0.2$ F and $K = 2.5$. The sensitivity $S_K^{P_0}$ of this realization is approximately -25 (the imaginary part is very small), which is about as high as can be tolerated. Further study of this configuration shows that, like the low-pass and high-pass noninverting realizations, it is suitable only for low-Q functions of 10 or less.

The second function given in (4.27) may be used to specify a tuning procedure similar to that used for the low-pass and bandpass filters, i.e., equal percentage changes in R_1, R_2, and R_3 (or equal percentage changes in C_1 and C_2) will change ω_c without affecting α, and changes in K will vary α without affecting ω_c. More details on tuning procedures are given in Sec. 4.5.

$j\omega$-axis-zeros Noninverting Filter

Frequently, filtering applications require the realization of a transfer function which has real frequency transmission zeros, i.e., zeros on the $j\omega$ axis. An example of such an application is the elliptic (or Cauer) filter. For the second-order case a network function of this type has the form

$$\frac{V_2(s)}{V_1(s)} = \frac{H(s^2 + b_0)}{s^2 + a_1 s + a_0} \qquad (4.30)$$

The magnitude characteristic for the case where the cutoff frequency and the peak magnitude are both normalized to unity has the form shown in Fig. 4.6. The values of the coefficients a_0, a_1, b_0, and H of (4.30) may be related to any two of the parameters K_p, K_s, and ω_s using the values given in Table 4.6. Comparing (4.30) with (4.4) for the network shown in Fig. 4.1, we see that in the passive portion of an active *RC* network capable of realizing (4.30), the numerator of its parameter $y_{13}(s)$ must be a multiple of $s^2 + b_0$. Thus there must be a twin-tee, parallel-ladder, or other similar zero-producing network connected between terminals 1 and 3. Such a condition is met by the network

Table 4.6 ELLIPTIC FUNCTION PARAMETERS[*]
[From top to bottom, a_1, a_0, b_0, $H(=K_s)$]

ω_s \ K_p	0.7	0.75	0.8	0.85	0.9	0.95	0.99
2.0	.597566	.672335	.761953	.87093	1.09079	1.28475	1.70530
	.748566	.807532	.889100	1.01055	1.21614	1.67671	3.39116
	7.46410	7.46393	7.46393	7.46393	7.46410	7.46394	7.46437
	.070208	.081143	.095295	.115081	.146639	.213409	.449766
1.8	.586497	.658788	.744765	.852101	.996903	1.22172	1.49664
	.761473	.821030	.903240	1.02526	1.23061	1.68388	3.25137
	5.93375	5.93377	5.93375	5.93377	5.93399	5.93377	5.93385
	.089828	.103773	.121775	.146865	.186645	.269588	.542449
1.6	.568640	.636848	.716947	.814969	.942467	1.12264	1.21673
	.780727	.840896	.923621	1.04564	1.24863	1.68414	3.01139
	4.55831	4.55832	4.55842	4.55832	4.55832	4.55832	4.55836
	.119892	.138356	.162086	.194981	.246530	.350990	.654022
1.4	.535956	.596787	.666375	.747996	.845981	.956021	.859430
	.811695	.872098	.954382	1.07401	1.26798	1.65969	2.61881
	3.33173	3.33166	3.33173	3.33167	3.33172	3.33167	3.33171
	.170533	.196320	.229155	.274010	.342509	.473248	.778164
1.3	.507505	.562111	.622959	.691325	.766598	.829058	.658076
	.835122	.894952	.975687	1.09135	1.27415	1.62319	2.34888
	2.76980	2.76979	2.76981	2.76979	2.76982	2.76980	2.76972
	.211054	.242333	.281803	.334915	.414008	.556729	.839569
1.2	.461178	.506162	.553873	.603117	.647985	.656811	.450447
	.867873	.925546	1.00200	1.10866	1.26987	1.55118	2.02432
	2.23597	2.23595	2.23597	2.23595	2.23591	2.23595	2.23595
	.271698	.310453	.358503	.421457	.511141	.659054	.896281
1.1	.372652	.401509	.428498	.450238	.457760	.420582	.244714
	.916613	.967014	1.03128	1.11609	1.23375	1.41020	1.63605
	1.71409	1.71408	1.71409	1.71405	1.71408	1.71409	1.71394
	.374317	.423125	.481308	.553478	.647782	.781571	.945011
1.05	.285907	.302274	.314810	.320161	.310931	.266561	.142026
	.951232	.991713	1.04130	1.10322	1.18267	1.28875	1.40382
	1.43865	1.43866	1.43866	1.43866	1.43867	1.43867	1.43868
	.462837	.516984	.579033	.651806	.739845	.850992	.966006

[*]D. J. Sticht and L. P. Huelsman, Direct Determination of Elliptic Network Functions, *International Journal of Computers and Electrical Engineering,* vol. 1, pp. 277–280, 1973.

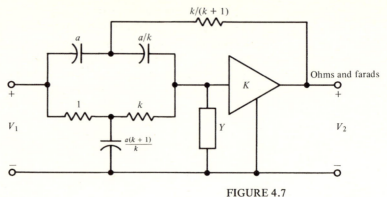

FIGURE 4.7
Second-order $j\omega$-axis-zeros filter.

configuration shown in Fig. 4.7.† For the case in which $a_0 > b_0$, $Y = 1/R$ and the transfer function is

$$\frac{V_2(s)}{V_1(s)} = \frac{K(s^2 + 1/a^2)}{s^2 + s[(K+1)/a][1/R + (2-K)/k] + [1 + (k+1)/R]/a^2} \tag{4.31}$$

In this relation, the constant k may be chosen as unity, or may be used to obtain a particular value of K or to reduce sensitivity. In the latter cases, however, a wider spread of element values results. For this network the following design relations may be used:

$$a = \sqrt{\frac{1}{b_0}} \qquad R = \frac{k+1}{a_0/b_0 - 1} \qquad H = K$$

$$K = 2 + \frac{k}{k+1}\left(\frac{a_0}{b_0} - 1 - \frac{a_1}{\sqrt{b_0}}\right) \tag{4.32}$$

For the case where $b_0 > a_0$ in (4.30), $Y = aCs$ and the transfer function is

$$\frac{V_2(s)}{V_1(s)} = \frac{K/[(k+1)C+1](s^2 + 1/a^2)}{s^2 + s\,(k+1)[C + (2-K)/k]/a[(k+1)C+1] + 1/a^2\,[(k+1)C+1]} \tag{4.33}$$

The following design equations may be used:

$$a = \sqrt{\frac{1}{b_0}} \qquad C = \frac{b_0/a_0 - 1}{k+1} \qquad H = \frac{a_0}{b_0}K$$

$$K = 2 + \frac{k}{k+1}\left(\frac{b_0}{a_0} - 1 - \frac{a_1\sqrt{b_0}}{a_0}\right) \tag{4.34}$$

†W. J. Kerwin and L. P. Huelsman, The Design of High-performance Active *RC* Bandpass Filters, *Proceedings of the IEEE International Convention Record,* part 10, pp. 74–80, March 1966; and W. J. Kerwin, An Active *RC* Elliptic Function Filter, *IEEE Region Six Conference Record,* vol. 2, pp. 647–654, April 1966.

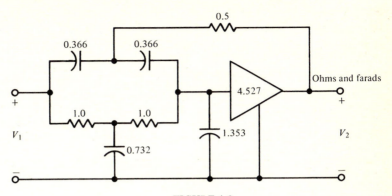

FIGURE 4.8
Example of second-order $j\omega$-axis-zeros filter.

As an example of the use of the above relations, consider an elliptic characteristic of the form shown in Fig. 4.6 in which $K_p = 0.8$ and $\omega_s = 2$. From Table 4.6 we find that the network function is

$$\frac{V_2(s)}{V_1(s)} = \frac{0.0953(s^2 + 7.464)}{s^2 + 0.762s + 0.889} \qquad (4.35)$$

The resulting network configuration, choosing $k = 1$ in (4.34), is shown in Fig. 4.8. The actual gain constant realized is $H = 0.539$; thus the peak transmission is $0.539/0.0953 = 5.656$ rather than the unity value shown in Fig. 4.6.

General Biquadratic Filter

For certain specialized filtering applications, especially those with critical phase requirements, a biquadratic network function in which the positions of the complex conjugate poles and zeros can be independently specified may be required. Such a function has the form

$$\frac{V_2(s)}{V_1(s)} = H \frac{s^2 + b_1 s + b_0}{s^2 + a_1 s + a_0} \qquad (4.36)$$

Although the $j\omega$-axis filter realization described in the preceding paragraphs can, in theory, be used to produce zeros off the $j\omega$ axis, the design equations become very involved. As an alternative approach we here present the circuit shown in Fig. 4.9, which uses two VCVSs, the first an inverting one with a gain of -1, the second a noninverting one with a gain of 2. The voltage transfer function for this filter is

$$\frac{V_2(s)}{V_1(s)} = \frac{2(Y_1 - Y_2)}{Y_3 - Y_4} \qquad (4.37)$$

The values of the admittances Y_i are found by dividing the numerator and denominator of (4.36) by the factor $s + c$, $c > 0$, and making partial fraction expansions of the resulting functions. Thus, for the numerator we obtain

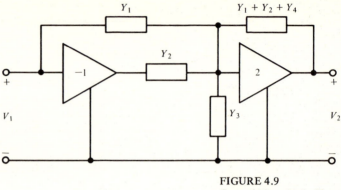

FIGURE 4.9
General biquadratic filter.

$$\frac{H(s^2 + b_1 s + b_0)}{s + c} = Hs + \frac{Hb_0}{c} + \frac{k_b s}{s + c} = 2(Y_1 - Y_2) \qquad (4.38)$$

where

$$k_b = \frac{H(c^2 + b_0 - b_1 c)}{-c} \qquad (4.39)$$

If k_b is positive, then we may identify

$$Y_1 = \frac{Hs}{2} + \frac{Hb_0}{2c} + \frac{1}{2/k_b + 2c/k_b s} \qquad Y_2 = 0 \qquad (4.40)$$

Thus, Y_2 is simply an open circuit, and Y_1 is realized as shown in Fig. 4.10a. If k_b is negative, then we may identify

$$Y_1 = \frac{Hs}{2} + \frac{Hb_0}{2c} \qquad Y_2 = \frac{1}{2/|k_b| + 2c/|k_b|s} \qquad (4.41)$$

with the realizations shown in Fig. 4.10b. A similar expansion of the denominator gives

$$\frac{s^2 + a_1 s + a_0}{s + c} = s + \frac{a_0}{c} + \frac{k_a s}{s + c} = Y_3 - Y_4 \qquad (4.42)$$

where

$$k_a = \frac{c^2 + a_0 - a_1 c}{-c} \qquad (4.43)$$

Y_3 and Y_4 are realized as shown in Fig. 4.10c and d. Usually, the value of c may be chosen so as to set either k_b or k_a in (4.39) and (4.43) to zero, and thus to simplify the resulting realization. As an example of the use of this network, consider the realization of the normalized all-pass (constant magnitude) transfer function

$$\frac{V_2(s)}{V_1(s)} = \frac{s^2 - 2s + 1}{s^2 + 2s + 1} \qquad (4.44)$$

Choosing $c = 1$ gives $k_a = 0$, and thus we obtain $Y_1 = (s + 1)/2$, $Y_2 = 2s/(s + 1)$, $Y_3 = s + 1$, and $Y_4 = 0$. The resulting filter is shown in Fig. 4.11.

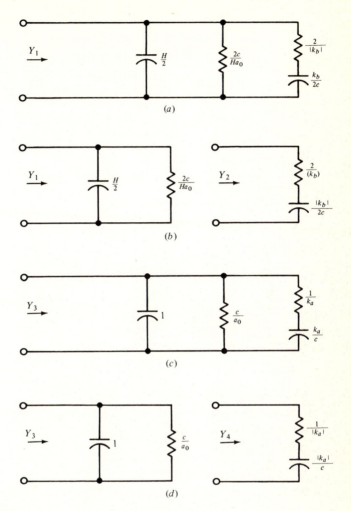

FIGURE 4.10
(*a*) Realization for Y_1 with $k_b > 0$; (*b*) realization for Y_1 and Y_2 with $k_b < 0$; (*c*) realization for Y_3 with $k_a > 0$; (*d*) realization for Y_3 and Y_4 with $k_a < 0$.

Low-pass Inverting Filter

All the active *RC* realizations discussed up to this point have been of the noninverting type requiring a positive-gain VCVS. It is also possible to design filters using an inverting (negative-gain) VCVS. In general, such realizations are not as practical as the noninverting ones. Here, however, for purposes of illustration, we shall consider one such circuit, the second-order low-pass filter. A realization is given by the circuit shown in Fig. 4.12. The numerators $n_{ij}(s)$ of the *y* parameters $y_{ij}(s)$ for the passive portion of this circuit which are required for analysis are

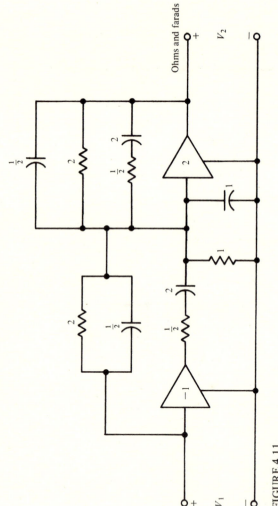

FIGURE 4.11
Example of biquadratic filter.

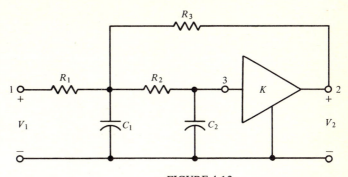

FIGURE 4.12
Second-order inverting low-pass filter.

$$n_{13}(s) = \frac{-1}{R_1 R_2 C_1} \qquad n_{23}(s) = \frac{-1}{R_2 R_3 C_1}$$

$$n_{33}(s) = C_2 \left[s^2 + s \left(\frac{1}{R_1 C_1} + \frac{1}{R_2 C_1} + \frac{1}{R_2 C_2} + \frac{1}{R_3 C_1} \right) + \frac{1}{R_2 C_1 C_2} \left(\frac{1}{R_1} + \frac{1}{R_3} \right) \right]$$

(4.45)

The denominators $d_{ij}(s)$ of the y parameters are

$$d_{ij}(s) = s + \frac{1}{C_1} \left(\frac{1}{R_1} + \frac{1}{R_2} + \frac{1}{R_3} \right) \qquad (4.46)$$

Substituting these results in the general expression given in (4.4), we obtain the voltage transfer function

$$\frac{V_2(s)}{V_1(s)}$$

$$= \frac{K/R_1 R_2 C_1 C_2}{s^2 + s[(1/C_1)(1/R_1 + 1/R_2 + 1/R_3) + 1/C_2 R_2] + (1/R_2 C_1 C_2)(1/R_1 + 1/R_3 - K/R_3)}$$

(4.47)

An examination of this voltage transfer function shows that complex conjugate poles are produced in this filter quite differently from the way they were produced in the noninverting ones. To see this, we insert the expressions of (4.45) into the denominator of (4.4) and compare the result with the denominator of (4.5). Thus we obtain

$$(s - \sigma_1)(s - \sigma_2) - \frac{K}{R_1 R_2 C_1 C_2} = s^2 + a_1 s + a_0 \qquad (4.48)$$

where the quantities of σ_1 and σ_2 are functions of the passive elements R_i and C_i. In order to produce complex conjugate poles, K must be negative, i.e., the VCVS must be an inverting one. The resulting loci of the poles as a function of K are shown in Fig. 4.13. From this we see that the real part of $S_K^{p_0}$ is zero, i.e., that the network is stable for all values of K.

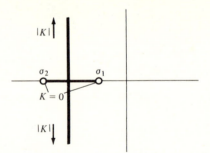

FIGURE 4.13
Locus of pole positions for the filter of
Fig. 4.12.

As an example of the determination of the values of the network elements required to realize a specific low-pass characteristic, consider the case where $R_1 = R_2 = R_3 = 4$ and $C_1 = C_2 = 1$ (ohms and farads). The voltage transfer function of (4.47) becomes

$$\frac{V_2(s)}{V_1(s)} = \frac{K/16}{s^2 + s + (2 - K)/16} \qquad (4.49)$$

The pole sensitivity is readily shown to be

$$S_K^{P_0} = 0 + j\frac{K/2}{K + 2} \qquad (4.50)$$

which for large values of K is equal to $0 + j0.5$. In general, large gain values are required. For example, for a Q of 5, the unnormalized denominator of the network function is $s^2 + s + 25$, and, from (4.49), $K = -398$. This value may be reduced by choosing different passive element values; for example, for $R_1 = 1, R_2 = 2, R_3 = 1/2, C_1 = 1$, and $C_2 = 1/4$ (ohms and farads), $K = -188$ for a Q of 5. Even at best, however, two or more orders of magnitude higher gains may be required than in the noninverting case, and lower multiplicative gain constants may result. Thus, despite the attractiveness of the low sensitivities, the problems of obtaining and stabilizing the higher gain values required, in general, make this type of filter less attractive than the noninverting type.

4.3 SECOND–ORDER INFINITE–GAIN– AMPLIFIER FILTERS

The active RC filters discussed in the preceding two sections of this chapter used low-gain amplifiers as their active element. In general, such an amplifier is most easily realized by using an operational amplifier and a pair of feedback resistors in one of the configurations shown in Figs. 1.6 and 1.21. In this section we present a different approach to the design of active RC filters, namely, the use of the entire passive RC network to provide the feedback around the operational amplifier. A general configuration for the second-order case is shown in Fig. 4.14. It is called a *multiple-feedback infinite-gain-amplifier filter*. The resulting transfer function is an inverting one, i.e., the dc gain is negative. The voltage transfer function for the circuit is

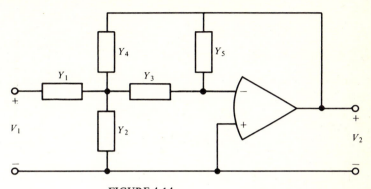

FIGURE 4.14
General multiple-feedback infinite-gain-amplifier filter.

$$\frac{V_2(s)}{V_1(s)} = \frac{-Y_1 Y_3}{Y_5(Y_1 + Y_2 + Y_3 + Y_4) + Y_3 Y_4} - \frac{1}{A}[(Y_3 + Y_5)(Y_1 + Y_2 + Y_4) + Y_3 Y_5]$$

(4.51)

where A is the open-loop gain of the operational amplifier. In general, this gain is very large, thus (4.51) may be approximated as

$$\frac{V_2(s)}{V_1(s)} = \frac{-Y_1 Y_3}{Y_5(Y_1 + Y_2 + Y_3 + Y_4) + Y_3 Y_4}$$

(4.52)

If the individual admittances Y_i are suitably chosen as either resistors or capacitors, this network function will yield a low-pass, high-pass, or bandpass characteristic. The details are given in the following paragraphs.

Low-pass Multiple-feedback Filter

The circuit configuration shown in Fig. 4.14 may be used to realize a second-order low-pass function having the general form given in (4.5) by choosing Y_1, Y_3, and Y_4 as resistors, and Y_2 and Y_5 as capacitors. The resulting network configuration is shown in Fig. 4.15. The voltage transfer function (for large values of the operational amplifier open-loop gain) is

$$\frac{V_2(s)}{V_1(s)} = \frac{-1/R_1 R_3 C_2 C_5}{s^2 + (s/C_2)(1/R_1 + 1/R_3 + 1/R_4) + 1/R_3 R_4 C_2 C_5}$$

(4.53)

The gain at direct current is $-R_4/R_1$. A design procedure is readily formulated in terms of the constants $a_0, a_1,$ and H $(H < 0)$ of (4.5) by letting

$$R_1 = R_3 = 1 \qquad R_4 = \frac{|H|}{a_0}$$

(4.54)

$$C_2 = \frac{2}{a_1} + \frac{a_0}{a_1|H|} \qquad C_5 = \frac{1}{C_2|H|}$$

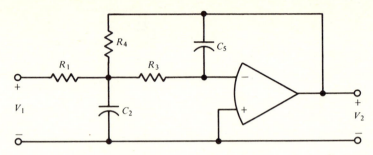

FIGURE 4.15
Low-pass multiple-feedback infinite-gain-amplifier filter.

Some other realizations for the cases of Table 4.1 are listed in Table 4.7. As an example of the design procedure, consider the realization of a normalized Butterworth filter (case 1 of Table 4.1) with a dc gain of -10. Inserting the values $a_0 = 1$, $a_1 = 1.41421$, and $H = -10$ in (4.54), we find $R_1 = 1.0$, $R_3 = 1.0$, $R_4 = 10$, $C_2 = 1.48492$, and $C_5 = 0.06734$ (ohms and farads). A comparison of this realization with the ones shown in Table 4.2 for the low-gain amplifier circuit shown in Fig. 4.2 readily illustrates one of the disadvantages of the infinite-gain realizations, namely that it is not usually possible to make elements of the same type have the same value. In addition, the spread of element values is usually quite large, although this latter disadvantage can be reduced somewhat by

Table 4.7 ELEMENT VALUES FOR THE LOW–PASS NETWORK OF FIG. 4.15 ($R_1 = R_3 = 1$) (ohms and farads)

Characteristic no. (from Table 4.1)	H	C_2	R_4	C_5
1	1.00000	2.12132	1.00000	0.47140
	2.00000	1.76777	2.00000	0.28284
	5.00000	1.55563	5.00000	0.12857
	10.00000	1.48492	10.00000	0.06734
2a	1.00000	2.46643	0.65954	0.40544
	2.00000	1.93466	1.31908	0.25844
	5.00000	1.61560	3.29771	0.12379
	10.00000	1.50925	6.59542	0.06626
2b	1.00000	2.82628	0.90702	0.35382
	2.00000	2.32410	1.81403	0.21512
	5.00000	2.02280	4.53508	0.09887
	10.00000	1.92236	9.07016	0.05202
3	1.00000	1.66667	0.33333	0.60000
	2.00000	1.16667	0.66667	0.42857
	5.00000	0.86667	1.66667	0.23077
	10.00000	0.76667	3.33333	0.13043

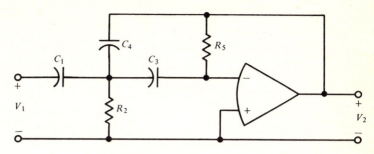

FIGURE 4.16
High-pass multiple-feedback infinite-gain-amplifier filter.

choosing a low value for dc gain. Even for low-gain circuits, however, the spread of element values may be a serious problem if the Q is even moderately high. For example, consider the case where $a_0 = 1$, $a_1 = 0.1$ ($Q = 10$), and $H = -1$ in (4.5). From the relations of (4.54) we find that $R_1 = 1.0$, $R_3 = 1.0$, $R_4 = 1.0$, $C_2 = 40.0$, and $C_5 = 0.025$ (ohms and farads). Obviously, the spread of capacitor values is excessive.

The infinite-gain multiple-feedback filter also has advantages compared with the low-gain one. One of these is its ability to exactly realize a specified dc gain. Another more significant advantage is its low sensitivity. The sensitivity to gain changes of the operational amplifier is practically zero. In addition, the sensitivities to changes in the passive elements are usually lower than the ones encountered in comparable low-gain filters.

High-pass Multiple-feedback Filter

The circuit configuration shown in Fig. 4.14 may be used to realize a second-order high-pass function having the general form given in (4.23) by choosing Y_2 and Y_5 as resistors and Y_1, Y_3, and Y_4 as capacitors. The resulting network configuration is shown in Fig. 4.16. The voltage transfer function (for large values of operational amplifier open-loop gain) is

$$\frac{V_2(s)}{V_1(s)} = \frac{-s^2 C_1/C_4}{s^2 + (s/R_5)(C_1/C_3 C_4 + 1/C_3 + 1/C_4) + 1/R_2 R_5 C_3 C_4} \tag{4.55}$$

The gain at infinite frequencies is $-C_1/C_4$. A design procedure is readily formulated in terms of the coefficients a_0, a_1, and H of (4.23) by letting

$$R_2 = \frac{a_1}{a_0(2 + 1/|H|)} \qquad R_5 = \frac{2|H| + 1}{a_1} \qquad C_1 = C_3 = 1 \qquad C_4 = \frac{1}{|H|} \tag{4.56}$$

Some other realizations for the cases of Table 4.3 are given in Table 4.8. As an example of the design procedure, consider the realization of a normalized maximally flat magnitude filter (case 1 of Table 4.3) with an infinite-frequency gain of -10. Inserting the values $a_0 = 1$, $a_1 = 1.41421$, and $H = -10$ in the relations of (4.56), we find

$R_2 = 0.673$, $R_5 = 14.851$, $C_1 = C_3 = 1$, $C_4 = 0.1$ (ohms and farads). Comments similar to those made for the low-pass infinite-gain-amplifier filter with respect to sensitivity and spread of element values also apply to realizations of this filter.

Bandpass Multiple-feedback Filter

There are several different choices of elements which may be used in the configuration shown in Fig. 4.14 to realize a second-order bandpass function having the general form given in (4.27). One of the most practical configurations is obtained by choosing Y_1, Y_2, and Y_5 as resistors, and Y_3 and Y_4 as capacitors. The resulting network configuration is shown in Fig. 4.17. The voltage transfer function (for high values of open-loop operational amplifier gain) is

$$\frac{V_2(s)}{V_1(s)} = \frac{-s/R_1 C_4}{s^2 + (s/R_5)(1/C_3 + 1/C_4) + (1/R_5 C_3 C_4)(1/R_1 + 1/R_2)} \quad (4.57)$$

The gain at resonance is $-R_5 C_3/[R_1(C_3 + C_4)]$. A design procedure is readily formulated in terms of the coefficients a_0, a_1, and H of (4.27) by letting

$$R_1 = \frac{1}{|H|} \quad R_2 = \frac{1}{2a_0/a_1 - |H|} \quad R_5 = \frac{2}{a_1} \quad C_3 = C_4 = 1 \quad (4.58)$$

As an example of the design procedure, consider the realization of a normalized bandpass filter with a Q of 10 and a gain at resonance of -100. Inserting the values $a_0 = 1$, $a_1 = 0.1$, and $H = -10$ in the relations of (4.58), we obtain $R_1 = 0.1$, $R_2 = 0.1$, $R_5 = 20$, and $C_3 = C_4 = 1$ (ohms and farads). Some other realizations for various values of Q and for different gain constants are listed in Table 4.9. Comments similar to those

Table 4.8 ELEMENT VALUES FOR THE HIGH–PASS NETWORK OF FIG. 4.16 ($C_1 = C_3 = 1$) (ohms and farads)

Characteristic no. (from Table 4.3)	H	R_2	C_4	R_5
1	1.00000	0.47140	1.00000	2.12132
	2.00000	0.56568	0.50000	3.53553
	5.00000	0.64282	0.20000	7.77817
	10.00000	0.67343	0.10000	14.84924
2a	1.00000	0.47521	1.00000	3.19061
	2.00000	0.57025	0.50000	5.31768
	5.00000	0.64801	0.20000	11.69889
	10.00000	0.67887	0.10000	22.33425
2b	1.00000	0.36591	1.00000	3.01308
	2.00000	0.43909	0.50000	5.02179
	5.00000	0.49897	0.20000	11.04795
	10.00000	0.52273	0.10000	21.09154

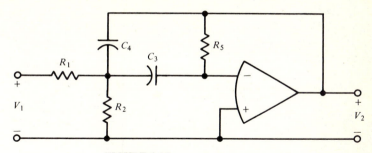

FIGURE 4.17
Bandpass multiple-feedback infinite-gain-amplifier filter.

made for the low-pass infinite-gain-amplifier filter with respect to sensitivity and spread of element values also apply to realizations of this filter.

Single-feedback Filters

A different approach to the direct use of infinite-gain amplifiers is the *single-feedback infinite-gain-amplifier filter,* in which only one feedback path is provided around the operational amplifier. This results in the network configuration shown in Fig. 4.18. It consists of two passive two-port networks with y parameters $y_{ij}^{(a)}$ and $y_{ij}^{(b)}$ interconnected by an operational amplifier. The voltage transfer function is

Table 4.9   ELEMENT VALUES FOR BANDPASS NETWORK OF FIG. 4.17 ($C_3 = C_4 = 1$) with a_0 of (4.27) equal to unity (ohms and farads)

$Q(= 1/a_1)$	H	R_1	R_2	R_5
2	1.00000	1.00000	0.33333	4.00000
	2.00000	0.50000	0.50000	4.00000
5	1.00000	1.00000	0.11111	10.00000
	2.00000	0.50000	0.12500	10.00000
	5.00000	0.20000	0.20000	10.00000
10	1.00000	1.00000	0.05263	20.00000
	2.00000	0.50000	0.05556	20.00000
	5.00000	0.20000	0.06667	20.00000
	10.00000	0.10000	0.10000	20.00000
20	1.00000	1.00000	0.02564	40.00000
	2.00000	0.50000	0.02632	40.00000
	5.00000	0.20000	0.02857	40.00000
	10.00000	0.10000	0.03333	40.00000

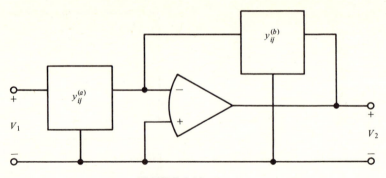

FIGURE 4.18
General single-feedback infinite-gain-amplifier filter.

$$\frac{V_2(s)}{V_1(s)} = \frac{-y_{12}{}^{(a)}(s)}{y_{12}{}^{(b)}(s) + (1/A)[y_{22}{}^{(a)}(s) + y_{11}{}^{(b)}(s)]} \quad (4.59)$$

where A is the open-loop gain of the operational amplifier. For large values of A, the voltage transfer function of (4.59) becomes

$$\frac{V_2(s)}{V_1(s)} = \frac{-y_{12}{}^{(a)}(s)}{y_{12}{}^{(b)}(s)} \quad (4.60)$$

If we now let $y_{12}{}^{(a)}(s) = n_{12}{}^{(a)}(s)/d_{12}{}^{(a)}(s)$ and $y_{12}{}^{(b)}(s) = n_{12}{}^{(b)}(s)/d_{12}{}^{(b)}(s)$, then the voltage transfer function has the form

$$\frac{V_2(s)}{V_1(s)} = \frac{-n_{12}{}^{(a)}(s)d_{12}{}^{(b)}(s)}{n_{12}{}^{(b)}(s)d_{12}{}^{(a)}(s)} \quad (4.61)$$

If we now require that the two passive networks have the same natural frequencies, i.e., that $d_{12}{}^{(a)}(s) = d_{12}{}^{(b)}(s)$, the resulting voltage transfer function is

$$\frac{V_2(s)}{V_1(s)} = \frac{-n_{12}{}^{(a)}(s)}{n_{12}{}^{(b)}(s)} \quad (4.62)$$

Thus, the numerators of the transfer admittances of the passive networks shown in Fig. 4.18 determine both the zeros *and the poles* of the voltage transfer function of a single-feedback infinite-gain-amplifier filter. Since passive *RC* networks can have transmission zeros anywhere in the complex frequency plane, the poles and zeros of the voltage transfer function given in (4.62) can be located anywhere in the complex frequency plane. Stability, of course, requires that the poles be restricted to the left half-plane. In Table 4.10, a collection of several of the passive networks which are most commonly used in single-feedback infinite-gain-amplifier filters is given, together with the necessary design information for them. In the table, network types 1 to 3 are used for producing transmission zeros at the origin and infinity. Of these, type 1 produces a single zero at the origin, type 2 has a numerator which is a constant and thus produces a

Table 4.10 PASSIVE *RC* NETWORKS FOR USE IN SINGLE-FEEDBACK INFINITE-GAIN-AMPLIFIER FILTERS

Type	Network	Transfer admittance	Element values (ohms and farads)
1		$y_{12} = \dfrac{-ks}{s+a}$	$k = 1/R \quad a = 1/RC$
2		$y_{12} = \dfrac{-k}{s+a}$	$k = \dfrac{1}{R_1 R_2 C} \quad a = \dfrac{1}{C}\left(\dfrac{1}{R_1} + \dfrac{1}{R_2}\right)$
3		$y_{12} = \dfrac{-ks^2}{s+a}$	$k = \dfrac{C_1 C_2}{C_1 + C_2} \quad a = \dfrac{1}{R(C_1 + C_2)}$
4		$y_{12} = \dfrac{-(s^2 + as + 1)}{s+a} \quad 1/2 < a < 2$	$R_1 = \dfrac{1}{2.5 - a} \quad R_2 = a - R_1$ $C_1 = 1 \quad C_2 = \dfrac{1}{R_1 R_2}$
5		$y_{12} = \dfrac{-(s+1)(s^2 + as + 1)}{(s - \sigma_1)(s - \sigma_2)} \quad a < 1$ $y_{12} x \dfrac{-(s^2 + as + 1)}{s+1}$	$C_1 = \dfrac{1}{R_1} = (2.5 - a)\dfrac{1+a}{2+a} \quad C_3 = \dfrac{1}{R_3} = \dfrac{C_1 C_2}{1+a}$ $C_2 = \dfrac{1}{R_2} = \dfrac{C_1}{C_1 - 1}$

transmission zero at infinity, and type 3 produces two transmission zeros at the origin. The remaining networks shown in the table, types 4 and 5, are used to produce complex conjugate zeros. Of these, type 4 is referred to as a *bridged-tee network*. As indicated by the specified range of the parameter a in the numerator of its transfer admittance, it may only be used for transmission zeros which are not located too close to the $j\omega$ axis, i.e., which are not too high in Q. For high-Q cases in which zeros close to the $j\omega$ axis are required, the *twin-tee network* shown as type 5 may be used. As indicated in the table, this network produces three transmission zeros, i.e., one complex conjugate pair and one negative real zero located at $s = -1$. The transfer admittance also has two negative real poles located at $s = \sigma_1$ and σ_2. The values of σ_1 and σ_2 are somewhat involved functions of the various resistors and capacitors of the network; however, in the high-Q case, as the transmission zeros approach the $j\omega$ axis (i.e., as the value of the numerator coefficient a becomes small), these values approach -1. Thus, the transmission zero at $s = -1$ is effectively canceled by one of the poles, and, with good accuracy, we may assume that the transfer admittance has the simplified approximate form shown in the table.

Realization of low-pass, high-pass, and bandpass filters using the configuration shown in Fig. 4.18 is readily accomplished by appropriate choices of the networks shown in Table 4.10. For a low-pass filter having the form given in (4.5), network A is a type 2 and network B is a type 4 (or a type 5). As an example of such a realization, for $a_0 = 1$, $a_1 = 1.414$, and H arbitrary, in the type 4 network we choose $a = 1.414$ and obtain $R_1 = 0.921$, $R_2 = 0.493$, $C_1 = 1$, and $C_2 = 2.22$ (ohms and farads). In the type 2 network we choose $R_1 = R_2 = 1/1.414H$ and $C = 2H$, where H may have any desired value. In a similar fashion, for a high-pass filter having the form given in (4.23), network A is a type 3 and network B is a type 4 (or a type 5). Finally, for a bandpass filter having the form given in (4.27), network A is a type 1 and network B is a type 4 (or a type 5).

The single-feedback infinite-gain-amplifier filter configuration has the disadvantage that more passive network elements are required than in other comparable realizations. In addition, the tuning procedure is usually more difficult due to the interaction, i.e., cancellation, of the denominators of the transfer admittances of the two passive networks. Despite these disadvantages, the single-feedback filter may usually be used to realize higher-Q network functions than can be realized with the multiple-feedback filter. In addition, larger values of H than are practical in the multiple-feedback case can be realized without producing an excessive spread of element values. The sensitivities of the two configurations are similar, the sensitivity with respect to the gain of the operational amplifier being essentially zero and the sensitivities to the passive elements being low.

General Biquadratic Filter

Frequently it is useful to be able to use an infinite-gain-amplifier filter configuration to realize a biquadratic network function in which the positions of the complex conjugate poles and zeros can be individually specified. The general form of such a function is given in (4.36). Here we present two techniques for realizing such a filter. The first method uses two single-input operational amplifiers in the configuration shown in Fig. 4.19. For this filter the voltage transfer function is

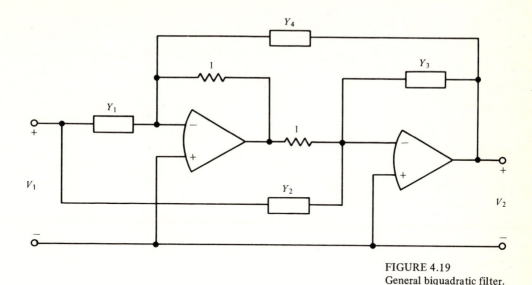

FIGURE 4.19
General biquadratic filter.

$$\frac{V_2(s)}{V_1(s)} = \frac{Y_1 - Y_2}{Y_3 - Y_4} \qquad (4.63)$$

The values of the admittances Y_i are found in a manner similar to that outlined for the biquadratic filter realization given in Sec. 4.2, namely, by dividing the numerator and denominator by the factor $s + c$, $c > 0$, and making partial fraction expansions. Thus for the numerator we obtain

$$Y_1 - Y_2 = \frac{H(s^2 + b_1 s + b_0)}{s + c} = Hs + \frac{Hb_0}{c} + \frac{k_b s}{s + c} \qquad (4.64)$$

where k_b is defined in (4.39). The only difference between this expression and the one given in (4.38) is the factor of 2 by which the term $Y_1 - Y_2$ is multiplied in that equation. Thus, the realization of Y_1 and Y_2 in (4.64) may be directly taken from Fig. 4.10a and b by deleting the factor of 2 that appears in the values of the elements. Similarly, expanding the denominator, we obtain an expression identical with the one given in (4.42), and thus the realizations given in Fig. 4.10c and d are directly applicable here. The value of c may, of course, be chosen so as to simplify the realization. As an example of the technique, to realize the function given in (4.44), setting $c = 1$, we obtain the realization shown in Fig. 4.20.

A second method of obtaining a biquadratic filter having the network function given in (4.36) is to use a single differential-input operational amplifier in the configuration shown in Fig. 4.21. For this filter the voltage transfer function is again given by (4.63), and the synthesis procedure for determining the admittances Y_i is identical with that described above for the two operational amplifier filter. In addition, any common terms in the expansion of the sums $Y_2 + Y_3$ and $Y_1 + Y_4$ may be

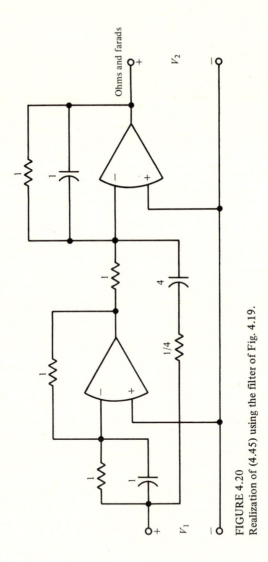

FIGURE 4.20
Realization of (4.45) using the filter of Fig. 4.19.

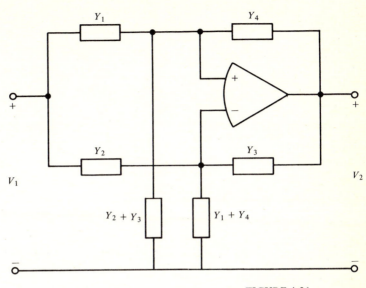

FIGURE 4.21
General biquadratic filter.

subtracted from both these sums to simplify the resulting network configuration. As an example of this type of filter, a realization of the all-pass function given in (4.44) (using $c = 1$) is shown in Fig. 4.22.

4.4 HIGH-*Q* FILTERS

The filters described in the preceding sections of this chapter are, in general, not suited for the realization of high-*Q* network functions. In this section, however, we describe two active *RC* filters which are well suited to realizing functions with *Q*'s whose values may go into the hundreds. The first such circuit is called a *state-variable filter*.[†] To see how such a filter works we may modify the second-order bandpass function given in (4.27) by multiplying numerator and denominator by the quotient $X(s)/s^2$, where $X(s)$ is an arbitrary function. Thus, assuming an inverting function, we obtain

$$\frac{V_2(s)}{V_1(s)} = \frac{-|H|s}{s^2 + a_1 s + a_0} = \frac{-|H|X(s)/s}{X(s) + a_1 X(s)/s + a_0 X(s)/s^2} \qquad (4.65)$$

This relation may now be written as the two separate equations

$$X(s) = V_1(s) - \frac{a_1 X(s)}{s} - \frac{a_0 X(s)}{s^2} \qquad V_2(s) = -|H|\frac{X(s)}{s} \qquad (4.66)$$

[†]W. J. Kerwin, L. P. Huelsman, and R. W. Newcomb, State-Variable Synthesis for Insensitive Integrated Circuit Transfer Functions, *IEEE Journal of Solid-State Circuits,* vol. SC-2, no. 3, pp. 87–92, September 1967.

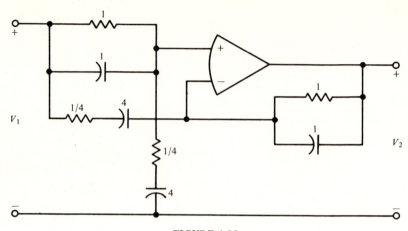

FIGURE 4.22
Realization of (4.45) using the filter of Fig. 4.21.

Since multiplying by $1/s$ in the (transformed) complex frequency domain corresponds with integration in the time domain, the equations of (4.66) are readily implemented by the analog-computer flowchart shown in Fig. 4.23. In this figure, the quantities $x(t) = L^{-1}[X(s)]$, $\int x(t)\,dt$, and $\int\int x(t)\,dt\,dt$ are called the state variables.[†] The operations of integration and summation called for in the flowchart are readily implemented by the operational amplifier circuits shown in Figs. 1.12 and 1.28, respectively. Using these we obtain, in Fig. 4.24, a circuit realization of the flowchart. For the case where the open-loop gain of the operational amplifiers is considered infinite, the transfer function is

$$\frac{V_2(s)}{V_1(s)} = \frac{N(s)}{D(s)}$$

$$= \frac{-s(1/R_1 C_1)[(1 + R_6/R_5)/(1 + R_3/R_4)]}{s^2 + s(1/R_1 C_1)[(1 + R_6/R_5)/(1 + R_4/R_3)] + (R_6/R_5)(1/R_1 R_2 C_1 C_2)} \qquad (4.67)$$

This circuit may also be used to produce a low-pass or a high-pass function by taking the output from the points indicated in Fig. 4.24. For these outputs,

$$\frac{V_{\text{LP}}(s)}{V_1(s)} = \frac{(1/R_1 R_2 C_1 C_2)[(1 + R_6/R_5)/(1 + R_3/R_4)]}{D(s)}$$

$$\frac{V_{\text{HP}}(s)}{V_1(s)} = \frac{s^2[(1 + R_6/R_5)/(1 + R_3/R_4)]}{D(s)} \qquad (4.68)$$

[†]P. M. DeRusso, R. J. Roy, and C. M. Close, "State Variables for Engineers," Chap. 6, Wiley, New York, 1965.

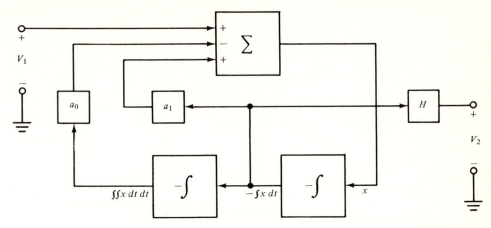

FIGURE 4.23
Analog-computer flowchart.

where $D(s)$ is defined in (4.67). Other biquadratic functions are easily produced by appropriate summing of the three indicated outputs.†

In the high-Q situations to which the state-variable filter is usually applied, the coefficient a_1 in (4.65) is very much smaller than a_0. In this case, to accurately design the filter, it is necessary to take account of the actual operational amplifier gains in the integrators. Thus, the integrators are characterized as

$$\frac{V_{out}(s)}{V_{in}(s)} = \frac{-K_i}{1 + (1 + K_i)sR_iC_i} \qquad i = 1, 2 \qquad (4.69)$$

which, of course, simply becomes $-1/sR_iC_i$ for large values of K_i. Using (4.69), the coefficient a_1 has the form

$$a_1 = \frac{1}{R_1C_1} \frac{1 + R_6/R_5}{1 + R_4/R_3} \frac{K_1}{1 + K_1} + \frac{1}{R_1C_1(1 + K_1)} + \frac{1}{R_2C_2(1 + K_2)} \qquad (4.70)$$

Choosing $R_1 = R_2 = R_5 = R_6 = C_1 = C_2 = 1$ so as to set a_0 of (4.65) to 1, we obtain $Q = 1/a_1$. For the high-Q (small a_1) case from (4.70) we find that the imaginary parts of the pole sensitivities are negligibly small and that the real parts are

$$\text{Re } S_{K_1}{}^{P_0} = \frac{\partial \sigma_0}{\partial K_1} \frac{K_1}{\sigma_0} \cong \frac{Q}{K_1} \qquad \text{Re } S_{K_2}{}^{P_0} = \frac{\partial \sigma_0}{\partial K_2} \frac{K_2}{\sigma_0} \approx \frac{Q}{K_2} \qquad (4.71)$$

where $p_0 = \sigma_0 + j\omega_0$ is the pole location. These sensitivity values are also clearly small, even for very large values of Q. Analysis of the other sensitivities for the state-variable

†Biquadratic functions may also be obtained by a slight modification of the original circuit. See P. E. Fleischer and J. Tow, Design Formulas for Biquad Active Filters Using Three Operational Amplifiers, *Proc. of the IEEE*, vol. 61, no. 5, pp. 662–663, May 1973.

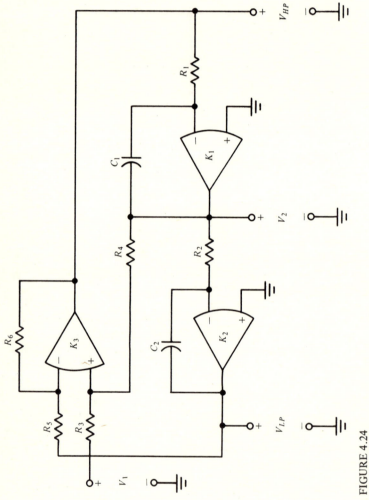

FIGURE 4.24
Circuit realization of the flowchart of Fig. 4.23.

filter shows them also to be low; thus this filter provides excellent stability even for very-high-Q realizations. The price that is paid for the excellent stability, of course, is the large number of active and passive elements used in the filter as compared with the number required in the realizations presented in the preceding sections of this chapter. Despite its complexity, the worth of this circuit is well attested to by the number of commercial companies who use it as the major element of their active *RC* filter product line.

A design procedure for the normalized state-variable filter is readily specified by setting all the R_i and C_i elements except R_4 to unity. Then, for the high-Q case,

$$R_4 = \frac{2}{a_1 - 1/K_1 - 1/K_2} \qquad (4.72)$$

Obviously, in the high-Q case, R_4 will be large. The passband gain at resonance is $-R_4/R_3$. As an example of the procedure, for $Q = 500$, with $K_i = 10^4$, we find that $R_4 = 1111.1$. For the low-Q case, from (4.70) we merely set $R_4 = (2/a_1) - 1$.

In high-Q high-frequency applications of the state-variable filter, the bandwidth of the operational amplifier may need to be considered. To determine the effect of such a parasitic, we may use a one-pole model for the operational amplifier by setting

$$K_1 = K_2 = \frac{K_0}{1 + s/p_n} \qquad (4.73)$$

where p_n is the magnitude of the normalized pole location, i.e., the actual magnitude divided by ω_c. For the high-Q case, and for all R_i and C_i except R_4 set to unity, we find that after some manipulation,

$$Q \approx \frac{1}{2/(1 + R_4) + 2/K_0 - 4/K_0 p_n} \qquad (4.74)$$

In general, p_n is proportional to the β of the operational amplifier transistors, and thus some degradation of performance may be expected unless $Q \ll K_0 p_n/4$. For example, for $K_0 = 10^4$ and $p_n = 1000 \text{ Hz}/f_c$, where $f_c = \omega_c/2\pi$, we require $Q f_c \ll 0.25 \times 10^7$. Thus for $f_c = 10 \text{ kHz}$, we have the limitation $Q \ll 250$.

A second high-Q active *RC* filter configuration is provided by the Tarmy-Ghausi circuit shown in Fig. 4.25.† It requires the use of three finite-gain amplifiers, of which two must have differential output. The amplifiers are realized as shown in Fig. 4.26, and the gains $-K_1$ and K_2 are defined as

$$-K_1 = \frac{V_3(s)}{V_1(s)} = -\frac{\alpha_1(1 + \alpha_2)}{1 + 2\alpha_2 + \alpha_1\alpha_2}$$

$$K_2 = \frac{V_3(s)}{V_2(s)} = \frac{1 + \alpha_1}{1 + 2\alpha_2 + \alpha_1\alpha_2} \qquad (4.75)$$

$$\alpha_1 = \frac{R_d}{R_c} \qquad \alpha_2 = \frac{R_b}{R_a}$$

†R. Tarmy and M. S. Ghausi, Very High-Q Insensitive Active *RC* Networks, *IEEE Transactions on Circuit Theory*, vol. CT-17, no. 3, pp. 358–366, August 1970.

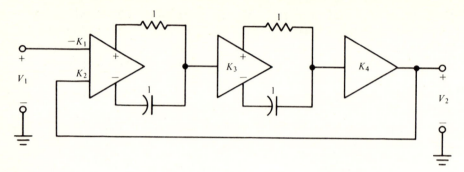

FIGURE 4.25
Tarmy-Ghausi high-Q circuit.

The voltage transfer function has the form

$$\frac{V_2(s)}{V_1(s)} = \frac{(1 - s)^2 [K_1 K_3 K_4/(1 + K_2 K_3 K_4)]}{s^2 + 2s[(1 - K_2 K_3 K_4)/(1 + K_2 K_3 K_4)] + 1} \qquad (4.76)$$

Due to the numerator factor $(1 - s)^2$, this filter deviates from ideal bandpass performance in that it has transmission at both low and high frequencies. This deviation is especially noticeable in the phase characteristic. The high-Q sensitivities are approximately

$$\text{Re } S_{\alpha_1}{}^{P_0} \approx \alpha_1 Q \qquad \text{Re } S_{\alpha_2}{}^{P_0} \approx 2\alpha_2 Q$$

In practice, the values of $\alpha_i = 1/Q$ are chosen so that these sensitivities are manageably small. A design procedure is readily formulated by noting that

$$Q = \frac{1}{1 - K_2 K_3 K_4} \qquad (4.77)$$

In practice, choosing $K_2 = 1 - 1/Q$, $K_3 = 1$, and $K_4 = 1 + \epsilon$, where ϵ can be varied to provide tuning, works well. The gain at resonance is only unity, however, and this is a major disadvantage of the circuit. On the other hand, the value of Q is not dependent (as a first approximation) on the bandwidth of the operational amplifiers. As an illustration of this, using the one-pole model for the operational amplifiers given in (4.73), we find that

$$Q \approx \frac{1}{1 - K_2 K_3 K_4 + 3/K_0}$$

whereas, in the state-variable filter, the dependence of Q on p_n was more strongly evident. It should be noted, however, that the addition of a high-gain stage to the Tarmy-Ghausi filter to make it more nearly comparable with the state-variable one would, of course, present an additional bandwidth restriction on its performance.

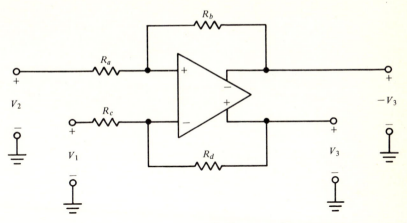

FIGURE 4.26
Amplifier with differential input and output.

4.5 CASCADE REALIZATION OF HIGH–ORDER FILTERS–TUNING

In the preceding sections of this chapter we have considered the realization of second-order filters using various finite- and infinite-gain-amplifier circuits. Many filtering applications, however, require filters of higher than second order, either to provide greater stop-band attenuation and sharper cutoff at the edge of the passband in the low-pass or high-pass case, or to provide a broad passband with some special transmission characteristic in the bandpass case. In this and the following sections we discuss some means of obtaining these high-order filters.

One of the simplest approaches to the realization of high-order filters is to factor the specified network function into quadratic factors, and realize each of these by a separate circuit using the second-order realizations given in Secs. 4.1 to 4.4. Since these circuits all have an operational amplifier as an output element, their output impedance is low (in theory, zero), and thus a simple cascade of such second-order realizations may be made without interaction occurring between the individual stages. As a result, the overall voltage transfer function is simply the product of the individual transfer functions. An advantage of this approach is that each of the filters may be tuned separately, a point of considerable practical importance when high-order network functions are to be realized. The success of the cascade method, of course, depends on the use of operational amplifiers which have as low an output impedance as possible. The use of high impedance normalization levels for the passive elements, which minimizes loading on the interior operational amplifier output stages, is also significant in reducing interaction. Some of the details of cascade realizations for various types of high-order filters follow.

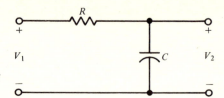

FIGURE 4.27
Realization of the first-order factor of (4.81).

In the nth-order case, the general low-pass voltage transfer function is

$$\frac{V_2(s)}{V_1(s)} = \frac{H}{s^n + a_{n-1}s^{n-1} + \cdots + a_1 s + a_0} \qquad (4.78)$$

Such a function has a magnitude characteristic which decreases at the rate of $20n$ dB/decade outside the passband. If n is even, (4.78) may be written in the form

$$\frac{V_2(s)}{V_1(s)} = \prod_{i=1}^{n/2} \frac{H_i}{s^2 + a_{i1}s + a_{i0}} \qquad (4.79)$$

and the individual quadratic functions $H_i/(s^2 + a_{i1}s + a_{i0})$ may be synthesized by any of the second-order low-pass realizations given in the preceding sections. The realizations may be individually impedance-normalized to provide convenient element values. The constants H_i are, of course, arbitrary, unless we want to realize H exactly, in which case the product of the H_i must equal H. If n is odd, (4.78) may be written

$$\frac{V_2(s)}{V_1(s)} = \frac{1}{s - \sigma_0} \prod_{i=1}^{(n-1)/2} \frac{H_i}{s^2 + a_{i1}s + a_{i0}} \qquad (4.80)$$

The quadratic factors are realized as before, and the first-order factor $1/(s + \sigma_0)$ may be realized (within a multiplicative constant) by the circuit shown in Fig. 4.27, which has the voltage transfer function

$$\frac{V_2(s)}{V_1(s)} = \frac{1/RC}{s + 1/RC} \qquad (4.81)$$

This first-order network must, of course, be isolated so as not to interact with the other second-order realizations.

The determination of the coefficients a_i of (4.78), and thus of the coefficients a_{i1} and a_{i0} of (4.79), is done according to the specific magnitude or phase characteristic desired. Some frequently used choices are given in Table 4.11, which is an extension of Table 4.1. As an example of the cascade technique, consider the realization of a fifth-order frequency-normalized low-pass filter having a maximally flat magnitude (Butterworth) characteristic which is 3 dB down at the passband edge of 1 rad/s. From Table 4.11 we see that the network function has the form

$$\frac{V_2(s)}{V_1(s)} = \frac{H}{(s + 1)(s^2 + 0.618s + 1)(s^2 + 1.618s + 1)} \qquad (4.82)$$

Table 4.11a DENOMINATOR POLYNOMIALS FOR LOW–PASS NETWORK FUNCTIONS OF THE FORM $s^n + a_{n-1}s^{n-1} + \cdots + a_1s + a_0$

Case (Table 4.1)	Order	a_0	a_1	a_2	a_3	a_4	a_5
1	3	1.00000	2.00000	2.00000			
	4	1.00000	2.61313	3.41421	2.61313		
	5	1.00000	3.23607	5.23607	5.23607	3.23607	
	6	1.00000	3.86370	7.46410	9.14162	7.46410	3.86370
2a	3	0.71569	1.53489	1.25291			
	4	0.37905	1.02545	1.71687	1.19739		
	5	0.17892	0.75252	1.30957	1.93737	1.17249	
	6	0.09476	0.43237	1.17186	1.58976	2.17184	1.15918
2b	3	0.49131	1.23841	0.98834			
	4	0.27563	0.74262	1.45392	0.95281		
	5	0.12283	0.58053	0.97440	1.68882	0.93682	
	6	0.06891	0.30708	0.93934	1.20214	1.93083	0.92825
3	3	15	15	6			
	4	105	105	45	10		
	5	945	945	420	105	15	
	6	10,395	10,395	4725	1260	210	21

Table 4.11b ROOTS OF DENOMINATOR POLYNOMIALS FOR LOW–PASS NETWORK FUNCTIONS

Case (Table 4.1)	Order	Roots		
1	3	$-0.50000 \pm j0.86603$	-1.00000	
	4	$-0.38268 \pm j0.92388$	$-0.92388 \pm j0.38268$	
	5	$-0.30902 \pm j0.95106$	$-0.80902 \pm j0.58778$	-1.00000
	6	$-0.25882 \pm j0.96593$	$-0.70711 \pm j0.70711$	$-0.96593 \pm j0.25882$
2a	3	$-0.31323 \pm j1.02193$	-0.62646	
	4	$-0.17535 \pm j1.01626$	$-0.42334 \pm j0.42095$	
	5	$-0.11196 \pm j1.01156$	$-0.29312 \pm j0.62518$	-0.36232
	6	$-0.07765 \pm j1.00846$	$-0.21214 \pm j0.73824$	$-0.28979 \pm j0.27022$
2b	3	$-0.24708 \pm j0.96600$	-0.49417	
	4	$-0.13954 \pm j0.98338$	$-0.33687 \pm j0.40733$	
	5	$-0.08946 \pm j0.99011$	$-0.23420 \pm j0.61192$	-0.28951
	6	$-0.06218 \pm j0.99341$	$-0.16988 \pm j0.72723$	$-0.23206 \pm j0.26618$
3	3	$-1.83891 \pm j1.75438$	-2.32218	
	4	$-2.89621 \pm j0.86723$	$-2.10379 \pm j2.65742$	
	5	$-3.35196 \pm j1.74266$	$-2.32467 \pm j3.57102$	-3.63674
	6	$-4.24836 \pm j0.86751$	$-3.73571 \pm j2.62627$	$-2.51593 \pm j4.49267$

Table 4.11c FACTORS OF DENOMINATOR POLYNOMIALS FOR LOW–PASS NETWORK FUNCTIONS OF THE FORM $(s + a)(s^2 + a_{11}s + a_{10})(s^2 + a_{21}s + a_{20})(s^2 + a_{31}s + a_{30})$

Case (Table 4.1)	Order	a	a_{11}	a_{10}	a_{21}	a_{20}	a_{31}	a_{30}
1	3	1.00000	1.00000	1.00000				
	4		1.84777	1.00000	0.76536	1.00000		
	5	1.00000	0.61803	1.00000	1.61804	1.00000		
	6		0.51764	1.00000	1.93185	1.00000	1.41421	1.00000
2a	3	0.62646	0.62646	1.14244				
	4		0.84667	0.35641	0.35072	1.06351		
	5	0.36231	0.58625	0.47677	0.22393	1.03578		
	6		0.57959	0.15699	0.42429	0.59004	0.15530	1.02300
2b	3	0.49417	0.49417	0.99421				
	4		0.67374	0.27940	0.27907	0.98650		
	5	0.28951	0.46839	0.42929	0.17893	0.98833		
	6		0.46413	0.12472	0.33973	0.55769	0.12438	0.99075
3	3	2.32218	3.67781	6.45943				
	4		4.20758	11.48780	5.79242	9.14014		
	5	3.64674	4.64935	18.15631	6.70391	14.27248		
	6		5.03186	26.51402	7.47142	20.85282	8.49672	18.80113

The resulting realization, impedance-normalized to produce all equal-valued elements, is shown in Fig. 4.28.

High-pass Filters

In the nth-order case, the general high-pass voltage transfer function has the form

$$\frac{V_2(s)}{V_1(s)} = \frac{Hs^n}{s^n + a_{n-1}s^{n-1} + \cdots + a_1 s + a_0} \qquad (4.83)$$

If n is even, this may be put in the form

$$\frac{V_2(s)}{V_1(s)} = \prod_{i=1}^{n/2} \frac{H_i s^2}{s^2 + a_{i1}s + a_{i0}} \qquad (4.84)$$

The separate quadratic factors are easily realized by using any of the second-order high-pass circuits presented in the preceding sections. For n odd, (4.83) may be factored as

$$\frac{V_2(s)}{V_1(s)} = \frac{s}{s - \sigma_0} \prod_{i=1}^{(n-1)/2} \frac{H_i s^2}{s^2 + a_{i1}s + a_{i0}} \qquad (4.85)$$

The quadratic factors are realized as before, and the first-order factor $s/(s + \sigma_0)$ is realized by the circuit shown in Fig. 4.29, which has the transfer function

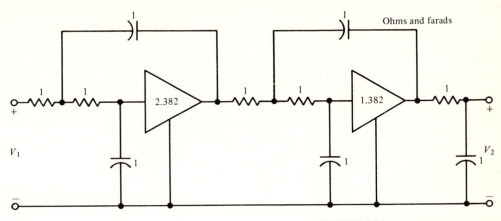

FIGURE 4.28
Fifth-order low-pass filter realization.

$$\frac{V_2(s)}{V_1(s)} = \frac{s}{s + 1/RC} \qquad (4.86)$$

Some frequently used cases defining the values of the coefficients of the high-order high-pass functions are given in Table 4.12, which is an extension of Table 4.3.

Bandpass Filters

In the *n*th-order case, the general bandpass voltage transfer function has the form

$$\frac{V_2(s)}{V_1(s)} = \frac{Hs^{n/2}}{s^n + a_{n-1}s^{n-1} + \cdots + a_1 s + a_0} \qquad (4.87)$$

where *n* can only be even. The factored form of this function is

$$\frac{V_2(s)}{V_1(s)} = \prod_{i=1}^{n/2} \frac{H_i s}{s^2 + a_{i1}s + a_{i0}} \qquad (4.88)$$

The separate quadratic factors are easily realized by any of the second-order bandpass circuits presented in the preceding sections. An alternative decomposition of a bandpass

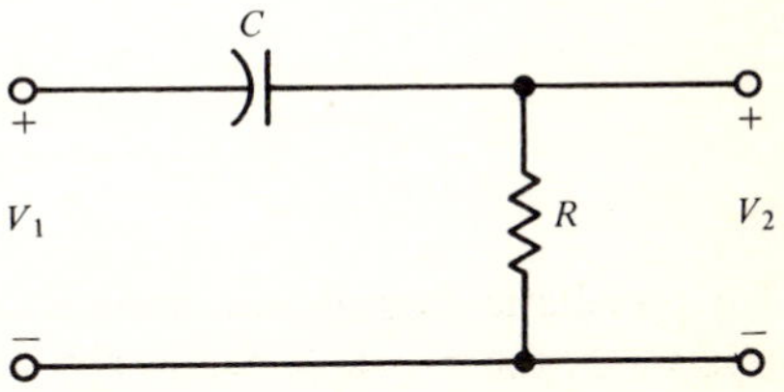

FIGURE 4.29
Realization of the first-order factor of (4.86).

Table 4.12a DENOMINATOR POLYNOMIALS FOR HIGH–PASS NETWORK FUNCTIONS OF THE FORM $s^n + a_{n-1}s^{n-1} + \cdots + a_1 s + a_0$

Case (Table 4.3)	Order	a_0	a_1	a_2	a_3	a_4	a_5
1	3–6	Same as low-pass (Table 4.11a)					
2a	3	1.39725	1.75063	2.14463			
	4	2.63717	3.15892	4.52940	2.70531		
	5	5.58899	6.55303	10.82793	7.31919	4.20580	
	6	10.55268	12.23240	22.91877	16.77626	12.36628	4.56263
2b	3	2.03537	2.01164	2.52063			
	4	3.62805	3.45684	5.27494	2.69426		
	5	8.14153	7.62716	13.74958	7.93310	4.72645	
	6	14.51233	13.47109	28.02078	17.44587	13.63210	4.72645

Table 4.12b ROOTS OF DENOMINATOR POLYNOMIALS FOR HIGH–PASS NETWORK FUNCTIONS

Case (Table 4.3)	Order	Roots		
1	3–6	Same as low-pass (Table 4.11b)		
2a	3	$-0.27418 \pm j0.89451$	-1.59628	
	4	$-1.18777 \pm j1.18108$	$-0.16489 \pm j0.95556$	
	5	$-0.61484 \pm j1.31128$	$-0.10809 \pm j0.97660$	-2.76005
	6	$-1.84584 \pm j1.72114$	$-0.35955 \pm j1.25121$	$-0.07590 \pm j0.98578$
2b	3	$-0.24852 \pm j0.97163$	-2.02358	
	4	$-1.20569 \pm j1.45785$	$-0.14144 \pm j0.99684$	
	5	$-0.54554 \pm j1.42542$	$-0.09052 \pm j1.00181$	-3.45427
	6	$-1.86087 \pm j2.13442$	$-0.30458 \pm j1.30397$	$-0.06277 \pm j1.00269$

Table 4.12c FACTORS OF DENOMINATOR POLYNOMIALS FOR HIGH–PASS NETWORK FUNCTIONS OF THE FORM $(s + a)(s^2 + a_{11}s + a_{10})(s^2 + a_{21}s + a_{20})(s^2 + a_{31}s + a_{30})$

Case (Table 4.3)	Order	a	a_{11}	a_{10}	a_{21}	a_{20}	a_{31}	a_{30}
1	3–6	Same as low-pass (Table 4.11c)						
2a	3	1.59628	0.54334	0.87531				
	4		2.37553	2.80573	0.32977	0.94028		
	5	2.76004	1.22968	2.09749	0.21618	0.96544		
	6		3.69168	6.36953	0.71910	1.69491	0.15180	0.97752
2b	3	2.02358	0.49705	1.00582				
	4		2.41138	3.57912	0.28288	1.01370		
	5	3.45427	1.09108	2.32938	0.18104	1.01181		
	6		3.72173	8.01881	0.60917	1.79300	0.12555	1.00933

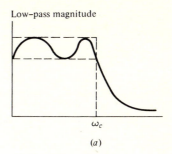

Low–pass magnitude

ω_c

(a)

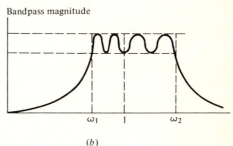

Bandpass magnitude

ω_1 1 ω_2

FIGURE 4.30
Low-pass-to-bandpass transformation.

(b)

network function into a product of low-pass and high-pass functions is also possible; for example, for $n = 4$, (4.88) could be written

$$\frac{V_2(s)}{V_1(s)} = \frac{H_1}{s^2 + a_{11}s + a_{10}} \frac{H_2 s^2}{s^2 + a_{21}s + a_{20}} \qquad (4.89)$$

which could be realized as a cascade connection of second-order low-pass and high-pass filters.

The determination of the coefficients of high-order bandpass functions is most easily done by using a transformation of the complex frequency variable. If we let p be the new complex frequency variable, then the transformation is defined as

$$s = \frac{p^2 + 1}{p} \qquad (4.90)$$

where s is the original complex frequency variable. Such a transformation, when applied to a low-pass network function, produces a bandpass network function with a center frequency of unity. The bandwidth of the low-pass function becomes the bandwidth of the bandpass function, and the band-edge frequencies of the bandpass function have 1 rad/s as their geometric mean. Thus, applying (4.90) (with $s = j\omega$) to the low-pass characteristic shown in Fig. 4.30a produces the bandpass characteristic shown in Fig. 4.30b, in which the indicated frequencies satisfy the relations

$$\omega_2 - \omega_1 = \omega_c \qquad \omega_1 \omega_2 = 1 \qquad (4.91)$$

Some transformations of the second- and third-order low-pass characteristics defined in Tables 4.1 and 4.11 are given in Table 4.13. If the transformation defined by (4.90) is applied to a high-pass function, a band elimination characteristic (notch filter) results. Relations similar to those given in (4.91) also apply to this case.

Elliptic Filters

Filters with transmission zeros on the $j\omega$ axis, originally introduced in Sec. 4.2, may also be required to be of greater than second order. In the general case, for n even, the transfer function of such a filter has the factored form

$$\frac{V_2(s)}{V_1(s)} = \prod_{i=1}^{n/2} \frac{H_i(s^2 + b_i)}{s^2 + a_{i1}s + a_{i0}} \tag{4.92}$$

For n odd, the function has the form

$$\frac{V_2(s)}{V_1(s)} = \frac{\displaystyle\prod_{i=1}^{(n-1)/2} H_i(s^2 + b_i)}{(s+a)\displaystyle\prod_{i=1}^{(n-1)/2}(s^2 + a_{i1}s + a_{i0})} \tag{4.93}$$

Specifically, the third-order elliptic filter will have the form

$$\frac{V_2(s)}{V_1(s)} = \frac{H(s^2 + b_1)}{(s+a)(s^2 + a_{11}s + a_{10})} \tag{4.94}$$

while the fourth-order filter has the form

$$\frac{V_2(s)}{V_1(s)} = \frac{H(s^2 + b_1)(s^2 + b_2)}{(s^2 + a_{11}s + a_{10})(s^2 + a_{21}s + a_{20})} \tag{4.95}$$

Table 4.13a **DENOMINATOR POLYNOMIALS OF BANDPASS NETWORK FUNCTIONS OF THE FORM** $s^4 + a_1 s^3 + a_2 s^2 + a_1 s + 1 = (s^2 + a_{11}s + a_{10})(s^2 + a_{21}s + a_{20})$

Low-Pass (Table 4.1)	BW	a_2	a_1	a_{11}	a_{10}	a_{21}	a_{20}
1	0.5	2.25000	0.70711	0.41602	1.42922	0.29108	0.69968
	0.2	2.04000	0.28285	0.15142	1.15218	0.13142	0.86792
	0.1	2.01000	0.14121	0.07321	1.07330	0.06821	0.93170
	0.05	2.00250	0.07071	0.03598	1.03599	0.03473	0.96528
2a	0.5	2.37905	0.71281	0.44442	1.65585	0.26839	0.60392
	0.2	2.06065	0.28512	0.15684	1.22260	0.12828	0.81793
	0.1	2.01516	0.14256	0.07485	1.10564	0.06770	0.90445
	0.05	2.00379	0.07128	0.03654	1.05148	0.03475	0.95103
2b	0.5	2.27563	0.54887	0.33488	1.56498	0.21398	0.63898
	0.2	2.04410	0.21954	0.11957	1.19609	0.09997	0.83606
	0.1	2.01103	0.10977	0.05734	1.09364	0.05243	0.91437
	0.05	2.00276	0.05489	0.02806	1.04577	0.02683	0.95623

Table 4.13b DENOMINATOR POLYNOMIALS OF BANDPASS NETWORK FUNCTIONS OF THE FORM $s^6 + a_1 s^5 + a_2 s^4 + a_3 s^3 + a_2 s^2 + a_1 s + 1 =$
$(s^2 + a_{11}s + a_{10})(s^2 + a_{21}s + a_{20})(s^2 + a_{31}s + a_{30})$

Low-Pass (Table 4.1)	BW	a_3	a_2	a_1	a_{11}	a_{10}	a_{21}	a_{20}	a_{31}	a_{30}
1	0.5	2.12500	3.50000	1.00000	0.30328	1.54171	0.50000	1.00000	0.19672	0.64863
	0.2	0.80800	3.08000	0.40000	0.10864	1.18911	0.20000	1.00000	0.09136	0.84097
	0.1	0.40100	3.02000	0.20000	0.05216	1.09046	0.10000	1.00000	0.04783	0.91704
	0.05	0.20013	3.00500	0.10000	0.02554	1.04425	0.05000	1.00000	0.02446	0.95762
2a	0.5	1.34237	3.38372	0.62645	0.19548	1.66029	0.31323	1.00000	0.11774	0.60230
	0.2	0.50689	3.06139	0.25058	0.06902	1.22646	0.12529	1.00000	0.05627	0.81536
	0.1	0.25129	3.01534	0.12529	0.03292	1.10756	0.06264	1.00000	0.02972	0.90288
	0.05	0.12538	3.00383	0.06264	0.01606	1.05242	0.03132	1.00000	0.01526	0.95019
2b	0.5	1.04975	3.30960	0.49417	0.15259	1.61489	0.24708	1.00000	0.09449	0.61923
	0.2	0.39927	3.04953	0.19767	0.05417	1.21283	0.09883	1.00000	0.04466	0.82451
	0.1	0.19816	3.01238	0.09883	0.02590	1.10138	0.04942	1.00000	0.02352	0.90795
	0.05	0.09889	3.00309	0.04942	0.01265	1.04948	0.02471	1.00000	0.01206	0.95285

Values of these coefficients for various values of the constants K_p, K_s, and ω_s, defined in Fig. 4.6, are given in Table 4.14. This table supplements Table 4.6. As an example of its use, consider the design of an elliptic filter with a maximum range of 0.95 to 1.0 in the passband from 0 to 1 rad/s, and an attenuation of at least 15 dB for all frequencies above 1.10 rad/s. From Fig. 4.6 we see that $K_p = 0.95$ and $\omega_s = 1.10$. From Table 4.6 we find that a second-order elliptic filter has $K_s = 0.782$ (-2.14 dB), from Table 4.14a that a third-order elliptic filter has $K_s = 0.394$ (-8.09 dB), and from Table 4.14b that a fourth-order filter has $K_s = 0.140$ (-17.08 dB). Obviously a fourth-order filter meets the specifications, and thus the required network function is

$$\frac{V_2(s)}{V_1(s)} = \frac{0.140(s^2 + 1.291)(s^2 + 4.350)}{(s^2 + 1.027s + 0.786)(s^2 + 0.126s + 1.052)} \qquad (4.96)$$

Two of the filters shown in Fig. 4.7 may be used to realize (within a multiplicative constant) this network function.

Tuning

In practice, to complete the actual construction of an active RC filter, a final tuning or adjustment of the element values and the gain may be required to compensate for element tolerance and parasitic effects. This becomes especially important in high-order cascaded realizations and in the high-Q case. First let us consider the tuning of a low-pass second-order filter having a network function of the form given in (4.17). Plots of the magnitude as a function of normalized frequency ω/ω_c readily show that the magnitude at low frequencies, say $\omega \leqslant 0.1\omega_c$, is constant, independent of α, and equal to H_c. Similarly, at the high frequency of $10\omega_c$, the magnitude characteristic is independent of α and is -40 dB below the values of H_c. Thus, the adjustment of the element values which determine H_c and ω_c is most easily done at these values. The tuning to correctly realize α is most easily done at the ω_α frequency, defined as

$$\omega_\alpha = \omega_c\sqrt{1 - \alpha^2/2} \qquad (4.97)$$

at which the peak magnitude occurs (for $\alpha < \sqrt{2}$). Some iterations between the three adjustments described above may be required, depending on the particular filter configuration which is being tuned. The same three steps may be used to tune high-pass second-order filters. In this case, the frequencies are the reciprocals of those referred to above.

The other basic second-order filter function to be considered here is the bandpass one having the form given in (4.27). Here the peak response occurs at ω_c, and thus the adjustment of element values to provide this frequency is one of the tuning steps. For the remaining steps, we note that the bandwidth (BW) is the difference between the frequencies at which the magnitude characteristic is down 3 dB from its peak value. These frequencies are

$$f_1 = \frac{f_c}{2Q}(-1 + \sqrt{1 + 4Q^2}) \qquad f_2 = \frac{f_c}{2Q}(1 + \sqrt{1 + 4Q^2}) \qquad (4.98)$$

Table 4.14a   THIRD–ORDER ELLIPTIC FUNCTION PARAMETERS OF (4.94) (from top to bottom, a, a_{11}, a_{10}, b_1, H, and K_s)

K_p \ ω_s	1.05	1.10	1.15	1.20	1.30	1.40	1.60
0.99	3.00155	2.38167	2.04962	1.84049	1.59035	1.44703	1.29096
	.085439	.164793	.239930	.308389	.423881	.514156	.641363
	1.17110	1.27550	1.35412	1.41484	1.50033	1.55605	1.62210
	1.20541	1.37031	1.53363	1.69962	2.04551	2.41363	3.22359
	2.91611	2.21688	1.80968	1.53210	1.16647	.932875	.649595
	.835656	.702859	.585336	.488077	.346305	.254634	.150824
0.95	1.39312	1.16920	1.05238	.978047	.887854	.834076	.773178
	.128759	.208146	.267535	.313860	.382084	.429629	.491552
	1.09401	1.12638	1.14386	1.15422	1.16538	1.17039	1.17419
	1.20541	1.37031	1.53359	1.69962	2.04548	2.41363	3.22359
	1.26436	.961054	.784827	.664187	.505756	.404446	.281626
	.550613	.393746	.298822	.235598	.158103	.113422	.066000
0.90	.989843	.850207	.776613	.729373	.671004	.635840	.595349
	.131719	.197936	.243978	.278588	.327809	.361387	.404241
	1.04501	1.05130	1.05182	1.05044	1.04612	1.04183	1.03480
	1.20541	1.37031	1.53360	1.69962	2.04550	2.41363	3.22359
	.858124	.652271	.532635	.450785	.343166	.274453	.191108
	.408593	.279160	.207846	.162349	.108001	.077236	.044840
0.85	.796463	.692067	.636652	.600774	.556216	.529099	.497668
	.125887	.182370	.220415	.248520	.287999	.314599	.348306
	1.01490	1.00922	1.00267	.996551	.986391	.978507	.967482
	1.20541	1.37031	1.53361	1.69962	2.04549	2.41362	3.22359
	.670582	.509696	.416235	.352254	.268216	.214500	.149361
	.330211	.221527	.163808	.127519	.084606	.060434	.035059
0.80	.672056	.588202	.543408	.514269	.477902	.455659	.429765
	.117935	.167009	.199450	.223182	.256259	.278406	.306340
	.993886	.981243	.970727	.962023	.948671	.938912	.925798
	1.20541	1.37031	1.53361	1.69962	2.04549	2.41362	3.22359
	.554123	.421193	.343958	.291087	.221642	.177253	.123426
	.277706	.184503	.135945	.105649	.069994	.049969	.028976
0.75	.580701	.510790	.473303	.448768	.418123	.399259	.377272
	.109462	.152592	.180768	.201218	.229618	.248517	.272306
	.978189	.960951	.947897	.937548	.922206	.911278	.896878
	1.20541	1.37031	1.53361	1.69962	2.04549	2.41362	3.22359
	.471237	.358197	.292535	.247550	.188505	.150742	.104965
	.238726	.157652	.115915	.089987	.059570	.042510	.024646
0.70	.508377	.448812	.416747	.395725	.369369	.353124	.334131
	.100986	.139154	.163859	.181717	.206412	.222806	.243387
	.965965	.945461	.930629	.919159	.902442	.890734	.875469
	1.20541	1.37033	1.53361	1.69962	2.04549	2.41363	3.22360
	.407390	.309657	.252888	.214008	.162958	.130318	.090743
	.207883	.136715	.100376	.077873	.051520	.036758	.021308

Table 4.14b FOURTH–ORDER ELLIPTIC FUNCTION PARAMETERS OF (4.95) [from top to bottom, $a_{11}, a_{10}, b_1, a_{21}, a_{20}, b_2$, and $H(= K_S)$]

K_p \ ω_s	1.05	1.075	1.1	1.15	1.20	1.25	1.3
	1.24184	1.36998	1.43445	1.48461	1.49416	1.48971	1.48067
	1.76639	1.64271	1.52732	1.35343	1.23106	1.14128	1.07349
	1.15362	1.22234	1.29092	1.42978	1.57242	1.71971	1.87203
0.99	.073511	.104589	.132384	.179552	.218090	.250180	.277321
	1.09961	1.12737	1.14901	1.18196	1.20610	1.22469	1.23958
	3.31266	3.85083	4.34993	5.29789	6.22434	7.15325	8.09589
	.503140	.389504	.309376	.209067	.150187	.112478	.086921
	1.11694	1.15719	1.17084	1.17132	1.16143	1.14961	1.13807
	1.17034	1.05721	.974214	.860463	.785950	.733037	.693369
	1.15363	1.22235	1.29093	1.42979	1.57244	1.71971	1.87203
0.97	.082382	.109829	.132841	.169869	.198774	.222152	.241517
	1.06041	1.07297	1.08228	1.09550	1.10463	1.11140	1.11664
	3.31252	3.85097	4.34995	5.29782	6.22442	7.15325	8.09588
	.315149	.233744	.182127	.120698	.086036	.064240	.049553
	1.00616	1.02456	1.02740	1.01897	1.00707	.994981	.983063
	.949099	.853237	.785841	.695368	.637238	.595833	.563827
	1.15362	1.22235	1.29090	1.42980	1.57240	1.71971	1.87203
0.95	.081657	.106240	.126428	.158254	.182889	.202523	.218409
	1.04043	1.04722	1.05200	1.05833	1.06243	1.06526	1.06699
	3.31238	3.85097	4.34973	5.29772	6.22421	7.15328	8.09613
	.245492	.180326	.139862	.092248	.065716	.049015	.037702
	.903747	.911065	.909212	.897844	.885261	.873655	.863732
	.800608	.718662	.662419	.587723	.539585	.505447	.480059
	1.15363	1.22235	1.29093	1.42979	1.57243	1.71971	1.87204
0.925	.078638	.100616	.118456	.146316	.167498	.184309	.198090
	1.02465	1.02747	1.02921	1.03102	1.03182	1.03207	1.03216
	3.31249	3.85095	4.34992	5.29794	6.22440	7.15331	8.09589
	.198567	.145089	.112302	.073994	.052617	.039221	.030239
	.825168	.827693	.822969	.811096	.799091	.788335	.779239
	.709059	.636774	.587138	.521773	.479903	.450211	.428090
	1.15363	1.22235	1.29041	1.42981	1.57240	1.71971	1.87203
0.9	.075144	.095172	.111155	.136283	.155159	.170095	.182294
	1.01374	1.01413	1.01394	1.01299	1.01180	1.01058	1.00947
	3.31240	3.85096	4.34581	5.29794	6.22422	7.15326	8.09589
	.169268	.123463	.095443	.062763	.044651	.033285	.025661
	.708408	.706199	.700475	.688058	.676966	.667721	.659911
	.598882	.537875	.496416	.442241	.407509	.383057	.364750
	1.15363	1.22235	1.29092	1.42978	1.57243	1.71971	1.87203
0.85	.068333	.085454	.099089	.120059	.135776	.148163	.158248
	.999163	.996486	.993986	.989660	.986076	.983077	.980558
	3.31250	3.85096	4.34987	5.29786	6.22434	7.15335	8.09589
	.133093	.096781	.074682	.049106	.034894	.026019	.020059
	.621079	.617140	.611017	.599431	.589557	.581212	.574306
	.532447	.478486	.442216	.394408	.363967	.342428	.326332
	1.15363	1.22235	1.29041	1.42982	1.57240	1.71971	1.87204
0.8	.062131	.077101	.088880	.107103	.120623	.131234	.139842
	.989514	.984974	.981121	.974761	.969765	.965686	.962302
	3.31250	3.85096	4.34613	5.29772	6.22423	7.15327	8.09589
	.110293	.080095	.061824	.040580	.028851	.021504	.016576

Table 4.14*b* FOURTH–ORDER ELLIPTIC FUNCTION PARAMETERS OF (4.95) [from top to bottom, $a_{11}, a_{10}, b_1, a_{21}, a_{20}, b_2,$ and $H(=K_S)$] (Continued)

K_ρ \ ω_s	1.05	1.075	1.1	1.15	1.20	1.25	1.3
0.75	.550687	.546069	.540306	.529453	.520419	.513005	.506861
	.487251	.438112	.405011	.361820	.334160	.314640	.300027
	1.15363	1.22235	1.29093	1.42979	1.57243	1.71971	1.87204
	.056514	.069762	.080214	.096136	.107970	.117237	.124738
	.982530	.976711	.971880	.964177	.958218	.953427	.949476
	3.31251	3.85096	4.34992	5.29791	6.22439	7.15327	8.09589
	.093955	.068176	.052571	.034540	.024536	.018289	.014096
0.7	.491207	.486393	.480919	.470946	.462799	.456159	.450655
	.454281	.408647	.377979	.337967	.312352	.294272	.280747
	1.15363	1.22235	1.29092	1.42979	1.57244	1.71972	1.87203
	.051378	.063179	.072459	.086558	.097012	.105183	.111795
	.977203	.970443	.964919	.956218	.949563	.944251	.939898
	3.31247	3.85102	4.34989	5.29795	6.22448	7.15335	8.09607
	.081317	.058963	.045465	.029861	.021211	.015810	.012188

where $Q = f_c/\text{BW} = f_c/(f_2 - f_1)$. As described in connection with the low-pass-to-bandpass transformation earlier in this section, the relation $f_1 f_2 = f_c^2$ must hold. As in the tuning procedure for the low-pass case, some iteration between the tuning steps described above may be required. In the high-Q case, the relations of (4.98) may be approximated as

$$f_1 = f_c - \frac{\text{BW}}{2} \qquad f_2 = f_c + \frac{\text{BW}}{2} \qquad (4.99)$$

4.6 GYRATOR FILTERS AND NEGATIVE–IMPEDANCE CONVERTERS

In this section we present two interesting applications of operational amplifiers. These are the gyrator and the negative-impedance converter. The former has been found to be especially valuable for the realization of a wide range of filter circuits.

Gyrator Filters

Before active *RC* filters became popular, practically all filtering was done using passive *RLC* filters. The realization of such filters is well documented, and extensive sets of tables which cover various applications may be found in the literature. Some examples of these are given in Table 4.15. Passive *RLC* realizations have certain advantages over active *RC* ones, the most important being their low sensitivity. Their main disadvantage, of course, is that they require the use of inductors, which, as was pointed out at the beginning of this chapter, are a rather undesirable element. In this section we present a method of designing active *RC* filters in which the low sensitivity of the passive *RLC* realization is retained, but the inductors are eliminated. The elimination process is accomplished by using a device called a gyrator, which electrically converts a capacitor to an inductor.

Table 4.15 DOUBLE-R TERMINATED LOW–PASS PASSIVE LC LADDER REALIZATIONS

Case (Table 4.1)	Order	C_1/L_1'	L_2/C_2'	C_3/L_3'	L_4/C_4'	C_5/L_5'	L_6/C_6'
1	2	1.4142	1.4142				
$R = 1$	3	1.0000	2.0000	1.0000			
$R' = 1$	4	0.7654	1.8478	1.8478	0.7654		
	5	0.6180	1.6180	2.0000	1.6180	0.6180	
	6	0.5176	1.4142	1.9319	1.9319	1.4142	0.5176
2a	2	1.5132	0.6538				
$R = 2$	3	2.9431	0.6503	2.1903			
$R' = \frac{1}{2}$	4	1.8158	1.1328	2.4881	0.7732		
	5	3.2228	0.7645	4.1228	0.7116	2.3197	
	6	1.8786	1.1884	2.7589	1.2403	2.5976	0.7976
2b	2	2.5721	0.4702				
$R = 3$	3	4.9893	0.4286	3.8075			
$R' = \frac{1}{3}$	4	3.0355	0.7929	3.7589	0.5347		
	5	5.3830	0.4915	6.6673	0.4622	3.9944	
	6	3.1307	0.8287	4.1451	0.8467	3.8812	0.5475
3	2	1.5774	0.4226				
$R = 1$	3	1.2550	0.5528	0.1922			
$R' = 1$	4	1.0598	0.5116	0.3181	0.1104		
	5	0.9303	0.4577	0.3312	0.2090	0.0718	
	6	0.8377	0.4116	0.3158	0.2364	0.1480	0.0505

Unprimed Network

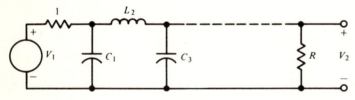

Primed Network

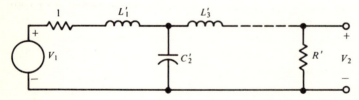

NOTES:
1 Either of the above configurations may be used to realize the indicated network function.
2 Normalized *high-pass* networks for cases 1, 2a, and 2b of Table 4.3 may be derived from the above network configurations (primed or unprimed) by making the following transformations of elements (values are given in henrys and farads).

3 Normalized *bandpass* (center frequency of 1 rad/s) networks may be derived from the above network configurations (primed and unprimed) by making the following transformations of elements (values are given in henrys and farads, BW is the desired bandwidth in radians per second).

A *gyrator* is a two-port circuit which ideally is defined by the y-parameter matrix

$$
\begin{bmatrix} I_1(s) \\ I_2(s) \end{bmatrix} = \begin{bmatrix} 0 & \dfrac{-1}{R} \\ \dfrac{1}{R} & 0 \end{bmatrix} \begin{bmatrix} V_1(s) \\ V_2(s) \end{bmatrix} \qquad (4.100)
$$

where R is positive and real and is called the *gyration resistance*. In the square matrix of (4.100), since $y_{12}(s) \neq y_{21}(s)$, obviously the gyrator is a nonreciprocal network element. A circuit symbol for a gyrator is shown in Fig. 4.31. If an admittance $Y_2(s)$ is connected at port 2 of the gyrator as shown in the figure, then the impedance $Z_{\text{in}}(s)$ seen at the terminals of port 1 is

$$
Z_{\text{in}}(s) = Y_2(s)R^2 \qquad (4.101)
$$

Thus, if $Y_2(s) = sC$ (a capacitor), the input impedance $Z_{\text{in}}(s) = sCR^2$, an inductor of value CR^2 henrys. Clearly, a gyrator and a single capacitor may be used to replace an inductor.[†] In practice the two values of R given in (4.100) may differ. This poses no problem in the simple inductor simulation illustrated in Fig. 4.31, since, from (4.101), it is only the product of these two quantities which determines the conversion ratio. It is more important to keep the $y_{11}(s)$ and $y_{22}(s)$ elements of the square matrix of (4.100) identically zero in order to provide the highest possible Q in the simulated inductor.

Now let us consider some methods of realizing gyrators. Many circuit realizations have been proposed. One of the best of the ones which directly use operational amplifiers is that shown in Fig. 4.32.[‡] For this circuit the y-parameter matrix is

$$
\begin{bmatrix} \dfrac{1/K_1}{(1 + 1/K_1)R_1} & \dfrac{-1}{(1 + 1/K_1)R_1} \\ \dfrac{R_3 + (1/K_1 - 1/K_2)(R_4 + R_3)}{(1 + 1/K_1)(R_4 + R_4/K_2 + R_3/K_2)} & \dfrac{1/K_1}{(1 + 1/K_1)R_2} \end{bmatrix} \qquad (4.102)
$$

[†]A more detailed discussion of a gyrator may be found in Ref. 1, chap. 5.
[‡]R. H. S. Riordan, Simulated Inductors using Differential Amplifiers, *Electronics Letters,* vol. 3, no. 2, pp. 50–51, February 1967.

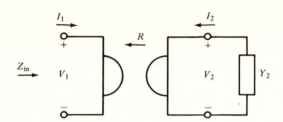

FIGURE 4.31
A gyrator with a load admittance.

where the K_i are the open-loop gains of the operational amplifiers. If we set $K_i = K$ and $R_i = R$, (4.102) simplifies to

$$\frac{1}{(1 + 1/K)R}\begin{bmatrix} 1/K & -1 \\ \dfrac{1}{(1 + 2/K)} & 1/K \end{bmatrix} \quad (4.103)$$

In the limit as the value of K becomes large, this reduces to the y-parameter matrix given in (4.100). As an example of the use of gyrators, consider the third-order high-pass filter shown in Fig. 4.33a. This has a normalized maximally flat magnitude voltage transfer characteristic which is 3 dB down (from its infinite-frequency value) at 1 rad/s. An equivalent realization for this filter, using gyrators, is shown in Fig. 4.33b.

The gyrator circuit shown in Fig. 4.32 has a limitation on its use in that one of the terminals of the input terminal pair at which the simulated inductance is realized is connected to ground. This means that this circuit can only be used to realize an inductor which has one terminal grounded. Passive RLC circuits, however, will in general include inductors which are not grounded. These are frequently referred to as *floating inductors*. One approach to the realization of such inductors is the cascade connection of two gyrators and a capacitor as shown in Fig. 4.34a. Assuming ideal gyrators, this circuit has the characteristics of an ideal floating inductor of value CR^2 as shown in Fig. 4.34b.[†] The gyrator shown in Fig. 4.32, however, is not suitable for use in this circuit because the port at which the capacitor is connected does not have one of its terminals at ground potential. As an alternative, the circuit shown in Fig. 4.35 may be used, as it does have one terminal of each of its ports at ground potential.[‡] For large values of open-loop operational amplifier gain, the y parameters of this circuit are again given by (4.100). Thus this circuit may be used in the cascade connection shown in Fig. 4.34 to realize a floating inductor. Because the gyrator circuit of Fig. 4.35 achieves the zero values of the parameters $y_{11}(s)$ and $y_{22}(s)$ by the subtraction of resistances of equal nominal value, aligning this circuit so as to produce an ideal floating inductor is considerably more critical than aligning the circuit of Fig. 4.32. Thus its use may be correspondingly less

---

[†]A. G. J. Holt and J. Taylor, Method of Replacing Ungrounded Inductances by Grounded Gyrators, *Electronics Letters*, vol. 1, no. 4, p. 105, June 1965.
[‡]G. J. Deboo, Application of a Gyrator-type Circuit to Realize Ungrounded Inductors, *IEEE Transactions on Circuit Theory*, vol. CT-14, no. 1, pp. 101–102, March 1967.

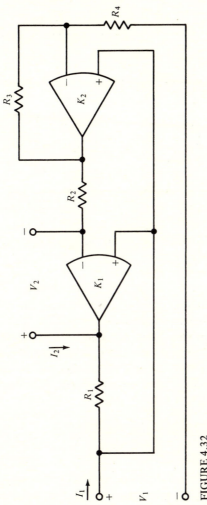

FIGURE 4.32
Circuit realization for a gyrator.

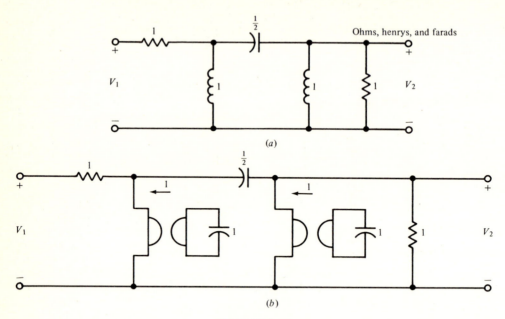

(a)

(b)

FIGURE 4.33
Realization of a third-order high-pass filter using gyrators.

satisfactory in critical applications such as a narrowband filter configuration in which such a floating inductor must be resonated with a capacitor. Many other ways of treating the floating inductor problem have been discussed in the literature.† One of these is the use of generalized impedance converters; this is discussed in the following section.

When careful attention has been paid to the design of the gyrator circuit, inductors with stabilities comparable to those obtained from ferrite inductors, and with Q factors in

† H. J. Orchard and D. F. Sheahan, Inductorless Bandpass Filters, *IEEE Journal of Solid-State Circuits*, vol. SC-5, no. 3, pp. 108–118, June 1970.

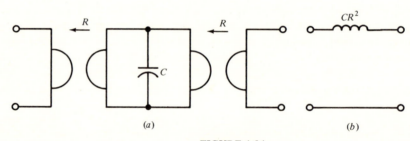

(a)   (b)

FIGURE 4.34
Using gyrators to realize a floating inductor.

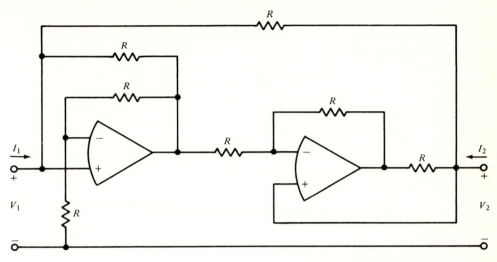

FIGURE 4.35
Circuit realization of a gyrator.

excess of 1000 have been simulated by gyrator-capacitor circuits. Filters constructed using such simulated inductors have been found to have stability, sensitivity, and performance characteristics directly comparable with those of filters using passive *RLC* elements in frequency ranges of up to 100 kHz.

Negative-impedance Converters

One of the more interesting devices to come out of the study of active *RC* circuits is the *negative-impedance converter* (NIC). Basically this is a two-port device which is defined by the transmission parameter matrix†

$$
\begin{bmatrix} V_1 \\ I_1 \end{bmatrix} = \begin{bmatrix} 1 & 0 \\ 0 & \dfrac{-1}{K} \end{bmatrix} \begin{bmatrix} V_2 \\ -I_2 \end{bmatrix} \tag{4.104}
$$

where K is called the gain of the NIC. If a load impedance $Z_2(s)$ is connected to port 2 of such a device, as shown in Fig. 4.36, the input impedance is $Z_{\text{in}} = -Z_2(s)/K$. Thus, negative-valued impedances of any type may be produced through the use of an NIC. A circuit realization for an NIC is shown in Fig. 4.37. The transmission parameters are those given in (4.104), and the NIC gain $K = R_a/R_b$. Due to the fact that the NIC is an active device, certain precautions must be observed with respect to the types of load impedances

†More formally, this is referred to as a current-inversion negative-impedance converter (INIC). A voltage-inversion negative-impedance converter (VNIC) may also be defined by interchanging the "11" and "22" elements in (4.104). See Ref. 1, chap. 4.

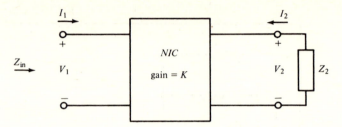

FIGURE 4.36
An NIC with a load impedance.

connected to its ports to prevent instability. In terms of the circuit shown in Fig. 4.37, this requires that the load resistance connected across the terminals of the input (left) port (this is sometimes referred to as the OCS or *open-circuit-stable* port) be greater than the value of any resistance connected to the terminals of the output (right) port (the SCS or *short-circuit-stable* port).†

NICs may be directly used in connection with passive RC elements to produce specific filter characteristics. As an example of this, the circuit shown in Fig. 4.38 provides a bandpass characteristic having the form

†This is true only for the case in which $R_a = R_b$. More generally it is required that $R_1 R_b > R_2 R_a$, where R_1 is the resistance connected to the input port and R_2 is that connected to the output port.

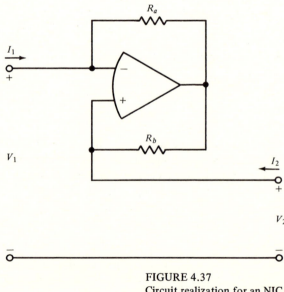

FIGURE 4.37
Circuit realization for an NIC.

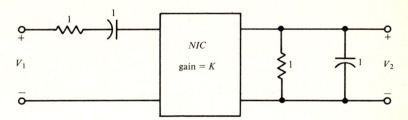

FIGURE 4.38
Realization of a bandpass filter using an NIC.

$$\frac{V_2(s)}{V_1(s)} = \frac{-Ks}{s^2 + (2 - K)s + 1} \quad (4.106)$$

Such realizations, however, have not proved useful except for special applications. The sensitivities in general are high, and although they may not be any higher than the ones encountered in the low-gain-amplifier noninverting realizations of Secs. 4.1 and 4.2, the NIC filters have the additional disadvantage that they cannot be cascaded without the use of additional isolating circuitry. One area in which NICs have found application, however, is the realization of other active elements. As an example of this, the circuit shown in Fig. 4.39 provides a realization for an ideal gyrator. Two NICs are required (one of these is used to produce the negative-valued resistor).

4.7 GENERALIZED IMPEDANCE CONVERTER FILTERS

In the last section we described a method of converting passive *RLC* filters to active *RC* form by replacing the inductors by gyrator-capacitor simulations. One of the problems encountered in this approach was seen to be the difficulty of simulating floating, i.e. ungrounded, inductors, such as those encountered in low-pass and bandpass passive *RLC* filters. In this section we present an alternative approach to obtaining low-sensitivity active *RC* filters from passive *RLC* ones, by using a device called a generalized impedance converter. To understand how the method works, consider the low-pass second-order

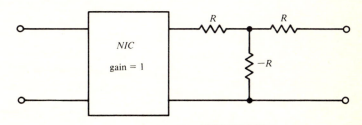

FIGURE 4.39
Circuit realization for a gyrator.

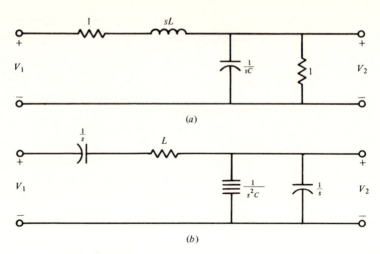

FIGURE 4.40
Impedance transformation by $1/s$ of a low-pass second-order filter.

filter shown in Fig. 4.40a. The elements of this are specified in terms of their impedances $Z(s)$. The voltage transfer function for this filter is

$$\frac{V_2(s)}{V_1(s)} = \frac{1}{s^2 LC + s(L + C) + 2} \qquad (4.107)$$

Now let us make an impedance normalization of the elements of this filter. Since the transfer function is dimensionless, such a normalization will not affect it. Specifically we will choose $1/s$ as the normalization. The individual passive impedances transform as

$$R \rightarrow R/s \qquad sL \rightarrow L \qquad 1/sC \rightarrow 1/s^2 C \qquad (4.108)$$

Thus, resistors are transformed to capacitors, inductors to resistors, and capacitors to elements which have impedances proportional to $1/s^2$ and are symbolized by four bars as shown in the transformed network in Fig. 4.40b. This latter element is called a *frequency-dependent negative resistor* (FDNR), since its impedance $Z_{\text{FDNR}}(s) = 1/s^2 C$ has the property

$$Z_{\text{FDNR}}(s) \Big|_{s=j\omega} = \frac{1}{s^2 C} \Big|_{s=j\omega} = \frac{-1}{\omega^2 C} \qquad (4.109)$$

which is purely real (and thus resistive), negative, and obviously a function of frequency, hence the name.† The voltage transfer function for the circuit shown in Fig. 4.40b is given by (4.107), just as that for the circuit of Fig. 4.40a is. As a result of the impedance transformation, however, the troublesome floating inductor of Fig. 4.40a has been

---

†L. T. Bruton, Network Transfer Functions Using the Concept of Frequency-dependent Negative Resistance, *IEEE Transactions on Circuit Theory,* vol. CT-16, no. 3, pp. 406–408, August 1969.

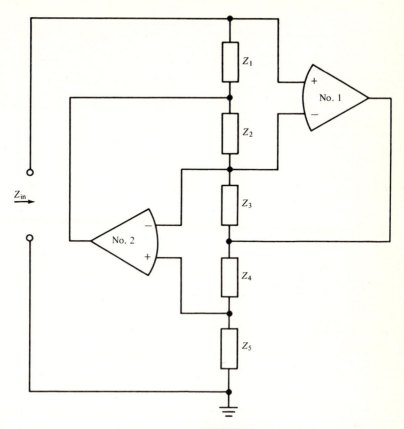

FIGURE 4.41
Generalized impedance converter (GIC).

removed, leaving in its place a floating resistor which is, of course, easy to realize. The FDNR, which is now the only active element to be considered, is a grounded element; thus by the impedance transformation, we have substituted a grounded active element for an ungrounded one, simplifying the realization procedure.

To realize a ground FDNR we may use a *generalized impedance converter* (GIC). Such a circuit is shown in Fig. 4.41. For this configuration, the input impedance $Z_{in}(s)$ is

$$Z_{in}(s) = \frac{Z_1(s)Z_3(s)Z_5(s)}{Z_2(s)Z_4(s)} \qquad (4.110)$$

If any two of the impedances $Z_1(s)$, $Z_3(s)$, or $Z_5(s)$ are chosen as capacitors, and the remaining ones as resistors, than an FDNR is realized.† It may be shown that the circuit of

---

†Similarly, if $Z_2(s)$ or $Z_4(s)$ is chosen as a capacitor, and the remaining elements as resistors, an inductor is realized. Thus, we see that the gyrator is a special case of a generalized impedance converter.

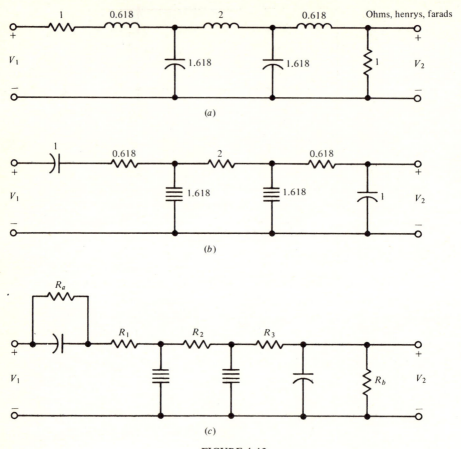

FIGURE 4.42
Steps in the realization of a fifth-order low-pass filter.

Fig. 4.41 is unconditionably stable (assuming ideal operational amplifiers) for all combinations of $Z_1(s)$ through $Z_5(s)$ which contain not more than two capacitors.[†]

As an example of the application of FDNR elements, consider the realization of a fifth-order maximally flat magnitude low-pass filter with equal resistance terminations. From Table 4.15a we obtain the realization shown in Fig. 4.42a. After an impedance transformation of $1/s$, the circuit has the form shown in Fig. 4.42b, which requires two FDNRs. This network must be slightly modified to provide a resistive path from the FDNRs to ground, since, as shown in Fig. 4.41, with $Z_1(s)$ and $Z_3(s)$ chosen as capacitors, such a path is required to provide input bias current to the noninverting terminal of operational amplifier 2. In theory such a resistive path could be provided by

[†]L. T. Bruton, Nonideal Performance of Two-amplifier Positive-impedance Converters, *IEEE Transactions on Circuit Theory*, vol. CT-17, no. 4, pp. 541–549, November 1970.

placing a large shunt resistor to ground from the input terminals of either of the FDNR realizations in Fig. 4.42*b*. Such an element, however, corresponds with a shunt inductor to ground in the *RLC* prototype filter, and this would produce zero transmission at zero frequency. An alternative modification which avoids this problem is to use two resistors R_a and R_b as shown in Fig. 4.42*c*. To select the values of these components we note that the dc gain of the filter shown in Fig. 4.42*a* is 0.5. Thus, for the circuit shown in Fig. 4.42*c*,

$$\lim_{\omega \to 0} \frac{V_2}{V_1}(j\omega) = \frac{R_b}{R_a + R_b + R_1 + R_2 + R_3} = \frac{R_b}{R_a + R_b + 3.236} \qquad (4.111)$$

By choosing R_a and R_b large with respect to the other resistors, and by requiring that $R_b = R_a + 3.236$, the desired criteria are met. For example, $R_a = 10$ and $R_b = 13.236$ works well. The complete filter, frequency-normalized to provide a cutoff frequency of 1 kHz, is shown in Fig. 4.43. An impedance normalization of 1000 has also been used. Note that the GIC elements have been chosen so as to obtain a nonunity conversion ratio, thus permitting the use of all equal-valued capacitors in the realization.

Just as the use of gyrators described in Sec. 4.6 works efficiently when all the inductors of a given passive *RLC* realization are grounded, as in a high-pass filter, so the use of FDNRs works well when all the capacitors are gounded, as in the low-pass case. An extension of the FDNR approach may be used, however, when floating FDNRs are to be realized, as occurs in the case of bandpass filters. To understand how this extension works, consider the GIC configuration shown in Fig. 4.44, in which a passive network defined by the transmission parameters

$$\begin{bmatrix} A(s) & B(s) \\ C(s) & D(s) \end{bmatrix} \qquad (4.112)$$

is connected in cascade with two identical GICs. The overall transmission parameters for the cascade are

$$\begin{bmatrix} 1 & 0 \\ 0 & \dfrac{Z_2(s)Z_4(s)}{Z_1(s)Z_3(s)} \end{bmatrix} \begin{bmatrix} A(s) & B(s) \\ C(s) & D(s) \end{bmatrix} \begin{bmatrix} 1 & 0 \\ 0 & \dfrac{Z_1(s)Z_3(s)}{Z_2(s)Z_4(s)} \end{bmatrix}$$

$$= \begin{bmatrix} A(s) & B(s)\dfrac{Z_1(s)Z_3(s)}{Z_2(s)Z_4(s)} \\ C(s)\dfrac{Z_2(s)Z_4(s)}{Z_1(s)Z_3(s)} & D(s) \end{bmatrix} \qquad (4.113)$$

This set of parameters is identical with those of a network derived from the original network by making an impedance transformation of $Z_1(s)Z_3(s)/Z_2(s)Z_4(s)$ on each element. In practice, rather than using the cascade shown in Fig. 4.44 to separately realize each floating FDNR, a combination of such elements can be simultaneously realized, thus reducing the total number of GICs that are required.

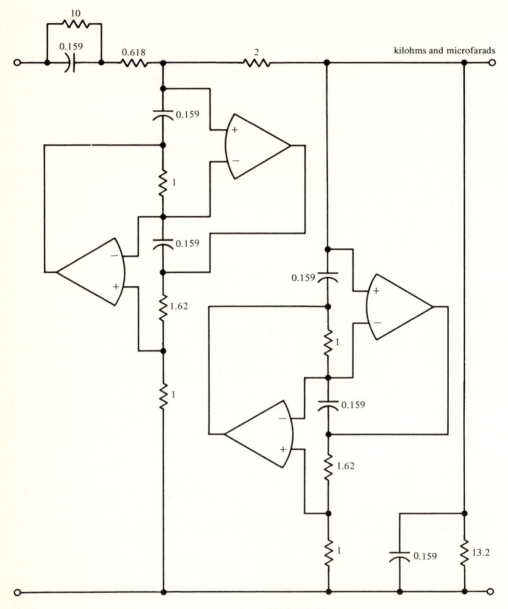

FIGURE 4.43
Final realization for the fifth-order low-pass filter.

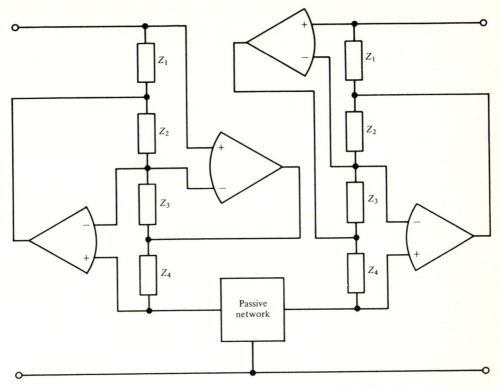

FIGURE 4.44
Realization of a floating FDNR.

As an example of the application of GIC techniques in the general case, consider the fourteenth-order resistance-terminated bandpass filter configuration shown in Fig. 4.45*a*.† After all the elements have been impedance-transformed by $1/s$, it has the form shown in Fig. 4.45*b*. Now let us consider the realization of some of the separate elements of this circuit. The series FDNR-capacitor combinations at the ends of the filter are most easily realized using the GIC circuit shown in Fig. 4.41, in which $Z_1(s) = 1/s$, $Z_2(s) = Z_4(s) = Z_5(s) = 1$, and $Z_3(s) = R_s + 1/sC_1$. Inserting these values in (4.110) we obtain

$$Z_{in}(s) = \frac{1}{s^2 C_1} + \frac{R_s}{s} \qquad (4.114)$$

which provides the input configuration for GIC *A* shown in Fig. 4.45*b*. Similarly, choosing $Z_1(s) = 1/s$, $Z_2(s) = Z_3(s) = Z_4(s) = 1$, and $Z_5(s) = R_L + 1/sC_7$ provides the output configuration of GIC *G*. The grounded FDNRs of value $1/C_2$ and $1/C_6$ shown in

†L. T. Bruton and D. Treleaven, Active Filter Design Using Generalized Impedance Converters, *EDN*, pp. 68–75, Feb. 5, 1973.

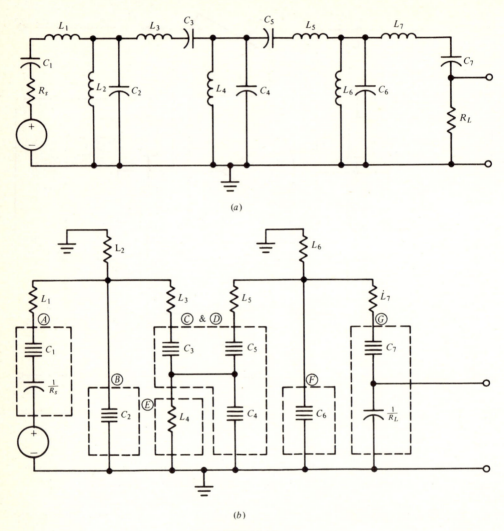

(a)

(b)

FIGURE 4.45
Steps in the realization of a fourteenth-order bandpass filter.

GICs B and F are realized by setting $Z_1(s) = Z_3(s) = 1/s$, $Z_2(s) = Z_4(s) = 1$, and $Z_5(s)$ to $1/C_2$ and $1/C_6$, respectively. The "T" configuration of two floating and one grounded FDNRs of values $1/C_3$, $1/C_4$, and $1/C_5$ and the grounded resistor of value L_4 shown in the center of the network is realized by impedance-transforming each of these elements by s^2, then setting $Z_1(s) = Z_3(s) = 1$, and $Z_2(s) = Z_4(s) = 1/s$ in the circuit shown in Fig. 4.44 for GICs C and D. Thus the "T" of FDNRs is realized by a "T" of resistors of values $1/C_3$, $1/C_4$, and $1/C_5$. The above procedure transforms the resistor of value L_4 to an element with impedance $s^2 L_4$ connected to ground. This is easily realized by the GIC of

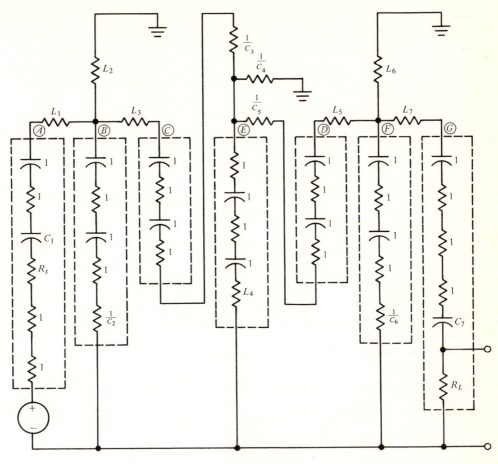

FIGURE 4.46
Final realization for the fourteenth-order bandpass filter.

Fig. 4.41, in which $Z_1(s) = Z_3(s) = 1$, $Z_2(s) = Z_4(s) = 1/s$, and $Z_5(s) = L_4$. The overall circuit is shown in Fig. 4.46. In this figure the GICs are indicated schematically by showing their impedances $Z_i(s)$. The realization of the fourteenth-order bandpass filter thus requires 14 capacitors and 7 GICs. It should be noted that the direct realization of each of the floating FDNRs in Fig. 4.45*b* by a separate cascade of two GICs and a resistor would be considerably less efficient, resulting in a circuit containing 24 capacitors and 11 GICs.

REFERENCES AND BIBLIOGRAPHY

1 HUELSMAN, L. P.: "Theory and Design of Active *RC* Circuits," McGraw-Hill, New York, 1968.

2 HUELSMAN, L. P., (ed.): "Active Filters: Lumped Distributed, Integrated, Digital, and Parametric," chaps. 2, 3, and 4, McGraw-Hill, New York, 1970.

3 MITRA, S. K.: "Analysis and Synthesis of Linear Active Networks," Wiley, New York, 1969.

4 NEWCOMB, R. W.: "Active Integrated Circuit Synthesis," Prentice-Hall, Englewood Cliffs, N.J., 1968.

5 SPENCE, R.: "Linear Active Networks," Interscience-Wiley, London, 1970.

6 SU, K. L.: "Active Network Synthesis," McGraw-Hill, New York, 1965.

7 TEMES, G. C., and S. K. MITRA: "Modern Filter Theory and Design," Interscience-Wiley, New York, 1973.

PROBLEMS

4.1 *(Sec. 4.1)* Derive the y parameters given in (4.6) and (4.7) for the low-pass second-order active RC circuit shown in Fig. 4.2.

4.2 *(Sec. 4.1)* Prove that the locus shown in Fig. 4.13, as defined by the roots of the denominator polynomial of (4.8), is a circle.

4.3 *(Sec. 4.1)* Determine which of the realizations for a Butterworth low-pass characteristic (case 1) given in Table 4.2 has the lowest sensitivity $|S_K^{P_0}|$ as defined in (4.12).

4.4 *(Sec. 4.1)* Repeat Prob. 4.3 for the realizations for the Chebychev 1/2-dB ripple low-pass characteristic.

4.5 *(Sec. 4.1)* Repeat Prob. 4.3 for the realizations for the Chebychev 1-dB ripple low-pass characteristic.

4.6 *(Sec. 4.1)* Repeat Prob. 4.3 for the realizations for the Thompson low-pass characteristic.

4.7 *(Sec. 4.1)* Find the sensitivity $S_x^{P_0}$ similar to that defined in (4.12) for each of the passive elements, R_1, R_2, C_1, and C_2, for the first active RC circuit realization given in Table 4.2.

4.8 *(Sec. 4.1)* Find the values of the passive network elements for the circuit shown in Fig. 4.2 such that the network function

$$\frac{V_2(s)}{V_1(s)} = \frac{H}{s^2 + 0.2s + 1}$$

is realized using the integer amplifier gain $K = 2$.

4.9 *(Sec. 4.1)*

   (*a*) Construct plots of Q, Re $S_K^{P_0}$, and Im $S_K^{P_0}$ versus K over a range $1 \leqslant K \leqslant 3$ for the first active RC circuit realization given in Table 4.2.

   (*b*) Repeat for the case where $R_2 = 10$ and $C_2 = 0.1$ over a range of $1.1 \leqslant K \leqslant 2.1$.

4.10 *(Sec. 4.2)* Determine which of the realizations for a maximally flat magnitude high-pass characteristic given in Table 4.4 has the lowest sensitivity $|S_K^{P_0}|$ as defined in (4.12).

4.11 *(Sec. 4.2)* Find the sensitivity $S_x^{P_0}$ similar to that defined in (4.12) for each of the passive elements R_1, R_2, C_1, and C_2 for the first active RC high-pass filter realization given in Table 4.4.

4.12 *(Sec. 4.2)*
(a) Find the sensitivity $S_K^{P_0}$ for the first $Q = 10$ active RC bandpass realization given in Table 4.5.
(b) Find the sensitivities $S_x^{P_0}$ for each of the passive elements.

4.13 *(Sec. 4.2)* If possible, find an alternative set of design relations to those given in (4.29), for the case in which $R_1 = C_1 = C_2 = 1$.

4.14 *(Sec. 4.2)*
(a) Find a realization for an elliptic magnitude filter having a characteristic similar to that shown in Fig. 4.6, in which $K_p = 0.9$ and $\omega_s = 1.2$, using the circuit shown in Fig. 4.7 with $k = 1$.
(b) Repeat using the circuit shown in Fig. 4.9 with $c = 1$.

4.15 *(Sec. 4.2)*
(a) Use the biquadratic filter shown in Fig. 4.9 to realize a bandpass filter having a Q of 10, and determine the sensitivity $S_K^{P_0}$ for the gain of each of the VCVSs.
(b) Realize the same function using the filter shown in Fig. 4.5 (choosing $C_2 = 1$) and the element values specified by (4.29), determine the sensitivity $S_K^{P_0}$, and compare the result of 1 percent changes in the gains in the two realizations.

4.16 *(Sec. 4.2)* Using the passive element values defined in the example following (4.50) and an appropriate frequency normalization, find a Butterworth realization for a low-pass inverting filter, and find the sensitivity $S_K^{P_0}$.

4.17 *(Sec. 4.3)* The multiple-feedback infinite-gain-amplifier filter shown in Fig. 4.15 is used to realize a normalized maximally flat magnitude low-pass characteristic (case 1, Table 4.1) with a dc gain of -1.586. Find the passive element values and their sensitivities $S_x^{P_0}$, and compare these latter with the ones obtained for the corresponding noninverting low-gain-amplifier filter in Prob. 4.7.

4.18 *(Sec. 4.3)*
(a) Determine the actual network function obtained using the element values given as the first case in Table 4.7, if the open-loop gain of the operational amplifier is -1000.
(b) Repeat for a gain of -100.

4.19 *(Sec. 4.3)* The multiple-feedback infinite-gain-amplifier bandpass filter shown in Fig. 4.17 is used to realize a $Q = 10$, $\omega_c = 1$, $H = 2.728$ network function. Find the passive element values and their sensitivities $S_x^{P_0}$, and compare these latter with the ones obtained for the corresponding noninverting low-gain-amplifier filter in Prob. 4.12.

4.20 *(Sec. 4.3)* Find a single-feedback infinite-gain-amplifier filter which realizes the transfer function

$$\frac{V_2(s)}{V_1(s)} = \frac{s^2 + 1}{s^2 + 2s + 1}$$

4.21 (Sec. 4.3)

    (*a*) The biquadratic infinite-gain-amplifier filter shown in Fig. 4.19 is used to realize a bandpass function with $Q = 10$, $\omega_c = 1$, and $H = 2.728$. Find the passive element values and their sensitivities $S_x^{P_0}$, and compare these latter with the ones obtained in Probs. 4.12 and 4.19.

    (*b*) Repeat for the filter shown in Fig. 4.21.

4.22 (Sec. 4.4)

    (*a*) Find a state-variable bandpass realization having the form shown in Fig. 4.24, with a center frequency of 1 kHz and a Q of 100. Assume infinite open-loop operational amplifier gains.

    (*b*) Repeat for operational amplifier open-loop gains of −1000.

4.23 (Sec. 4.4) Repeat Prob. 4.22*a* and *b* for the Tarmy-Ghausi circuit shown in Fig. 4.25.

4.24 (Sec. 4.4) Find the sensitivities $S_x^{P_0}$ of the passive elements in the state-variable realization found in Prob. 4.22*a*.

4.25 (Sec. 4.4) Repeat Prob. 4.24 for the Tarmy-Ghausi circuit realization found in Prob. 4.23*a*.

4.26 (Sec. 4.4) Using the passive element values found in Prob. 4.22*a*, find the actual pole locations (dominant and parasitic) realized with operational amplifier open-loop gains of 1000 and pole locations of 10 kHz.

4.27 (Sec. 4.4) Repeat Prob. 4.26 for the Tarmy-Ghausi circuit realization found in Prob. 4.23*a*.

4.28 (Sec. 4.5) Design a sixth-order maximally flat magnitude low-pass active *RC* filter with a −3-dB cutoff frequency of 10 kHz using the low-gain-amplifier circuit shown in Fig. 4.2.

4.29 (Sec. 4.5) Repeat Prob. 4.28 using the multiple-feedback infinite-gain-amplifier circuit shown in Fig. 4.15.

4.30 (Sec. 4.5) Repeat Prob. 4.28 using the single-feedback infinite-gain-amplifier circuit shown in Fig. 4.18.

4.31 (Sec. 4.5) Design a sixth-order bandpass active *RC* filter with a center frequency of 1 kHz and with a bandwidth of 0.2 kHz in which there is 1/2-dB ripple. Use low-gain amplifier circuits for complex conjugate pole pairs with $Q < 10$, and infinite-gain-amplifier single-feedback circuits for pole pairs with $Q > 10$.

4.32 (Sec. 4.5) Design an elliptic filter in which the passband is 0–1 kHz and the tolerance is 15 percent of the maximum magnitude. The stop band should have a tolerance no greater than 5 percent, starting from a stop-band frequency of 1.15 kHz. Use the low-gain-amplifier circuit shown in Fig. 4.7.

4.33 (Sec. 4.5) Repeat Prob. 4.32 using the biquadratic low-gain-amplifier circuit shown in Fig. 4.9.

4.34 (Sec. 4.5) Repeat Prob. 4.32 using the biquadratic infinite-gain-amplifier circuit shown in Fig. 4.19.

4.35 (Sec. 4.5) Repeat Prob. 4.32 using the biquadratic infinite-gain-amplifier circuit shown in Fig. 4.21.

4.36 (Sec. 4.5)

    (*a*) Find the network function for a fourth-order band elimination filter having a maximally flat magnitude characteristic in the passband, a (geometric) center frequency of 1 rad/s, and a bandwidth (between −3-dB frequencies) of 1 rad/s.

(*b*) Find a realization for this filter using the biquadratic low-gain-amplifier circuit shown in Fig. 4.9.

4.37 *(Sec. 4.6)* Design a second-order maximally flat magnitude high-pass active *RC* filter with a −3-dB frequency of 1 kHz, using a gyrator to simulate the inductor. Find the sensitivities $S_x^{P_0}$ for each of the passive elements and for the gyration factor R^2.

4.38 *(Sec. 4.6)* Design a fourth-order maximally flat magnitude high-pass active *RC* filter with a −3-dB frequency of 10 kHz, using gyrators to simulate the inductors.

4.39 *(Sec. 4.6)* Design a fourth-order maximally flat magnitude low-pass active *RC* filter with a −3-dB frequency of 10 kHz, using gyrators to simulate the (floating) inductors.

4.40 *(Sec. 4.6)* Find the voltage transfer functions for the circuits shown in Fig. P4.40, and find element values that will realize a voltage transfer function with a maximally flat magnitude passband characteristic.

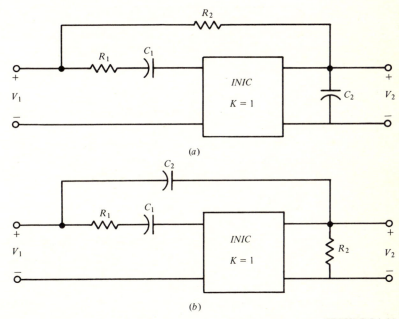

(*a*)

(*b*)

FIGURE P4.40

4.41 *(Sec. 4.7)* Design a second-order maximally flat magnitude low-pass active *RC* filter with a −3-dB frequency of 1 kHz, using a GIC. Find the sensitivities $S_x^{P_0}$ for each of the passive elements and for the impedances Z_i of the GIC.

4.42 *(Sec. 4.7)* Design a fourth-order maximally flat magnitude low-pass active *RC* filter with a −3-dB frequency of 10 kHz, using GICs. Compare the realization with that obtained in Prob. 4.40 for the number of passive and active elements.

5

ELECTRONIC SWITCHING CIRCUITS

5.1 INTRODUCTION

We refer to continuously variable quantities in instrumentation, computing, control, and communication systems as *analog variables*; the term is historically derived from "analogies" between mechanical and electrical quantities in early analog computers. *Digital variables*, on the other hand, take countable sets of *discrete* states (0 or 1; ON or OFF; 1, 2, 3, 4, . . . ; etc.). *Analog comparators* (Sec. 3.2) are the basic *analog-to-digital* transducers, producing a digital (discrete-state) output as a function of analog inputs. Similarly, *switches* are the basis of all *digital-to-analog* transducers, which produce different continuously variable outputs given different discrete (digital) inputs. The many possible kinds of switching circuits (e.g., single-pole/single-throw, double-pole/double-throw, star switches connecting one point to one of several different points, and crossbar switches for matrix interconnections) can all be constructed from simple *on-off switches,* each controlled by a basic digital ON/OFF (0 or 1) decision. Accurate analog switches are needed for the following system-design purposes:

1 *General-purpose analog-data interconnections* in instrumentation, control, computer, and communication circuits
2 *Multiplexing* (timesharing) of devices such as analog-to-digital converters
3 *Timed sampling* of continuously changing variables, e.g., for analog-to-digital conversion
4 *Digital-to-analog conversion*; programming of test signals (Chap. 6)
5 *Integrator switching* in analog computers
6 *Accurate modulation and demodulation*; analog multiplication

5.2 IDEAL SWITCHES AND REAL SWITCHING CIRCUITS

Switches in automatic systems are most frequently operated by *electrical signals from digital logic circuits,* typically 0 V for digital 0 and +4 V for digital 1. The earliest automatic switches, still widely used, were *relays*, which afford good low-resistance contacts but switch far too slowly for many purposes; many relays also tend to bounce a few times between ON and OFF when they are turned ON. An *ideal electronic switch* would, in fact, have the following desirable properties:

1 *Zero forward impedance* Z_{ON} when the switch is ON.
2 *Infinite back impedance* Z_{OFF} when the switch is OFF.
3 *Zero switching time* between ON and OFF conditions.
4 No *offset voltages or currents* introduced by the switch into the signal circuit.
5 *Zero crosstalk* from the switch-control circuit into the signal circuit.
6 These properties should be obtained *without critical circuit adjustments.*

Electronic switching circuits approximating these desirable properties are not easy to design. In general, compromises will be necessary.

To turn an electrical signal circuit ON or OFF, one can insert an electronic switch *in series with the signal current,* or the switch may *shunt the signal current to ground* (Fig. 5.1*a*). Real electronic switches do have a finite forward resistance R_{ON} when the switch is ON, and also a finite back impedance (leakage impedance) Z_{OFF} representing both resistive and capacitive leakage when the switch is OFF (Fig. 5.1*b*). The leakage resistance R_{OFF} is quite large (greater than 50 MΩ) for most electronic switches, but even a few picofarads of leakage capacitance C_{OFF} can produce serious leakage at high frequencies. A very effective method of minimizing such leakage effects is to *combine shunt and series switches* into tee-network transfer impedances (Fig. 5.1*c*). Given the forward impedance $Z_{ON} = R_{ON}$ of the closed shunt switch and the leakage impedance

$$Z_{OFF} = \frac{R_{OFF}}{j\omega R_{OFF}C_{OFF} + 1} \tag{5.1}$$

the tee-network short-circuit transfer impedance (Sec. 1.5) of the switching network in Fig. 5.1*c* is

$$Z = R + Z_{OFF} + \frac{RZ_{OFF}}{Z_{ON}} = R + \frac{R_{OFF}}{j\omega R_{OFF}C_{OFF} + 1}\left(1 + \frac{R}{R_{ON}}\right) \tag{5.2}$$

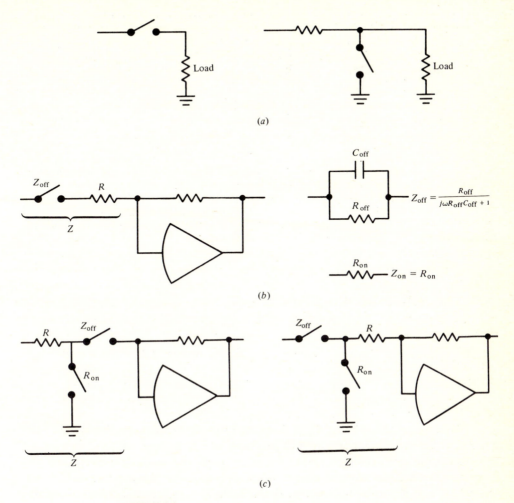

FIGURE 5.1
(*a*) Series and shunt switches; (*b*) a simple series switch; (*c*) the large tee-network transfer impedance formed by combination series-shunt switching.

We see that capacitive as well as resistive leakage is effectively shunted to ground. For $R = 10$ kΩ, $R_{ON} = 50$ Ω, $R_{OFF} = 50$ MΩ, $C_{OFF} = 5$ pF, $|Z|$ exceeds 10^4 MΩ at direct current and 60 MΩ at 100 kHz. To switch even higher frequencies, the resistance R can be replaced by a third switch, so that

$$Z = 2Z_{OFF} + \frac{Z_{OFF}^2}{Z_{ON}} \qquad (5.3)$$

(see also Fig. 5.3*b* and *d* and Ref. 1).

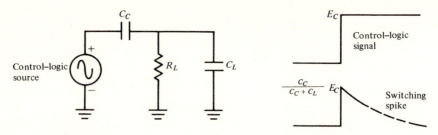

FIGURE 5.2

Turn-on spike (grounded control-signal source, grounded load) due to a switch capacitance C_C. The dashed lines indicate that the spike will be modified when the switch actually turns ON. Until then (neglecting nonlinear circuit effects), the spike has the form of a decaying exponential with initial amplitude $E_C C_C/(C_C + C_L)$ and time constant $R_L(C_C + C_L)$. For a purely capacitive load ($R_L = \infty$), the exponential becomes a step (based on Ref. 1).

Electronic switching devices such as diodes and bipolar transistors can introduce not only spurious ON and OFF switch impedances, but also *offset voltages and currents* (Secs. 5.3 and 5.4).

If the switch-control circuit and the switched (signal) circuit have common dc impedances, a portion of the switching waveform may appear in the signal circuit, where switching steps give rise to control-voltage–dependent voltage offsets or *pedestals*.

For fast switching, switch-control voltages are usually fast-rising steps or pulses derived from digital logic circuits. Even small coupling capacitances C_C between the control circuit and the switched (signal) circuit *differentiate* these steep-sided waveforms and create *switching spikes* and/or steps in the signal circuits (Fig. 5.2). Switching spikes may have the full control-voltage step amplitude and constitute very annoying interference which can trip signal-circuit comparators and may even overload amplifiers in low-voltage signal circuits. In capacitive signal-load circuit, switching spikes are integrated into steps (Sec. 5.8).

Reducing the coupling capacitance C_C will make switching spikes shorter but not smaller. Spike amplitudes can be reduced by increasing the signal load capacitance or by otherwise reducing the frequency response of the signal circuit, if this is compatible with signal-bandwidth requirements.

5.3 DIODE SWITCHING CIRCUITS

Figure 5.3 shows a variety of *diode switches (diode-bridge switches)*. Figure 5.4 presents simple diode equivalent circuits in the ON and OFF conditions. Diode switches are perhaps the simplest electronic switches and can be quite fast. It is best to employ modern fast-recovery, low-leakage silicon-junction diodes in matched monolithic sets. The leakage capacitance C_{OFF} in Fig. 5.4 can be less than 2 pF, R_{OFF} is greater than 100

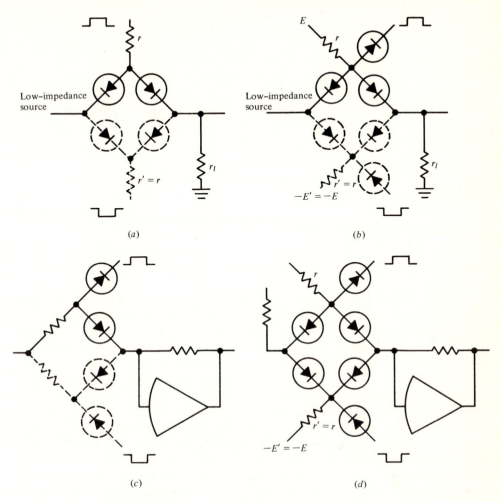

(a)

(b)

(c)

(d)

FIGURE 5.3
Diode bridge switches; r_L can be an operational amplifier input resistor. Circuits (b), (c), and (d) combine shunt and series switching and turn OFF quickly. In (a), (b), and (c), half of each circuit can be used to switch positive input voltages (solid lines) or negative input voltages (dashed lines). But this interferes with pedestal-free control-current returns, and the offset-voltage and leakage cancellation makes the full bridge desirable even with input voltages of one sign. Circuit (d) does *not* permit half-bridge operation and requires careful balancing of E/r and $-E/r$ to minimize offset in the ON condition (based on Ref. 1).

MΩ, and R_{ON} will be 25 to 50 Ω. Unfortunately, diode characteristics (see also Chap. 4) imply an *offset voltage* of about 0.6 V when the diode is ON and an *offset current* I_0 when the diode is OFF. The offset voltage rises with temperature by 2 to 3 mV/$^\circ$C and can be canceled in matched diode-pair circuits. The small leakage current, typically 0.1 to 2 nA at room temperature, doubles every 7 to 10°C.

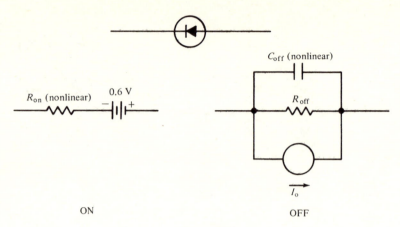

FIGURE 5.4
Diode models in the OFF and ON states.

To turn each diode switch in Fig. 5.3 fully OFF (i.e., to turn *all* series diodes OFF) requires control voltages *exceeding the largest signal-voltage excursion in absolute value.* Again, to keep a diode-bridge switch ON, the control circuits must supply a *current* at least *as large as the largest signal current required to flow through the switch.* Otherwise, one of the series diodes could turn OFF, and the bridge would act as a voltage limiter in the manner discussed in Chap. 3. The resulting control-current requirement can be quite large if the switch must pass high-frequency currents charging capacitors in sample-hold circuits (Sec. 5.8). The six-diode switches of Fig. 5.3b and d combine shunt and series switching in the manner discussed in Sec. 5.2. The symmetrical bridge circuits in Fig. 5.3, moreover, can reduce the effective diode leakage current by a factor of at least 3 if matched bridge diodes are fabricated on the same monolithic chip. Unfortunately, switching spikes from the push-pull control circuit will not cancel accurately because of the nonlinear nature of the diode capacitances. One may try to use trimmer capacitors to vary the control-step rise and fall times on one side of the bridge.

The *turnoff time* of a diode switch will depend on the time constant of the circuit which charges the OFF diode capacitance or capacitances until each diode is forward-biased. In the case of the six-diode switch of Fig. 5.3b, the turnoff time constant is of the order of $C_{OFF}R_{ON}$, so that six-diode switches can turn OFF very quickly. The *turn-on time* will depend on the resistance in the signal circuit.

5.4 BIPOLAR–TRANSISTOR SWITCHES (REF. 1)

Switch-Transistor Characteristics. Inverted Connection

The collector characteristics of a bipolar transistor (Fig. 5.5) indicate that a relatively small change of the transistor base current will switch either *p-n-p* or *n-p-n* transistors between *saturation* (ON condition) and *cutoff* (OFF condition); note that a saturated

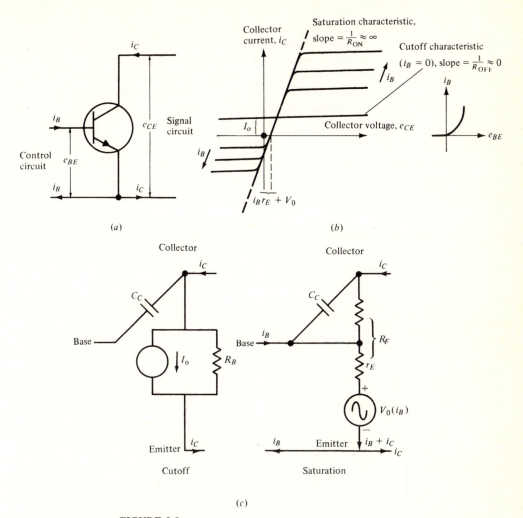

FIGURE 5.5
(*a*) *n-p-n* transistor switch; (*b*) transistor characteristics; (*c*) simplified equivalent circuits (based on Ref. 1).

transistor will pass current in both directions. In the *n-p-n* circuit of Fig. 5.5, a sufficiently *positive* base current i_B will cause saturation, and the transistor is cut off for zero or negative i_B.

Actual transistor collector characteristics show that (Ref. 1)

1 The reciprocal slopes $\Delta e_{CE}/\Delta i_C = R$ of the saturation and cutoff characteristics, respectively, yield *forward resistances* $R = R_{ON}$ between 10 and 100 Ω and *back resistances* $R = R_{OFF}$ between 1 and 500 MΩ, much as in the case of junction-diode switches.

2 The saturation and cutoff characteristics in Fig. 5.5*b* do not pass exactly through the coordinate origin, as ideal switch characteristics would. The resulting offset errors are interpreted as follows: the saturation characteristic is displaced horizontally by a positive *saturation offset voltage* $i_B r_E + V_0(i_B)$ of 10 to 500 mV; and the cutoff characteristic is displaced vertically by a positive *leakage current* I_0 between 0.02 and 50 nA at 25°C.

This situation is summarized by the simplified *equivalent circuits* of Fig. 5.5*c*. If an *n-p-n* or *p-n-p* junction transistor is *inverted*, i.e., if one interchanges the roles of emitter and collector, the device still operates as a transistor whose characteristics are shaped like those in Fig. 5.5*b*, although they are *compressed vertically*. More specifically, experimental and theoretical investigation of the normal and inverted transistor characteristics reveals that:

1 The current amplification factor $\beta = \partial i_C/\partial i_B$ is decreased by a factor of 5 to 100 in the inverted connection, and $-\log_e \alpha = -\log_e (\partial i_C/\partial i_E)$ is decreased by about a factor of 10. The inverted connection is, therefore, not suitable for amplification or for power switching with small control currents.
2 The switch forward resistance (reciprocal slope of the saturation characteristic) is only slightly higher in the inverted connection.
3 *The inverted transistor connection typically decreases the offset voltage* $V_0 + i_B r_E$ *by a factor of at least 10.*

NOTE: In earlier (germanium, silicon-alloy) switching transistors, the leakage-current offset in the OFF state (as well as the offset voltage in the ON state) was substantially reduced by the inverted connection. This is *not* true for modern silicon switching transistors, where what little leakage is left is due mainly to surface leakage and charge generation rather than to diffusion in the junction.

Transistors suitable for analog switching are silicon epitaxial-planar types combining low saturation resistance and leakage with relatively small junctions for high switching speed. For typical silicon transistors of this kind (e.g., Motorola 2N2220, 2N2369, and MPS6512) used in the inverted connection at 25° C, we find:

Inverted-saturation voltage $i_B r_E + V_0 i_B < 1$ mV ($i_B = 0.5$ mA)
Inverted leakage current $I_0 < 0.1$ nA ($i_B \approx 0$)
Inverted-saturation resistance ("dynamic resistance") $R_{ON} < 25$ Ω ($i_B = 0.5$ mA)
Back resistance $R_{OFF} > 1$ to 100 MΩ

Since the emitter-base and collector-base junctions of a junction transistor behave essentially like junction diodes, a transistor switch exhibits nonlinear emitter-base and collector-base capacitances (typically 5 to 15 pF for 1-V reverse bias) and turnoff storage time (below 0.1 μs for switching transistors). Transistor capacitances and storage times are, approximately, inversely proportional to the quoted alpha-cutoff frequency f_α, which is, therefore, a useful measure of the attainable switching speed. Since the collector-base junction is larger than the emitter-base junction, an inverted transistor has a higher input capacitance than a normal transistor, but typical switching times are still well below 1 μs for switching transistors with $f_\alpha > 100$ MHz.

Circuit Design

To turn a transistor switch OFF, the base-control circuit must supply a voltage at least large enough to reverse-bias *both* the emitter-base junction and the collector-base junction. To turn the transistor switch ON, the base current must exceed the signal current divided by β_N (normal connection) or β_I (inverted connection).

Bipolar-transistor switching circuits (like diode switches but unlike FET switches) are controlled by current rather than voltage, so that *the designer must guard against impedances common to this control current and the signal circuit* if pedestal-type offsets in the signal circuit are to be avoided.

This isolation between control and signal circuits is achieved in the *transistor shunt switch* (simple bipolar-transistor chopper) of Fig. 5.6a, where the switching transistor returns the control current to ground and causes only a relatively small saturation offset. But control-current return poses a real difficulty for the *series-transistor switch* of Fig. 5.6b, *unless the signal-source impedance* R_S *is quite small compared to the load.*

The best way to return the series-transistor-switch control current is with a *transformer-floated circuit* as in Fig. 5.6c; the circuit shown employs two matched inverted transistors to cancel much of the saturation offset voltage. But such a transformer-coupled circuit cannot be turned ON for an indefinite time (since the transformer will not pass direct current) and is therefore mainly used for chopper operation. The *dc transformer circuit* of Fig. 5.6d solves this problem at the expense of some circuit complication.

Nonsaturating Transistor Switches

To avoid the extra time ("storage time") required to sweep carriers out of a saturated-transistor base-collector junction when the switch is turned OFF, two types of bipolar-transistor switches avoid saturation and utilize the "working" region of the transistor characteristics in Fig. 5.5b (i.e., the transistor works like an amplifier in the ON condition of the switch). The first such switch is the *switched current source* described in Sec. 6.3, and the second method involves *switching an amplifier stage* (Secs. 5.8 and 5.9). Such switches turn OFF within 10 to 30 ns even when they must supply large currents (10 to 100 mA) in the ON condition.

5.5 JUNCTION FIELD–EFFECT TRANSISTOR (JFET) SWITCHES

The *junction field-effect transistors* (JFETs) sketched in Fig. 5.7 contain "channels" of diffused p-type or n-type impurities which act, for *zero* gate-to-source voltage e_{GS}, *like a passive silicon resistor.* In this ON state, the JFET passes current in either direction, and *there is no voltage offset.* JFET forward resistances range between 3 and 100 Ω and

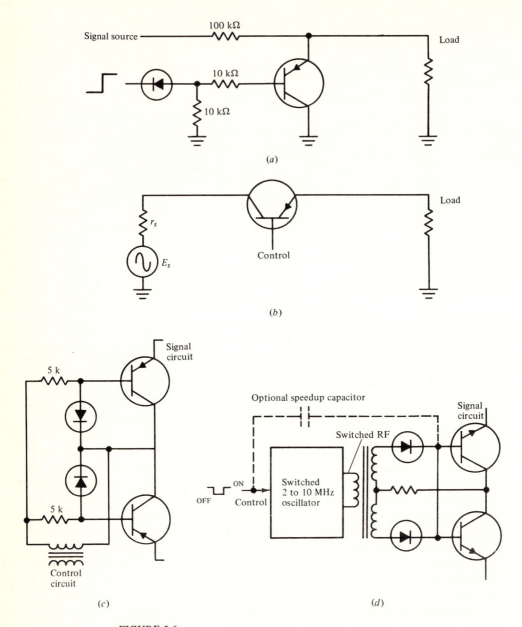

FIGURE 5.6
(*a*) Inverted-transistor shunt switch; (*b*) series switch; (*c*) matched inverted-transistor pair with floating control circuit; (*d*) dc transformer control for a transistor pair.

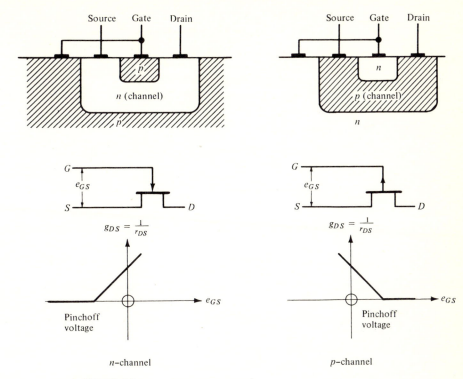

FIGURE 5.7
Junction field-effect transistors (JFETs) and their dc switching characteristics.
Note that switch forward resistances will depend on the control voltage e_{GS}.

increase slightly with temperature (0.1 to 0.7 percent/°C) and also with current (about 2 percent/mA).

Increasingly positive (for p-channel JFETs) or increasingly negative (for n-channel JFETs) e_{GS} will "pinch" the flow of carriers in the channel (electrostatic field effect) and decrease the channel conductance g_{DS} (Fig. 5.7) until the current ceases at the *pinchoff voltage*. In this OFF condition, there are small leakage currents (typically less than 0.01 to 0.3 nA at 25°C, doubling every 10°C) across the gate-source and gate-drain junctions; these junctions also act as nonlinear capacitances. Figure 5.8 shows simple equivalent circuits in the ON and OFF states.

Forward resistances of n-channel JFETs are slightly lower than those of p-channel JFETs of similar geometry. Unfortunately, JFET junction capacitances increase with decreasing R_F (larger junctions). High-priced units may have $R_F = 4$ Ω at 25°C with 20-pF junction capacitances. $R_F = 30$ Ω is more typical.

Figure 5.9 shows practical n-channel JFET switching circuits (reverse voltages and junctions for p-channel). To turn each switch in Fig. 5.9 OFF, *the negative driver-amplifier output must exceed the largest signal excursion at S by 8 to 12 V in absolute value.* To turn the JFET ON, *we would like to keep the gate-to-source voltage at*

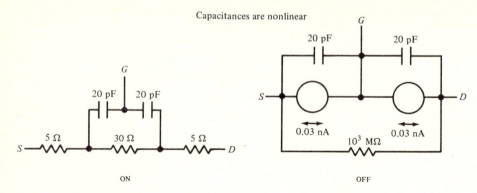

FIGURE 5.8
Simple ON and OFF equivalent circuits for a typical JFET switch. Note that gate-junction capacitances are nonlinear (highest at low junction voltages).

0 V. We must not make the gate so positive that the gate-to-channel junction (Fig. 5.7) is forward-biased; any changes of e_{GS} from 0 V will, moreover, "modulate" the switch conductance and disturb resistance matching in accurate computing circuits.

In Fig. 5.9a, the ON value of the driver-amplifier output exceeds the largest signal excursion by at least 0.6 V to turn the coupling diode OFF. The gate voltage will then follow the signal voltage at the source, unless signal excursions are fast compared to the time constant determined by R_C (typically 20 to 100 kΩ) and the circuit capacitances at G, so that e_{GS} is near zero at least for low signal frequencies. Unfortunately, lower R_C values load the signal circuit at S when the switch is ON, but low R_C also speeds up the turnon time of the switch.

The circuit of Fig. 5.9b serves to switch a voltage ON for a short time, during which the signal remains more or less constant (as in a multiplexer sampling successive signal sources, Sec. 5.7). R_C is omitted; e_{GS} is reduced to (approximately) zero by the positive charge transmitted by the coupling-diode capacitance as the driver output swings more positive than the diode turnoff voltage. One may employ a varactor diode (Ref. 5) or an external 10- to 100-pF capacitor to get the correct turnon charge. But the gate will not follow the signal when the switch is ON; large positive signal excursions may even pinch the FET OFF and get clipped.

In the improved circuit of Fig. 5.9c, the output stage of a CMOS switch driver shorts the JFET gate to the source through a few hundred ohms of MOSFET resistance (see also Sec. 5.6) when the JFET is ON, so that e_{GS} is kept neatly near zero.

Note that JFET switches will normally turn ON when the control-voltage supply is turned OFF or fails. If this creates problems, special precautions may be needed.

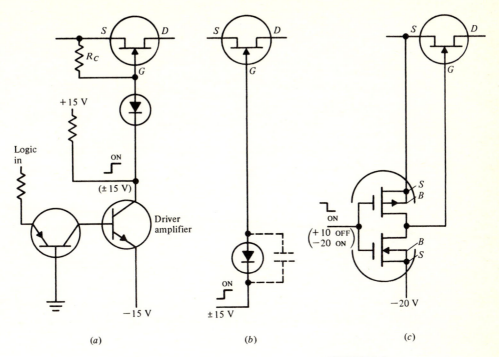

(a) (b) (c)

FIGURE 5.9
n-channel JFET switching circuits.

5.6 METALIZED–OXIDE–SEMICONDUCTOR FIELD–EFFECT TRANSISTOR (MOSFET) AND CMOS INTEGRATED CIRCUITS

The *enhancement-mode* insulated-gate FETs illustrated in Fig. 5.10 are OFF for zero gate-to-source voltage e_{GS}. The MOSFET conductance increases to turn ON as e_{GS} is made more positive (n-channel) or more negative (p-channel). MOSFETs with *depletion* characteristics similar to those in Fig. 5.7 also exist but are rarely used as switches.

Figure 5.11 shows simple equivalent circuits. With the gate electrode insulated by a thin silicon-oxide layer, gate leakage is essentially negligible, and gate-source and gate-drain capacitances are very low (<2 pF, Fig. 5.12). On the other hand, MOSFET forward resistances (75 to 800 Ω) tend to be larger than JFET resistances.

In MOSFET switches, the *substrate-to-channel junction must be back-biased for all signal-voltage excursions* to avoid surprises depending on the circuit design (the circuit may "latch up" with a forward-biased substrate junction until all power supplies are momentarily turned OFF). For signals below 0.6 V in absolute value, it may be possible to simply short substrate and source. Some MOSFETs come with an extra insulating

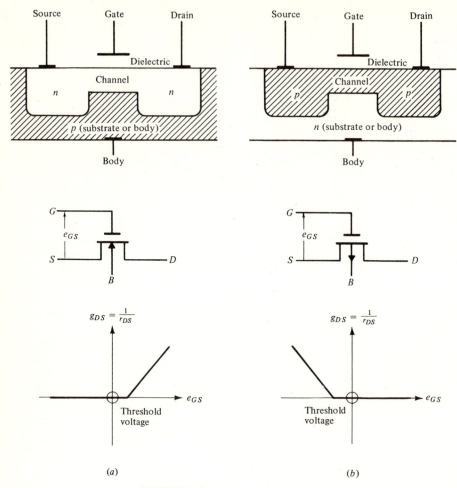

FIGURE 5.10
Enhancement-type MOSFETs and their dc switching characteristics.

diode in their substrate connection to prevent latch-up. Note also that the thin gate-insulation layer of many MOSFETs is easily damaged by electrostatic charges during circuit construction.

If their higher forward resistances can be tolerated, the design of MOSFET switches is much simpler than that of JFET switches because the insulated MOSFET gate can simply be overdriven in the ON and OFF conditions. Multiple MOSFET switches and MOSFET switch drivers are easily constructed as monolithic integrated circuits. In particular, complementary pairs of n-channel and p-channel MOSFETs (*CMOS integrated circuits*) are very useful as switches (Fig. 5.13) and also as switch-driving amplifiers (Fig. 5.9c).

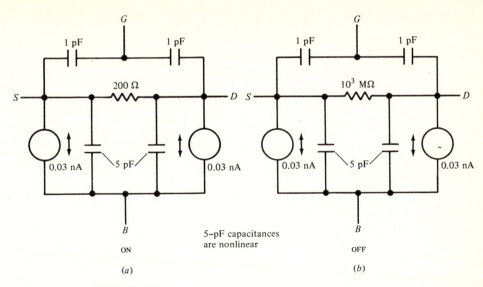

FIGURE 5.11
Simple MOSFET switch equivalent circuits. The substrate-junction capacitances are nonlinear (highest for low junction voltages).

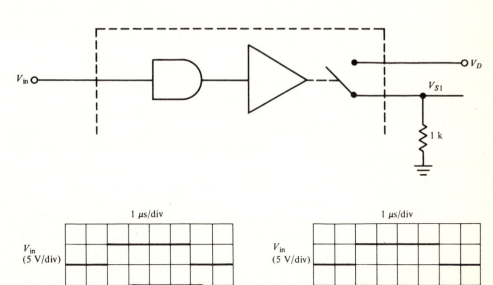

Switching waveforms for $V_D = +10$ V   *Switching waveforms for $V_D = 0$ V*

FIGURE 5.12
Switching performance of an integrated-circuit MOSFET switch/switch driver combination, showing the low switching delay and short switching spikes (Analog Devices, Inc.).

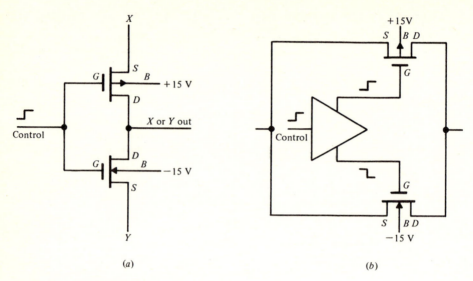

FIGURE 5.13
(*a*) A CMOS SPDT switch; (*b*) a CMOS SPST switch with complementary gate-control signals from a CMOS push-pull switch driver.

5.7 SOME IMPROVED SWITCHING CIRCUITS

Figure 5.14 illustrates a good technique for switching operational amplifier inputs with FET switches. Note that each circuit involves complementary shunt and series switching of the input resistor R to ground potential (ground or summing junction). This yields important advantages:

1. The voltage across each switch is practically zero in both ON and OFF states; this simplifies FET gate-drive requirements, especially with JFETs.
2. Combined series-shunt switching (Sec. 5.2) minimizes signal leakage through the FET junction capacitances.

In addition, the impedances seen by the source in the ON and OFF states are similar and can be accurately matched in Fig. 5.14*b*; this is important in the ladder circuit of Fig. 6.7.

Figure 5.15*a* shows a *multiplexer circuit* employing summing-point FET switches in such a way that *the value of* R_{ON} *cannot affect the dc output.* In the alternative multiplexer circuit of Fig. 5.15*b*, R_{ON} is negligible compared with the high input impedance of a follower amplifier.

Finally, the accurate shunt switch of Fig. 5.16 has its FET switch inside a high-gain operational amplifier feedback loop, so that the FET resistance R_{ON} is effectively divided by the loop gain (Ref. 1).

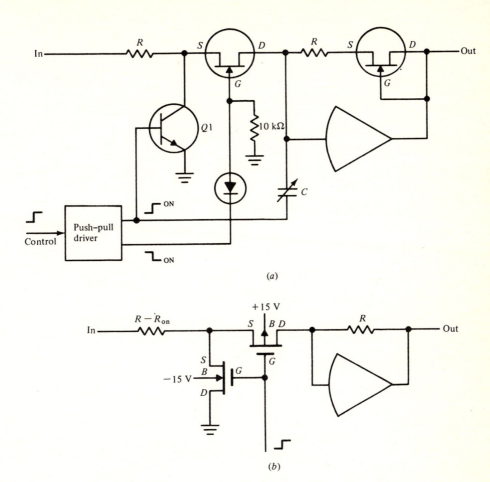

FIGURE 5.14

Summing-junction switching with FETs. In (*a*), variations of R_{ON} with temperature and current are compensated by a matched JFET in the feedback circuit. *C* is a trimmer for partial switching spike cancellation. The shunt switch is sometimes simply replaced by a pair of shunt limiter diodes limiting FET source excursions to ±0.6 V. In (*b*), the signal source (e.g., the ladder circuit of Fig. 6.7) sees practically equal switch impedances to ground in the OFF and ON states.

5.8 SAMPLE–HOLD CIRCUITS

Sample-hold (or *track-hold*) *circuits* (see Ref. 1) hold a timed sample of an analog input for analog-to-digital conversion or analog storage (delay). The output X_0 of the basic capacitor-storage track-hold circuit shown in Fig. 5.17*a* tracks the input X_1 while the switch is ON (*TRACK mode*). A HOLD command opens the switch, so that the capacitor

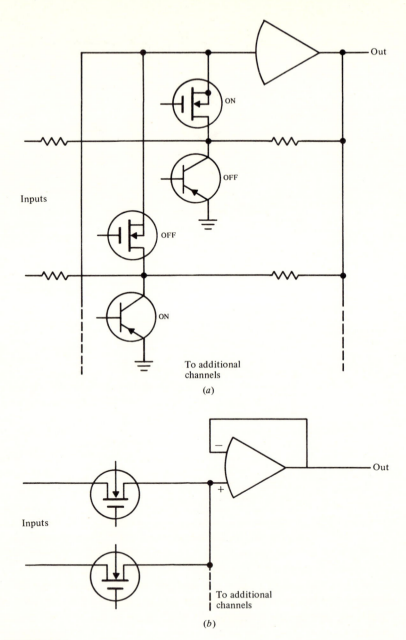

FIGURE 5.15
Multiplexer circuit designed to minimize the effects of FET switch forward resistances.

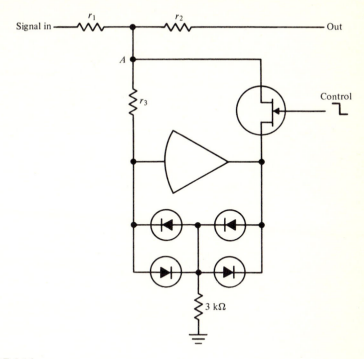

FIGURE 5.16
Tee network with an active shunt switch. Point A is returned to ground potential either through r_3 or through the FET resistance divided by the operational amplifier loop gain.

stores the last input value, and the output ideally remains constant (*HOLD mode*). The storage circuit can drive external loads without discharging the holding capacitor.

An ideal track-hold circuit would follow its input instantaneously and exactly in TRACK, and would hold instantaneously and exactly for an indefinite time after receiving a HOLD command. Figure 5.17*b* illustrates the response of an actual track-hold circuit. The *current capacity of amplifiers and switches* not only limits the maximum tracking rate, but also places a lower limit on the TRACK period required to catch up with specified input-signal excursions. For accurate tracking of high-frequency signal components too small to cause rate limiting, we require low phase shift in TRACK, i.e., sufficient bandwidth in the capacitor-charging input circuit or amplifier. The effect of the capacitor-charging time constant on small-signal phase shift can be compensated with suitable equalizing networks in the input amplifier. The charging time constant must, in any case, be less than one-seventh of the shortest contemplated TRACK period for 0.1 percent accuracy (about one-tenth for 0.01 percent) to allow for exponential step response.

The HOLD output, ideally constant, drifts because of capacitor leakage and dielectric absorption, switch and amplifier-input leakage, and amplifier voltage offset. A total error current i_E contributes the absolute drift error

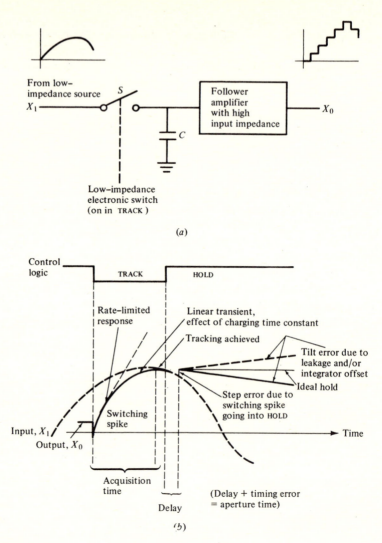

FIGURE 5.17

(a) Switched-capacitor sample-hold circuit; (b) typical response, showing error effects. Rate limits are determined by current capacities of switches and/or amplifiers and may differ for positive and negative rates. Switching spikes can be positive or negative. Timing error and switching spikes going into TRACK will not affect HOLD output. Errors due to dc offset in TRACK are not shown (based on Ref. 1).

$$e_H = \frac{|i_E|}{C} \tau_H \qquad (5.4)$$

where C is the holding capacitance and τ_H is the elapsed time in HOLD. An amplifier offset of e_D V referred to the input will be reproduced at the follower-amplifier output.

With electronic switching, the *switching spike* caused by differentiation of the HOLD control step in the switch capacitances charges the holding capacitor and contributes a spurious voltage step (Fig. 5.17b) whose amplitude is approximately

$$e_s = a_S E_C \frac{C_s}{C} \qquad (5.5)$$

where E_C is the control-step amplitude and C_s is the effective switch capacitance. a_S is a proportionality factor, which we may attempt to reduce through push-pull spike cancellation in symmetrical switching circuits, and/or by adding inverted spikes through a trimmer capacitor. With well-designed circuits, switching-spike errors are not serious in ordinary track-hold operation, but may accumulate in certain computations (Ref. 1).

Finally, switch-response delays in executing the HOLD command result in possibly serious *timing errors,* i.e., the sample-hold circuit actually holds a later input voltage value than desired.

The holding errors in (5.4) and (5.5) are seen to improve with *increasing* holding capacitance C, while *low* values of C require less current and *favor fast tracking.* The required compromise is best resolved with high-current solid-state amplifiers and switches, which permit the use of relatively large holding capacitances. In practice, i_D and e_D can be kept below 1 to 10 nA and 50 to 500 μV, respectively.

If one must track fast signals accurately *and* one also requires relatively long holding periods, it is best to employ a fast (low-C) sample-hold to track the signal and to sample the output of this fast sample-hold with a second, accurately holding (high-C), sample-hold after the input sample-hold has just settled into HOLD. Such *track-hold pairs* are, in any case, needed in applications requiring true zero-order-hold output (i.e., output of successive sample levels not separated by noisy TRACK periods, as in simple sample-holds).

Figure 5.18 shows an improved sample-hold circuit (see also Sec. 5.8), and Table 5.1 shows typical specifications.

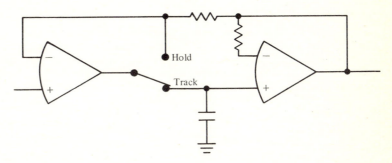

FIGURE 5.18
This commonly used sample-hold circuit improves accuracy and frequency response in the TRACK mode by enclosing the entire circuit in a feedback loop.

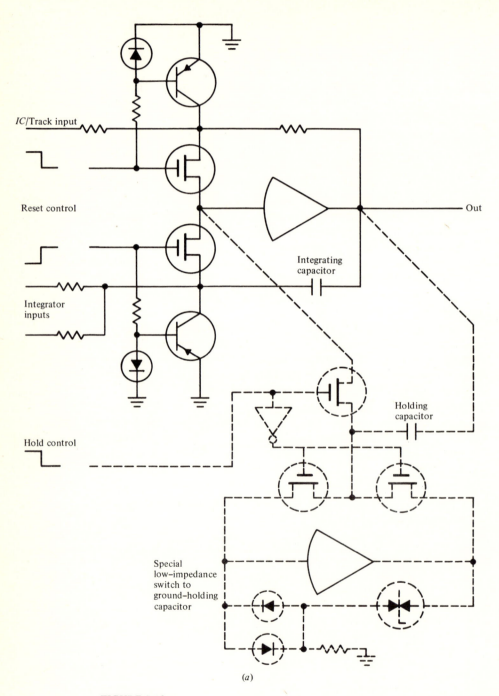

IC/Track input

Reset control

Integrator
inputs

Integrating
capacitor

Hold control

Holding
capacitor

Out

Special
low–impedance
switch to
ground-holding
capacitor

(*a*)

FIGURE 5.19

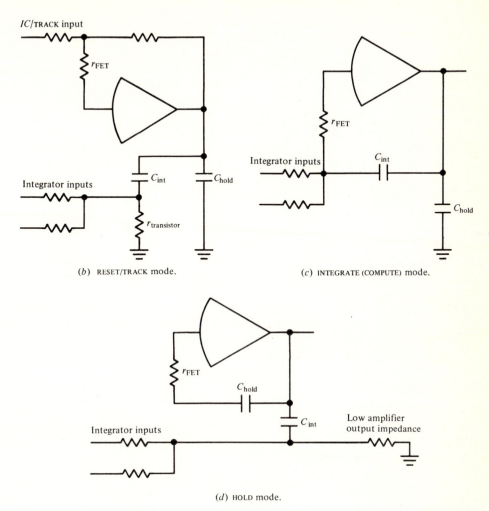

IC/TRACK input

r_{FET}

Integrator inputs

C_{int} C_{hold}

$r_{transistor}$

(*b*) RESET/TRACK mode.

r_{FET}

Integrator inputs

C_{int}

C_{hold}

(*c*) INTEGRATE (COMPUTE) mode.

r_{FET}

C_{hold}

C_{int}

Low amplifier
output impedance

Integrator inputs

(*d*) HOLD mode.

FIGURE 5.19 (*continued*)
Three-mode integrator-control circuit. Typical integrators of this type can hold
within 0.01 percent of half-scale (10 or 100 V) for at least 200 s/μF and switch
into COMPUTE or HOLD within 200 to 400 ns, with at most 50 to 100 ns
between integrators. An "active" switch uses operational amplifier feedback to
ground the holding capacitor through a very small impedance (0.01 to 1 Ω); this
minimizes tracking phase shift (Ref. 11).

5.9 INTEGRATOR MODE CONTROL AND RELATED SWITCHING CIRCUITS

The simplest scheme for resetting an integrator merely resets it to *zero* by shorting the integrating capacitor (say with a JFET switch); the desired initial output is then added to the output of the integrator proper. In spite of its simplicity, this resetting method is rarely used because the two voltages to be added must be scaled down and thus lose accuracy.

Figure 5.19 shows a *three-mode integrator control circuit* for a modern analog/hybrid computer (Refs. 1 and 11). In the RESET mode, the amplifier output follows the phase-inverted initial-condition (IC) input and charges the integrating and holding capacitors through the low feedback-amplifier output impedance. In the COMPUTE mode, the integrator inputs are integrated, and the holding-capacitor charge follows. In the HOLD mode, the circuit acts like an integrator with zero input and holds its last output voltage. We see that *switched integrators can also be used as sample-hold circuits.*

Figure 5.20 shows a *two-mode integrator* designed for very fast analog computation (Refs. 1, 9, and 12); without the integrand input, the circuit is useful as a sample-hold. In the RESET/TRACK mode, the switched unity gain amplifier acts as an impedance transformer, reducing the effective RESET/TRACK-network resistances to $0.2\ \Omega$, so that the effect of the 1- or 10-kΩ integrator inputs becomes negligible. Typical integrators of this type can hold within 0.1 percent of half-scale (10 V) for at least 100 s/μF and switch into COMPUTE or HOLD within 20 to 40 ns, with at most 5 to 10 ns between integrators. The impedance transformation also reduces phase shift in RESET/TRACK.

Switched Operational Amplifiers (Ref. 9)

Substitution of a resistance for the integrator capacitance C in Fig. 5.20 produces an *electronic two-input SPDT switch* with sign-inverted low-impedance output (switched phase inverter or summer). Figure 5.21 shows a more general type of switched operational amplifier implementing

Table 5.1   TYPICAL SPECIFICATIONS OF ±10-V SAMPLE–HOLD CIRCUITS USED FOR ANALOG–TO–DIGITAL CONVERSION*

Gain accuracy	0.01 to 0.05 percent of half-scale ±0.002 percent of half-scale/°C
Bandwidth in TRACK mode	To several megahertz, full signal, depending on application. Small-signal dynamic error should match gain accuracy at working frequencies; frequently not properly specified
Acquisition time (step response or settling time in TRACK mode)	2 to 10 μs within gain accuracy
Aperture time (switching time into HOLD)	50 to 500 ns
HOLD drift	0.1 to 2 mV
Noise	1 mV peak-to-peak (output)
Input-output crosstalk	0.01 to 0.04 percent of half-scale

*From G. A. Korn and T. M. Korn, *Electronic Analog and Hybrid Computers,* McGraw-Hill, New York, 1972.

$$X_0 = -\frac{Z_0(P)}{Z_1(P)}X_1 \quad \text{or} \quad X_0 = -\frac{Z_0'(P)}{Z_1'(P)}X_1' \quad P \equiv \frac{d}{dt} \quad (5.6)$$

In addition to switching-spike and switch-leakage errors, switched operational amplifiers will have transfer-function errors due to finite amplifier gain and follower and/or switch impedances. Such errors are readily calculated by the methods of Chap. 2. The circuit of Fig. 5.21, for example, is an ordinary operational amplifier with the switch OFF; with the switch ON, $X_0 = -(Z_0/Z_1)X_1 + e$. The error $e = e(t)$ is given by

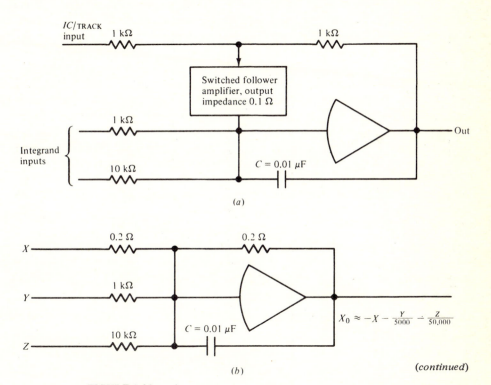

(a)

(b)

(continued)

FIGURE 5.20

(a) Two-mode integrator or sample-hold circuit; (b) equivalent circuit in the TRACK/IC mode; (c) switched follower amplifier. The switched follower not only has very low output impedance, but employs a switched emitter follower output stage to avoid saturation of the switching transistors, which pass 30 mA when ON. Output impedance is 0.1 Ω at 100 Hz, and output current is 30 mA. The integrator switches into COMPUTE/HOLD within 40 ns, with ±10-ns maximum difference between two integrators. Output impedance is 0.1 Ω at 100 Hz; output current is 30 mA. COMPUTE/HOLD drift in 10 ms (0.01 μF) is 2 mV ± 0.05 mV/°C. In RESET/TRACK with 0.01 μF, the integrator tracks ±10-V sine waves up to 40 kHz and follows a 10-V step within 0.1 percent in 10 μs; small-signal dynamic error is within 0.1 percent up to 10 kHz (no load), and temperature drift is ±100 μV/°C (University of Arizona and Burr-Brown Research Corporation, Tucson, Ariz.; based on Ref. 1).

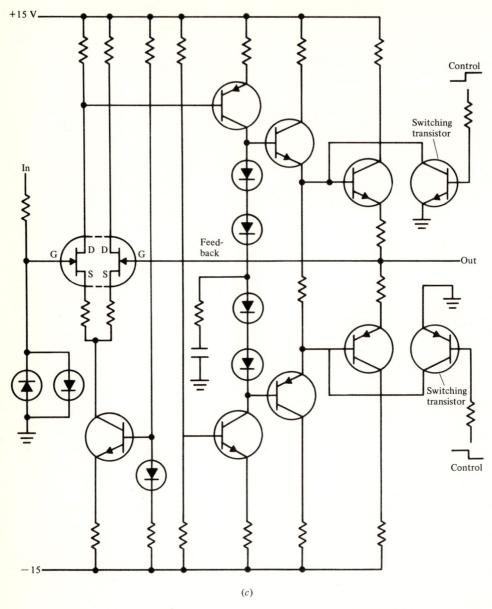

(c)

FIGURE 5.20   (*continued*)

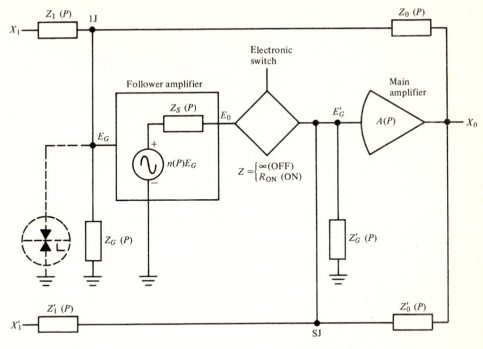

FIGURE 5.21
Switched operational amplifier (Ref. 9).

$$e \approx -\left(\frac{1}{A_{ON}\beta_{ON}}\frac{Z_0}{Z_1}X_1 + \frac{\beta_1}{\alpha\beta_{ON}}\frac{Z_0}{Z_0'}X_0'\right)$$

$$A_{ON} = \frac{\alpha A}{1 - A\beta_1} \qquad \beta_{ON} = \left(1 + \frac{Z_0}{Z_1} + \frac{Z_0}{Z_G}\right)^{-1} \qquad \beta_1 \approx \frac{Z_S + R_{ON}}{Z_0'}$$

(5.7)

at frequencies where $|A_{ON}\beta_{ON}| \gg 1$ and $|Z_S + R_{ON}| \ll |Z_0|,\ |Z_1'|,\ |Z_G'|$. Note that the second error term makes this one-switch circuit less practical than multiswitch circuits if low computing impedances are to be used.

If the switch in Fig. 5.21 is enclosed in the follower-amplifier feedback loop in the manner of Fig. 5.20, then $Z_S + R_{ON}$ in Eq. (5.7) is simply replaced by Z_S.

REFERENCES AND BIBLIOGRAPHY

1 KORN, G. A., and T. M. KORN: "Electronic Analog and Hybrid Computers," 2d ed., McGraw-Hill, New York, 1972.

2 HOESCHELE, D. G.: "Analog-to-Digital/Digital-to-Analog" Conversion Techniques, Wiley, New York, 1968.

3 SCHMID, H.: "Electronic Analog/Digital Conversions," Van Nostrand Reinhold, New York, 1970.

4 GRAEME, J. G., ET AL.: "Operational Amplifiers," McGraw-Hill, New York, 1971.

5 COHEN, J.: Solid-state Signal Switching, *EDN*, Nov. 15, 1972.

6 ——: Sample-and-Hold Circuits using FET Analog Gates, *EEE*, January 1971.

7 EVANS, A. D.: The Basics of Using FETs for Analog-Signal Switching, *EDN*, May 20, 1973.

8 ALTMAN, L.: CMOS Forges Monolithic Analog Gate, *Electronics*, Apr. 26, 1973.

9 KORN, G. A.: Performance of Operational Amplifiers with Electronic Mode Switching, *IEEETEC*, June 1963.

10 KEAY, C. S., and J. A. KENNEWELL: D/A Converter Switches Inputs with TTL Gates, *Electronics*, Dec. 8, 1969.

11 MARJANOVIC, S., ET AL.: Analog Section of Hybrid Computing System HRS-100, *Proceedings of the 7th AICA Conference*, Prague, 1973.

12 DOSTAL, J.: Three-mode Integrator, *Proceedings of the 7th AICA Conference*, Prague, 1973.

6

DIGITAL-TO-ANALOG AND
ANALOG-TO-DIGITAL CONVERTERS†

6.1 DIGITAL REPRESENTATION OF VOLTAGES;
SIMPLE CONVERSION CIRCUITS

Digital computing devices represent each variable or parameter by the discrete states of switching devices. As an example, the numbers $0, 1, 2, \ldots, 9$ could be represented by the 10 possible states (positions) of a 10-position star switch, or by 9 of the 16 possible combined states of 4 ON/OFF or SPDT switches.

The latter representation turns out to be practical and convenient. We therefore commonly represent a nonnegative voltage as a *binary fraction*

$$Y_D = \frac{1}{2}a_1 + \frac{1}{4}a_2 + \cdots + \frac{1}{2^n}a_n \qquad (6.1)$$

$(0 \leqslant Y_D < 1)$ of a *reference voltage* X_1 (for example, 10 V or 100 V), where $a_1, a_2, \ldots, a_n$ are *binary digits* (*bits*), each capable of taking the two values 0 and 1. The value

†Portions of Chap. 6 are based on material taken from Chap. 5 of G. A. Korn and T. M. Korn, "Electronic Analog and Hybrid Computers," 2d ed., McGraw-Hill, 1972, by permission of the authors and publisher.

Table 6.1 BINARY CODES REPRESENTING REAL FRACTIONS X BY n-BIT WORD: $(a_0, a_1, a_2, \ldots, a_{n-1})$*

Each binary digit a_k is either 0 or 1. a_0 is the most significant bit (MSB), and a_{n-1} is the least significant bit (LSB).

---

1 **Nonnegative fractions:** $X = \dfrac{1}{2}a_0 + \dfrac{1}{2^2}a_1 + \cdots + \dfrac{1}{2^n}a_{n-1}$, where $0 \leqslant X \leqslant 1 - \dfrac{1}{2^n}$.

EXAMPLE (3 bits):

Decimal	Binary		Decimal	Binary
0.000	000		$\frac{4}{8} = 0.500$	100
$\frac{1}{8} = 0.125$	001		$\frac{5}{8} = 0.625$	101
$\frac{2}{8} = 0.250$	010		$\frac{6}{8} = 0.750$	110
$\frac{3}{8} = 0.375$	011		$\frac{7}{8} = 0.875$	111

2 **Signed fraction (positive, negative, or zero).** The sign bit a_0 is 0 for $X \geqslant 0$ and 1 for $X < 0$.

(*a*) **Sign-and-magnitude code:** $X = (-1)^{a_0}\left(\dfrac{1}{2}a_1 + \dfrac{1}{2^2}a_2 + \cdots + \dfrac{1}{2^{n-1}}a_{n-1}\right)$, where

$\dfrac{1}{2^{n-1}} - 1 \leqslant X \leqslant 1 - \dfrac{1}{2^{n-1}}$. There are *two* binary representations of 0: 000 . . . and 100 . . . . This may cause complications, e.g., in statistical work.

(*b*) **One's-complement code:**

$$X = \left(\dfrac{1}{2^{n-1}} - 1\right)a_0 + \dfrac{1}{2^2}a_2 + \cdots + \dfrac{1}{2^{n-1}}a_{n-1}$$

where $\dfrac{1}{2^{n-1}} - 1 \leqslant X \leqslant 1 - \dfrac{1}{2^{n-1}}$. There are *two* binary representations of 0: 000 . . . and 111 . . . . *One obtains the code for* $-X$ *very simply by complementing each bit.* This code is used in some arithmetic units.

(*c*) **Two's-complement code:**

$$X = -a_0 + \dfrac{1}{2}a_1 + \dfrac{1}{2^2}a_2 + \cdots + \dfrac{1}{2^{n-1}}a_{n-1}$$

where $-1 \leqslant X \leqslant 1 - \dfrac{1}{2^{n-1}}$. It has a unique 0 and simple arithmetic. *To obtain code for* $-X$, *complement every bit and add 1 LSB.* This code is used in almost all minicomputers.

EXAMPLES (4-bit codes):

Decimal	Sign and magnitude				One's-complement				Two's-complement			
$+\frac{7}{8} = +0.875$	0	1	1	1	0	1	1	1	0	1	1	1
$+\frac{6}{8} = +0.750$	0	1	1	0	0	1	1	0	0	1	1	0
	. . .				. . .				. . .			
$+\frac{3}{8} = +0.375$	0	0	1	1	0	0	1	1	0	0	1	1
$+\frac{2}{8} = +0.250$	0	0	1	0	0	0	1	0	0	0	1	0
$+\frac{1}{8} = +0.125$	0	0	0	1	0	0	0	1	0	0	0	1
$+0$	0	0	0	0	0	0	0	0	0	0	0	0
-0	1	0	0	0	1	1	1	1				
$-\frac{1}{8} = -0.125$	1	0	0	1	1	1	1	0	1	1	1	1
$-\frac{2}{8} = -0.250$	1	0	1	0	1	1	0	1	1	1	1	0
$-\frac{3}{8} = -0.375$	1	0	1	1	1	1	0	0	1	1	0	1
	. . .				. . .				. . .			
$-\frac{6}{8} = -0.750$	1	1	1	0	1	0	0	1	1	0	1	0
$-\frac{7}{8} = -0.875$	1	1	1	1	1	0	0	0	1	0	0	1
-1									1	0	0	0

---

*From Korn, G. A.: *Minicomputers for Engineers and Scientists*, McGraw-Hill, New York, 1973.

(6.1) of the n-bit binary fraction is then given by the states of n binary switching devices (for example, n toggle switches or a flip-flop register); thus, 011 stands for the fraction $\frac{3}{8}$, 100 stands for $\frac{4}{8} = \frac{1}{2}$, etc. Many other binary *codes* relating the sequence $(a_1, a_2, \ldots, a_n)$ to a fraction are possible, but (6.1) is the most frequently employed coding scheme.

If we want to represent *negative as well as positive fractions* Y_D, we can add a *sign bit* a_0 and use the n-bit coding scheme defined by

$$Y_D = (1 - 2a_0)\left(\frac{1}{2}a_1 + \frac{1}{4}a_2 + \cdots + \frac{1}{2^{n-1}}a_{n-1}\right) \qquad (6.2)$$

which is similar to (6.1), except that a sign bit $a_0 = 1$ acts like a minus sign and reverses the sign of our fraction (*sign-and-magnitude code*). Unfortunately, we now have *two* codes for $Y_D = 0$ (000 ... and 100 ...).

A binary code more useful for digital counting and arithmetic is the *two's-complement code*, which has a unique zero 000 ... :

$$Y_D = -a_0 + \frac{1}{2}a_1 + \frac{1}{4}a_2 + \cdots + \frac{1}{2^{n-1}}a_{n-1} \qquad (6.3)$$

Table 6.1 further illustrates binary-fraction codes. Many other codes exist (see also Sec. 6.2).

A *digital-to-analog converter (DAC)* is a switched resistance network (possibly also including amplifiers) which produces the output voltage $X_1 Y_D$, where X_1 is an analog input (*reference voltage*) and Y_D is a fraction determined by the state of a set of switches, usually binary (ON/OFF or SPDT) switches. Thus, the simple binary-code DAC of Fig. 6.1 multiplies its analog input voltage X_1 by the n-bit nonnegative fraction

$$Y_D = \frac{1}{2}a_1 + \frac{1}{4}a_2 + \cdots + \frac{1}{2^n}a_n \qquad (6.4)$$

The kth switch is ON if the corresponding binary digit a_k is 1, so that $0 \leqslant Y_D < 1$. a_1 is the *most significant bit (MSB)* of Y_D, and a_n is the *least significant bit (LSB)*. The analog input X_1 may be a fixed reference voltage (*fixed-reference DAC*) or a variable (*multiplying DAC, or MDAC*). MDACs are more difficult to design and more expensive because switch-resistance effects, and possibly switch offset voltages, will vary with X_1.

Most frequently, a flip-flop register, which can be set by parallel bit lines from a digital computer, stores the bit values $a_0, a_1, a_2, \ldots$ and produces bit voltages (for example, 0 V for $a_i = 0$ and +4 V for $a_i = 1$) which keep the correct switches turned OFF and ON. In Fig. 6.2a, n-bit lines from a digital computer set a DAC register so that the output voltage X_0 is proportional to a desired "digital input" Y_D.

Figure 6.2b illustrates *double-buffered* DAC operation. It is often desirable to "update" two or more DAC outputs *simultaneously*, i.e., to switch these voltages simultaneously to new values. But most digital computers can output *only one number at a time*. In Fig. 6.2b, therefore, *buffer registers* no. 1 and no. 2 are loaded *successively* by the computer. Afterward, a common transfer pulse transfers the bit values in each buffer register to the corresponding DAC register, so that both DACs are updated together.

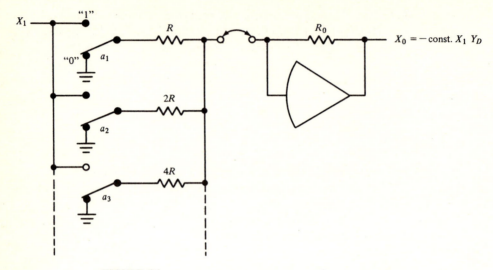

FIGURE 6.1
Simple digital-to-analog converter (DAC) for n-bit nonnegative binary numbers

$$Y_D = \frac{1}{2} a_1 + \frac{1}{4} a_2 + \cdots + \frac{1}{2^n} a_n$$

where $a_k = 0$ or 1 ($k = 1, 2, \ldots, n$). The kth switch is ON if the corresponding binary digit a_k is 1. The circuit has *constant source impedance*, so that it could drive a simple load resistance r_L instead of an operational amplifier summing point.

6.2 PRACTICAL DIGITAL–TO–ANALOG CONVERTERS; VOLTAGE–LADDER CIRCUITS

Conversion of Positive and Negative Numbers

The simple digital-to-analog converter of Fig. 6.1 was designed to convert *nonnegative* binary numbers. Such circuits will, however, correctly convert both negative and positive numbers expressed in the *sign-and-magnitude code* (Sec. 6.1) if we add a *sign-bit switch* which switches either the analog input X_1 or the output X_9 through a unity gain inverter whenever the *sign bit* a_0 is 1.

Figure 6.3 illustrates the conversion of a *positive or negative two's-complement–coded n-bit binary fraction*

$$Y_D = -a_0 + \frac{1}{2} a_1 + \frac{1}{4} a_2 + \cdots + \frac{1}{2^{n-1}} a_{n-1}$$

Note, first, that four-quadrant multiplication of the two's-complement–coded digital input Y_D and the analog input X_1 requires both $-X_1$ and X_1 as input voltages. The most

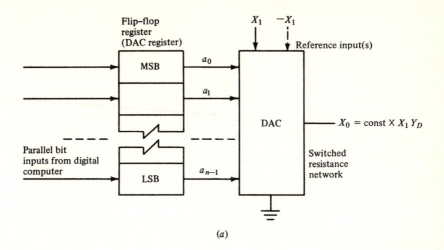

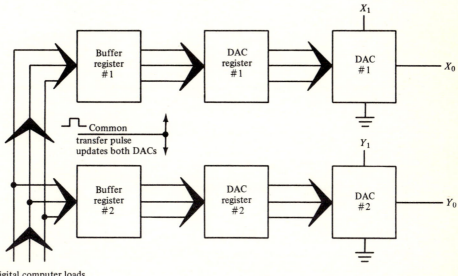

FIGURE 6.2

(a) DAC with flip-flop register; (b) two double-buffered DACs updated by the same transfer pulse. The common transfer pulse in b updates both DACs simultaneously.

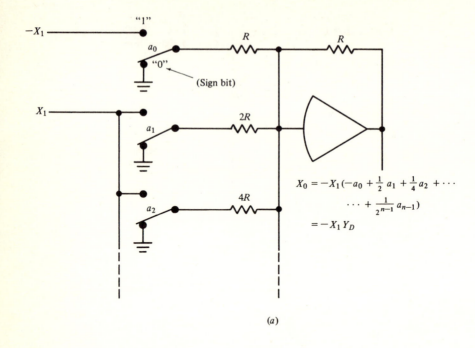

$$X_0 = -X_1\left(-a_0 + \frac{1}{2}a_1 + \frac{1}{4}a_2 + \cdots\right.$$
$$\left.\cdots + \frac{1}{2^{n-1}}a_{n-1}\right)$$
$$= -X_1 Y_D$$

(a)

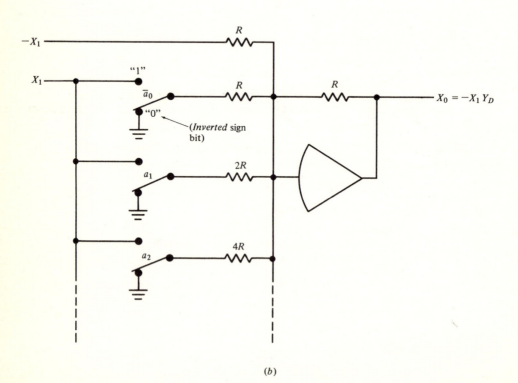

$X_0 = -X_1 Y_D$

(b)

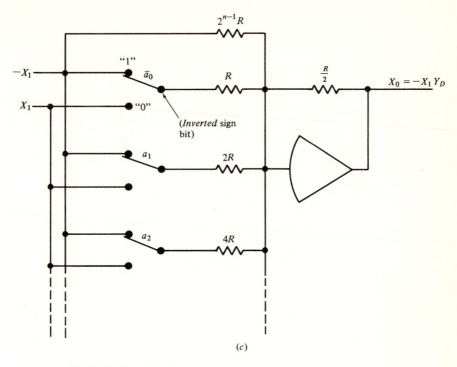

(c)

FIGURE 6.3
Three different methods for converting two's-complement-coded positive or negative binary fractions

$$Y_D = -a_0 + \frac{1}{2} a_1 + \frac{1}{4} a_2 + \cdots + \frac{1}{2^{n-1}} a_{n-1}$$

If X_1 is a fixed reference voltage, the scheme of Fig. 6.3b requires only switches passing current in one direction. All three schemes work just as well with ladder networks.

significant bit (MSB) will be the sign bit a_0, and the least significant bit (LSB) is a_{n-1}. Note also that $-1 \leqslant Y_D < 1$. Starting at -1, Y_D would increase as follows:

$$100 \cdots 00 = -1$$
$$100 \cdots 01 = -(1 - 2^{1-n})$$
$$100 \cdots 10$$
$$\cdots\cdots\cdots\cdots$$

$$\underline{111 \cdots 11 = -2^{1-n}}$$

(Two's-complement code)

$$\underline{000 \cdots 00 = 0}$$

$$000 \cdots 01 = 2^{1-n}$$
$$000 \cdots 10$$
$$\cdots\cdots\cdots\cdots$$

$$011 \cdots 11 = 1 - 2^{1-n}$$

The use of the inverted sign bit in Fig. 6.3b and c ($\bar{a}_0 = 0$ if $a_0 = 1$ and vice versa) is sometimes referred to as an *offset-binary code*. The scheme of Fig. 6.3c is used in some multiplying digital-to-analog converters. Note that the bit switches no longer multiply by $\bar{a}_0, a_1, a_2, \ldots$, but by $(2\bar{a}_0 - 1), (2a_1 - 1), (2a_2 - 1), \ldots$, so that Y_D is in the form

$$Y_D = \frac{1}{2}\left[-(2a_0 - 1) + \frac{1}{2}(2a_1 - 1) + \frac{1}{4}(2a_2 - 1) + \cdots + \frac{1}{2^{n-1}}(2a_{n-1} - 1) - \frac{1}{2^{n-1}}\right]$$

(6.5)

Table 6.2 **BINARY–NUMBER REPRESENTATION OF VOLTAGES**

Unsigned octal numbers*	Voltage
0000	0
0001	0.00244140625
0002	0.0048828125
0004	0.009765625
0010	0.01953125
0020	0.0390625
0040	0.078125
0100	0.15625
0200	0.3125
0400	0.625
1000	1.25
2000	2.5
4000	5
6000	7.5
7000	8.75
7400	9.375
7600	9.6875
7700	9.84375
7740	9.921875
7760	9.9609375
7770	9.98046875
7774	9.990234375
7776	9.9951171875
7777	9.99755859375
(10000)	(10)

*Each octal digit may be regarded as *a shorthand notation for three binary digits*, thus:

Octal	Binary	Octal	Binary
0	000	4	100
1	001	5	101
2	010	6	110
3	011	7	111

Tee-Network Digital-to-Analog Converters

If more than 5 or 6 bits are needed, the resistance $2^k R$ in the simple current-summing networks of Figs. 6.2 and 6.3 becomes uncomfortably large. One way to reduce the network impedance levels when an operational amplifier does the current summing (as in Figs. 6.2 and 6.3) is to replace each resistance $2^k R$ by a corresponding *tee-network transfer impedance* (Sec. 1.5). Figure 5.16 shows a tee-network bit-switching circuit designed for fast, calibration-free MDACs (Ref. 1).

While the requirement of a given static accuracy at low cost favors *high* DAC-network resistances to make switch resistances less critical, each DAC must settle to within the specified accuracy after switching, and the required settling time will be proportional to some circuit resistance (Fig. 6.12). The sum of this settling time and the (usually smaller) switching time constitutes the conversion time, whose reciprocal is the maximum conversion rate still permitting the specified accuracy. High conversion rates, and also low MDAC phase shift, favor *low* DAC-network resistances. Hence DAC design will require a compromise between accuracy, conversion rate and/or phase shift, and switch costs (see also Sec. 6.5).

Digital-to-analog converters employing *switched-capacitance networks* to decode digital data arriving in serial form (i.e., bit by bit) are described in Refs. 2 and 3.

Voltage-Ladder Digital-to-Analog Converters

The basic *binary voltage-ladder network* illustrated in Fig. 6.4 is equivalent to a voltage source of internal impedance R and source voltage

$$\frac{1}{2}a_1 + \frac{1}{2^2}a_2 + \cdots + \frac{1}{2^n}a_n$$

at the terminal Q, where each bit value a_k is either 0 or 1. This fact is easily proved by the successive Thévenin-equivalent circuits shown in Fig. 6.4b, c, and d. Figure 6.5 illustrates how binary voltage ladders can be connected for conversion of two's-complement–coded

---

Table 6.3 TYPICAL DIGITAL–TO–ANALOG CONVERTERS (BASED ON REF. 1)*

---

1 Low-cost: 8 to 12 bits, temperature range 0 to 50°C (or −50 to 150°C, more expensive); settling time 0.3 to 10 μs. Resistances 10 to 500 kΩ.

2 High-quality, designed for slow conversion: to 15 bits. 0 to 50°C; settling time 50 μs. Fixed-reference DACs for digital voltmeters, precision instruments; MDACs of this type serve as *digital attenuators* in hybrid computers; switch only *between* analog-computer runs. Available for ±100 V. Resistances 100 to 500 kΩ.

3 Fast, high-quality: 14 bits, 0 to 50°C; settling time 0.5 to 5 μs. For hybrid-computer applications requiring fast conversion *during* analog computer runs. Expensive. Available for ±100 V. Resistances 5 to 20 kΩ.

4 Very fast, low-resolution: 4 to 8 bits, −50 to 150°C settling times below 0.05 μs for communications applications. Resistances 1 to 10 kΩ.

---

*Accuracy within $\frac{1}{2}$ LSB; settling times to within static accuracy; do not include output-amplifier settling. Output range is ±10 V unless noted.

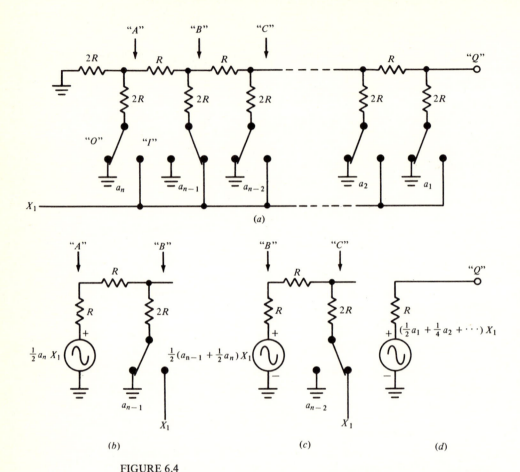

FIGURE 6.4
(a) A binary ladder network; (b) to (d) Thévenin-equivalent circuits for successive sets of ladder sections, showing how successive inputs are effectively divided by powers of 2. *Circuit c is the equivalent-voltage-source circuit for the complete ladder.* If such a ladder network is loaded with 2R, the impedance between each ladder switch and ground is 3R.

positive or negative numbers; the two circuits shown are, respectively, analogous to Fig. 6.3a and b. The design of a ladder circuit analogous to Fig. 6.3c is left as a problem.

Ladder-network DACs necessarily require SPDT switches but need fewer resistors than tee-network DACs; above all, binary ladders require *only two different resistance values of the same order of magnitude* and can, in fact, be made up of resistances R or 2R alone. This greatly simplifies close matching of film-resistor temperature coefficients; indeed, the entire ladder network can be deposited on a substrate by thick-film or thin-film techniques.

Figure 6.6 shows practical bipolar-transistor *bit switches* (SPDT switches) suitable for voltage-ladder switching. The ON resistances of such switches (5 to 15 Ω, Sec. 5.4) are within 0.01 percent of 200-kΩ ladder resistances 2R. With smaller ladder resistances,

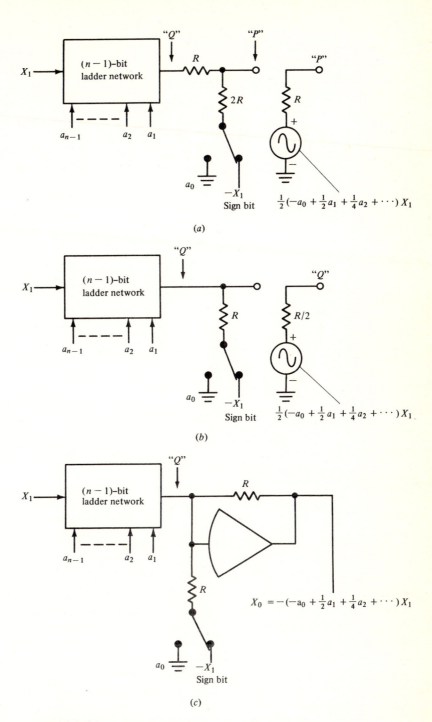

FIGURE 6.5

Ladder circuits for converting two's-complement–coded positive or negative quantities. Equivalent-voltage-source circuits are shown for (*a*) and (*b*).

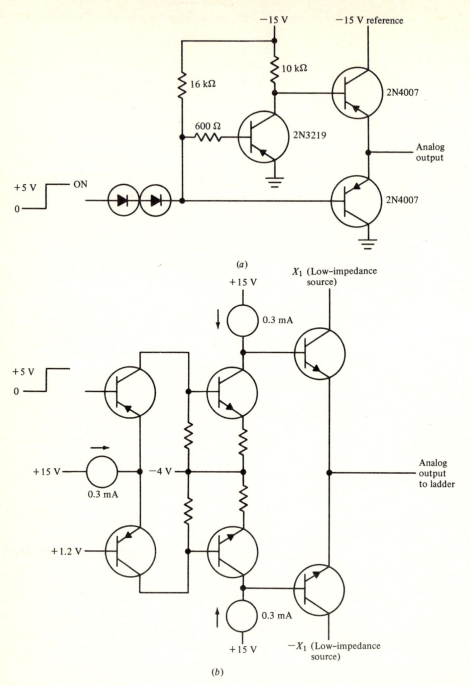

FIGURE 6.6

(a) Fixed-reference voltage-ladder switch; (b) a polarity-changing voltage-ladder switch for a low-cost (integrated-circuit) MDAC. Analog input voltages must be offset to correct for bipolar-transistor saturation voltage. This cannot be done much better than 0.1 percent of half-scale when the analog input varies, so that more accurate MDACs usually employ FET switches with trimmed ladder resistors (*Crystalonics, Inc., and Analog Devices, Inc.*).

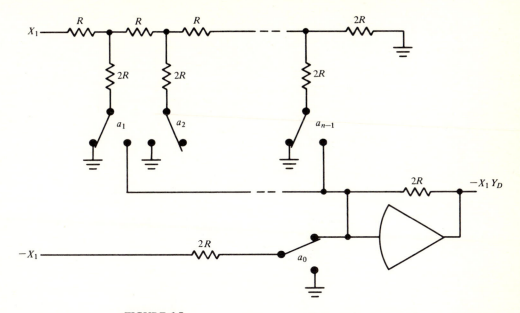

FIGURE 6.7
This "inverted" binary ladder works nicely with the FET/bipolar SPDT switch of Fig. 5.14. The $2R$ resistances in series with FET switches can be trimmed to account for the FET ON resistance, and the $2R$ feedback resistance can include a FET for temperature compensation as in Fig. 5.14a.

tighter accuracy specifications, and/or FET bit switches (Sec. 5.5), the $2R$ ladder resistances must be trimmed to account for the switch ON resistances. A common technique is to specify $2R$ resistances 30 Ω too low and then to select switching FETs with 30 $\Omega \pm 3$ Ω ON resistances.

The "inverted" binary ladder circuit of Fig. 6.7 permits low-level FET switching in the manner of Fig. 5.9. Note the compensation of FET ON resistance changes with temperature by a FET in the output-amplifier feedback circuit.

Figure 6.8 is an example of a voltage-ladder DAC converting a *binary-coded-decimal (BCD) code*, with

$$Y_D = \frac{1}{2}a_1 + \frac{1}{4}a_2 + \frac{1}{8}a_3 + \frac{1}{16}a_4 + \frac{1}{10}\left(\frac{1}{2}b_1 + \frac{1}{4}b_2 + \frac{1}{8}b_3 + \frac{1}{16}b_4\right)$$

$$+ \frac{1}{100}\left(\frac{1}{2}c_1 + \frac{1}{4}c_2 + \frac{1}{8}c_3 + \frac{1}{16}c_4\right) \qquad (6.6)$$

In such a BCD code, each *decimal* digit of Y_D is separately represented in binary form. Each group of 4 bits can, for instance, operate one decimal digit of a decimal-number display. The operation of such networks is readily analyzed with the aid of Thévenin's theorem in the manner of Fig. 6.4 (Prob. 6.6).

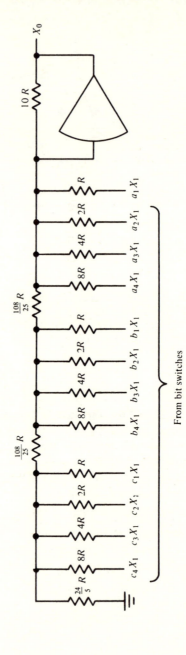

FIGURE 6.8

DAC circuit for converting binary-coded-decimal (BCD) numbers. Many other schemes exist.

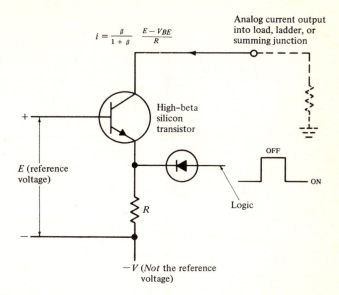

$$i = \frac{\beta}{1 + \beta} \frac{E - V_{BE}}{R}$$

Analog current output
into load, ladder, or
summing junction

High–beta
silicon
transistor

OFF

E (reference
voltage)

ON

R

Logic

−V (*Not* the reference
voltage)

FIGURE 6.9
A simple switched current source suitable for fixed-reference DACs. Monolithic integrated-circuit construction is possible. The V_{BE} temperature dependence should be compensated in the source of the reference voltage E, or through a reference current source in the manner of Fig. 6.10. Reverse the junctions and bias for negative current sources.

6.3 CURRENT-SWITCHING DACs

A *bipolar-transistor current switch* like that shown in Fig. 6.9 does not depend on transistor saturation to turn ON. Such a switch operates as a feedback-controlled current source in the "working region" of its characteristics when it is ON. To turn the switch OFF, the logic waveform opens an alternate current path through a diode or transistor ("current stealing") and cuts the current-source transistor OFF. Compared to voltage switching, current switching has two significant advantages:

1 With reasonably high current gain in the current-source feedback circuit, the output current does not depend on switch-transistor forward resistance.
2 Current switches turn OFF quickly because there is no need to get a transistor out of saturation.

Switched current sources can be used as operational amplifier inputs in DAC circuits similar to Figs. 6.2 and 6.3, or in current-ladder circuits like that shown in Fig. 6.11a which, like binary voltage ladders, employ only two different resistance values. Current switches are employed mainly in fixed-reference DACs, because the inherent high impedances of current sources tend to cause analog-input phase shift.

The output of the simple switched current sources in Fig. 6.9 depends, unfortunately, on transistor V_{BE} and thus on temperature, so that accurate current-

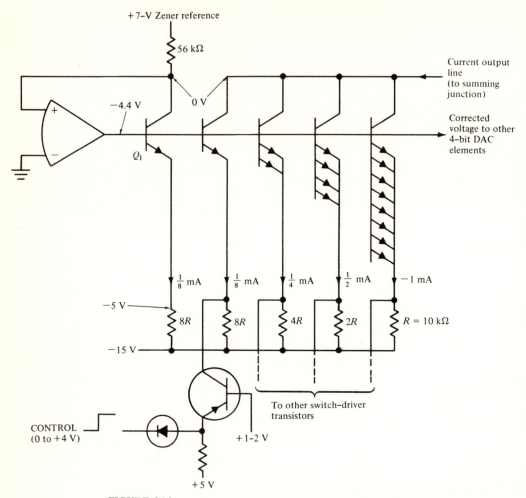

FIGURE 6.10
Trimmer-free four-bit current-switching DAC element for a 12-bit fixed-reference DAC (Fig. 6.11*b*). A feedback loop including the extra current source Q_1 sets all current-source base voltages to compensate for V_{BE} and β changes with temperature and power supply. The four current sources shown correspond to the most significant bits and are deposited on the same substrate as Q_1 for V_{BE} tracking within 500 μV between −50 and +120°C. Similarly good β tracking is achieved by doubling the emitter areas of successive current sources for equal current densities. This integrated circuit, using *n-p-n* current-source transistors, permits faster switching than earlier monolithic circuits using *p-n-p*'s (Analog Devices, Inc.).

switch DACs require some form of temperature compensation. In the ingenious current-switch network of Fig. 6.10, the ON current level in five closely matched transistors is determined by an operational amplifier feedback circuit. V_{BE} matching is close, since all transistors are deposited on the same monolithic integrated-circuit chip. Transistors designed to pass successive bit currents of $\frac{1}{8}$, $\frac{1}{4}$, $\frac{1}{2}$, and 1 mA are given successively doubled emitter areas, so that similarly good current-gain (β) tracking is obtained. Figure 6.11*b* shows how several 4-bit DAC elements of this type are combined into accurate 8- or 12-bit DACs; while only the first 4 bits are obtained from the same integrated circuit, the less significant bits need not be matched as accurately.

6.4 STATIC-ERROR SPECIFICATIONS

The *resolution* of a digital-to-analog converter is given by the number of bits. Conversion *accuracy* should be within the voltage limits determined by one-half of the least significant bit ($\frac{1}{2}$ LSB). The DAC analog output must always increase (or always decrease) with the digital input (*monotonicity*).

DAC static errors, which result from *network-resistance tolerances, switch resistances, switch offsets,* and *reference-voltage errors,* must be figured *on a worst-case basis for all digital inputs, working temperatures,* and *power-supply-voltage limits.* In general, the sign bit and the most significant bits will contribute relatively larger switch-resistance and switch-offset errors than the less significant bits. Reference 2 contains a number of carefully detailed worst-case design examples (see also Prob. 6.5).

6.5 A DISCUSSION OF DAC SETTLING TIME; SPIKES AND GLITCHES

To maintain a specified data rate (samples per second), a digital-to-analog converter must settle to within a specified accuracy in a specified time (*settling time*). Switched-resistance networks, like those used in DACs, involve distributed circuit capacitances between 2 and 30 pF and will exhibit exponential step responses approximated by those of the simple network in Fig. 6.12. We see that *better accuracy will require relatively longer settling times.* On the other hand, *DAC network resistances which are large compared to semiconductor-switch resistances favor static accuracy but increase the settling time constant.* Thus, for any given accuracy, faster conversion will require lower network impedances, which may have to be matched to individual bit switches at substantial expense.

Analog-input phase-shift errors in multiplying digital-to-analog converters (MDACs) are also improved by low network impedances and can be further reduced by equalizing capacitors in the input or feedback networks. Phase-shift errors within 0.1 percent up to 10 kHz are possible, but most commercially available MDACs are much slower.

Amplifier Settling Time

Most digital-to-analog converters involve an output amplifier (phase inverter or follower amplifier) whose step response will crucially affect the DAC settling time. The correct specification of DAC settling time must, then, refer to the time required for DAC *and*

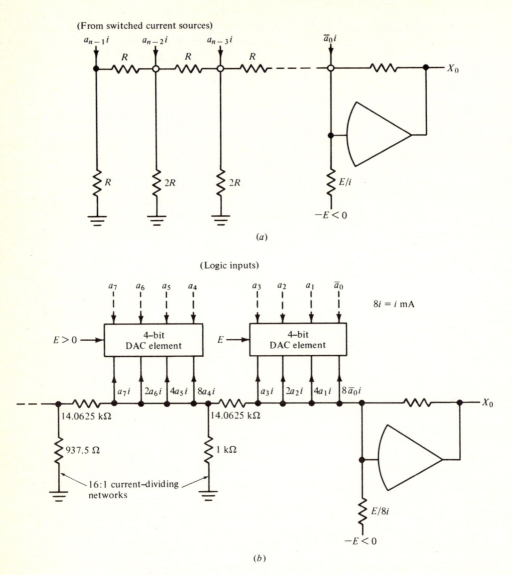

FIGURE 6.11
Fixed-reference DACs employing switched current sources. The sign-changing scheme of Fig. 6.11*b* permits the use of similar one-directional current sources. Each a_k is either 0 or 1, and E is a fixed reference voltage.

amplifier to settle within a specified accuracy after large or small steps. Amplifier step response involves a short *propagation time* and, at least after large steps, a *slewing period* in which the essentially overloaded amplifier supplies the maximum possible current to load and circuit capacitances. This is followed by a *linear transient* (often with a small overshoot), which must be controlled by careful amplifier equalization. Typical

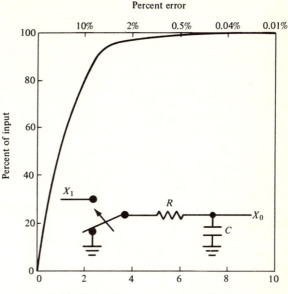

FIGURE 6.12
The step response

$$X_0 = X_1 (1 - e^{-1/RC})$$

of a switched RC circuit (time constant RC) indicates the settling times needed
for given accuracy.

DAC/amplifier settling times range from 30 ns (0.1 to 0.01 percent of half-scale) to
slower and more accurate DACs settling to within 0.01 percent of half-scale within 1 to 4
μs (see also Fig. 6.13).

Spikes and Glitches

All digital-to-analog converters will exhibit *switching spikes* due to switch capacitances
(Sec. 5.9). Spike amplitude will be reduced by slow output amplifiers, but fast DACs can
exhibit spike amplitudes much larger than one LSB. More serious transient disturbances
than capacitive spikes are the so-called *glitches,* which arise from the fact that many DAC
voltage changes involve operations of multiple bit switches with unequal turnon and
turnoff times. Since one bit switch may turn OFF more quickly than another turns ON,
the DAC output can momentarily take a completely false value during the switching
interval. Such multiple switch operations take place, for instance, on each carry when a
DAC is driven by a counter (Fig. 6.14). Glitches, like spikes, are reduced by faster
switches, but glitches cannot be completely eliminated in fast DACs. In applications
where a smooth DAC output waveform is crucial, as in the case of DACs driving

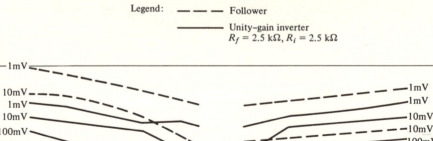

FIGURE 6.13
"V curves" showing settling times to within a specified number of millivolts for a fast FET-input operational amplifier (Analog Devices, Inc.; based on Ref. 16).

cathode-ray-tube displays, it is therefore necessary to follow the digital-to-analog converter with a sample-hold circuit ("deglitcher").

6.6 ANALOG–TO–DIGITAL CONVERSION

Analog-to-digital converters (ADCs, encoders) produce a digital output proportional to an analog input voltage X_A or (usually) to the ratio of an analog input X_A and an internal or externally supplied *reference voltage E*. As a rule, each conversion is started and timed by a *CONVERT command* in the form of a pulse or level change supplied by a switch, digital computer, timer, or sensing device. Each conversion will require a finite time, during which the input may or may not have to be "held" by a sample-hold circuit (Sec. 5.8). End of conversion is usually indicated by a level change in the output of a flip-flop *(DONE or CONVERSION COMPLETED flag)*. This is needed because the converter output should not be read or transferred to a computer before the conversion is finished.

The basic ingredients of all electronic ADCs are *analog comparators* (Sec. 3.2), whose binary output indicates the sign of an analog input sum or difference.

6.7 PARALLEL–COMPARATOR
AND CASCADE–ENCODER ADCs

The simplest *analog-to-digital converters* employ multiple parallel comparators, one for each digital output level (Fig. 6.15). This requires 2^n comparators for *n*-bit conversion and is, therefore, not practical if high resolution is needed. But parallel-comparator ADCs

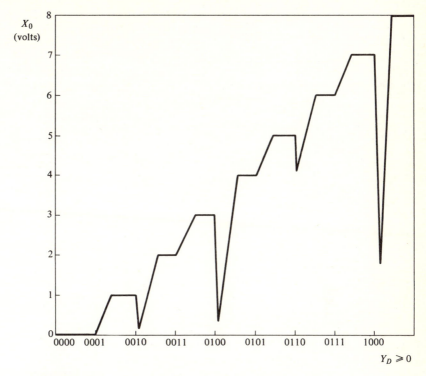

FIGURE 6.14
Output of a counter-driven fast digital-to-analog converter showing glitches associated with each counter carry. No capacitive switching spikes are shown (Analog Devices, Inc.; based on Ref. 16).

are the fastest of all, since all digits are produced simultaneously. Three- to four-bit converters of this type are available as monolithic integrated circuits with 5- to 20-ns conversion times.

A *cascade encoder* (Ref. 17) reduces the required number of comparators but produces binary digits *in succession*. Referring to Fig. 6.16a with $X_A > 0$, comparator 1 decides whether $X_A > E/2$. If $X_A > E/2$, comparator 2 decides whether $X_A > E/2 + E/4$; if $X_A < E/2$, comparator 2 decides whether $X_A > E/4$, etc. Cascade encoders can be fast: the converter shown in Fig. 6.16a requires one clock period per bit, but improved circuits separate the converter stages with analog delay lines one clock period (0.1 μs) long, so that the earlier stages can begin to work on the next analog sample before each conversion is completed (Ref. 3).

Figure 6.16b shows a ±10-V cascade encoder based on the precision absolute-value circuit of Fig. 3.26 (Refs. 18 and 19). The *first encoder stage* produces a digital 1 if $X_A < 0$ (sign bit) and also an analog output X_1 such that

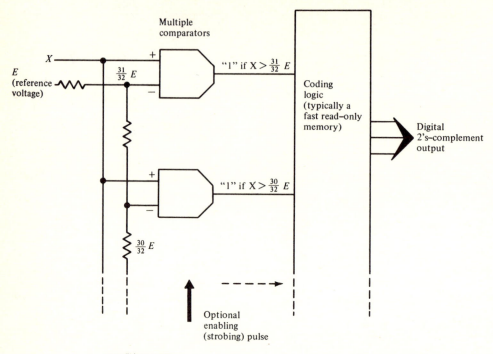

FIGURE 6.15
An "instantaneous" analog-to-digital converter employing multiple parallel comparators. Typical integrated-circuit comparators are simply differential amplifier stages with low accuracy (±100 to 300 mV) but high speed (5 to 20 ns). A timed strobing pulse can enable the comparators and will serve both as a sampling pulse and as the CONVERT command. The coding logic can be a monolithic read-only-memory (ROM) chip.

$$X_1 > 5\,\text{V} \quad \text{if} \quad \begin{cases} X_A > 5\,\text{V} \quad (X_A > 0) \\[2ex] X_A < -5\,\text{V} \quad (X_A < 0) \end{cases}$$

The *second encoder stage*, quite similar to the first, produces a digital 1 if $X_1 - 5\,\text{V} > 0$, and also a new analog output X_2 to be tested by the third encoder stage, and so on. The conversion time will be the time required for the signal to propagate through n stages, typically 0.2 μs per stage with fast operational amplifiers. Note that the code produced is not the usual binary code, but a "reflected" code (*Gray code*), which requires code conversion (Refs. 18 and 19). Many variations of this circuit exist (Refs. 3 and 19).

Cascade encoders are usually limited to about 8 bits because of error accumulation in successive stages.

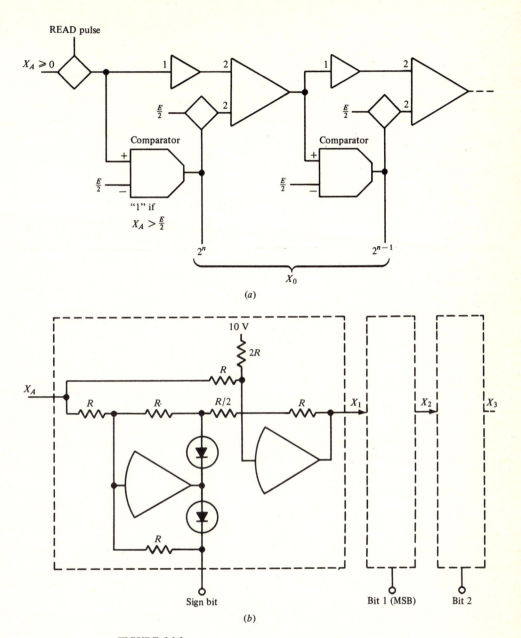

(a)

(b)

FIGURE 6.16
Cascade encoders employing (a) comparators and switches and (b) precision limiter absolute-value circuits to convert successive bits.

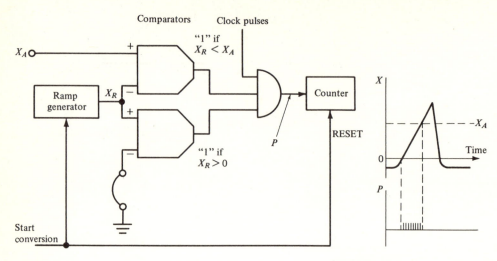

FIGURE 6.17
A simple ramp-comparator ADC.

6.8 RAMP–COMPARATOR CONVERTERS

A simple *ramp-comparator converter or analog-to-time-to-digital* converter gates clock pulses into a counter while an accurate ramp voltage varies between zero and the analog input $X_A > 0$ (Fig. 6.17). The counter, which serves as the ADC output register, is then read and reset to zero. The second comparator response can initiate a DONE (conversion completed) signal.

Either binary or decimal counters can be used. To read positive or negative voltages between −10 and +10 V, we can set the first comparator at −10 V instead of 0 V and employ a binary (two's-complement) counter initially reset to −10.00 instead of 0. The converter will then read the correct two's-complement code.

The two comparators and the output logic can be a single-chip integrated circuit, so that comparator levels are closely matched and will drift together with temperature changes.

6.9 INTEGRATING CONVERTERS

If a constant or variable voltage $X_A > 0$ (one could handle negative inputs by adding a known positive offset) is applied to the input of an electronic integrator initially reset to zero, the integrator output will take ERC/X_A s to reach the negative reference level $-E$. In *analog-to-frequency-converter ADCs* the integrator is then quickly reset to zero, and the process repeats (Fig. 6.18a). The average frequency of the reset pulses, measured by the pulse count during a specified *observation period T*, is approximately

$$\frac{N}{T} \approx \frac{1}{ERC} \bar{X}_A \qquad \bar{X}_A = \frac{1}{T} \int_0^T X_A(t)\, dt$$

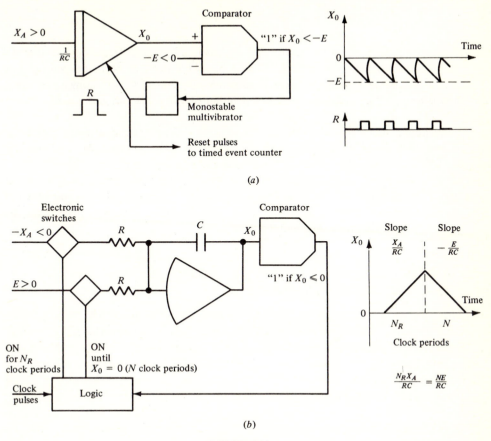

FIGURE 6.18
(a) Analog-to-frequency converter; (b) dual-slope converter.

where we have neglected the effect of the integrator-resetting time. This will be permissible if T_{RESET} is less than the desired measurement-error percentage of the integrator time constant RC, and $X_A \leqslant E$. In practice, T_{RESET} can be less than 10 μs, so that $RC = 10$ ms is large enough for 0.1 percent error (some of the reset-time error could be "calibrated out").

Reset-time errors could be reduced by employing two integrators, one integrating while the other one resets. A better voltage-to-frequency conversion scheme switches the input of a single integrator between X_A and $-X_A$ whenever the integrator output reaches $-E$ and $+E$. This method does away with reset switching and also compensates integrator dc offsets.

The *dual-slope integrating ADC* shown in Fig. 6.18b starts with its integrator output at 0 V and then integrates the analog input $-X_A < 0$ for N_R clock-pulse periods, or $T = N_R \, \Delta t$ seconds. Digital logic now switches the integrator input to a $+E$ reference, and comparator/logic circuits count N pulses while the integrator output returns to 0 V. Then

$$N = \frac{N_R}{E}\,\bar{X}_A \qquad \bar{X}_A = \frac{1}{T}\int_0^T X_A(t)\,dt$$

which is independent of the integrator RC and also of the clock frequency, so that no precision crystal oscillator is needed. Dual-slope ADCs serve in both inexpensive and precision (0.01 percent of half-scale $+\frac{1}{2}$ LSB) digital voltmeters, with conversion times between 0.1 and 50 ms. Bipolar operation is obtained with an input-voltage offset as in Sec. 6.8.

Integrating ADCs read the *time average* $\bar{X}_A$ of a possibly variable analog input X_A over a specified observation time T. *This is especially useful for rejecting periodic and nonperiodic high-frequency noise in low-level dc measurements.* In particular, line-frequency noise effects can be reduced by at least 50 dB if we make the observation interval T a multiple of the line-frequency period. It may also be possible to "synchronize" the observation time with other periodic noise. For *timed* measurements, integrating ADCs will require sample-hold circuits.

6.10 ANALOG–TO–DIGITAL CONVERTERS USING DIGITAL–TO–ANALOG FEEDBACK

For high accuracy and fast conversion independent of capacitor properties, one employs a *feedback digital-to-analog converter* which reconverts the digital output X_D to a voltage EX_D, where E is the *ADC reference voltage* (Fig. 6.19a). The feedback voltage EX_D is now compared with the analog input X_A. With $|X_A| \leqslant E$ and $|X_D| \leqslant 1$, we vary X_D during each conversion cycle until the *absolute analog error* $|e_A| = |EX_D - X_A|$ is sufficiently small, i.e., less than $\frac{1}{2}$ LSB:

$$X_D = \frac{X_A + e_A}{E} \qquad |e_A| < \frac{E}{2^n}$$

where n is the required number of bits, *including the sign bit*. Different schemes for varying X_D yield various compromises between circuit simplicity and maximum n-bit conversion time. We shall measure conversion time in clock periods, where each clock period must be sufficiently long to permit comparator, logic, and D/A converter switches to settle. Typical clock periods will be between 0.05 and 10 μs.

In the simplest scheme, the D/A converter register is a clock-driven 2^n-step counter which starts at the most negative value once per conversion cycle and counts up to produce a staircase D/A output analogous to the ramp voltage in Fig. 6.17a. X_D is read when the staircase voltage reaches X_A. *The maximum conversion time is 2^n clock periods*, since each staircase step will be just $2E/2^n$ V. Such converters are used in some inexpensive digital panel meters.

"Continuous" Converters

To reduce the average conversion time, the D/A converter register can be a reversible counter; we count up or down as required by the sign of the error $e_A = EX_D - X_A$ (*continuous* or *incremental converter, digital servo,* Fig. 6.19b). A small comparator dead space about $e_A = 0$ will prevent oscillation about the correct output. Incremental

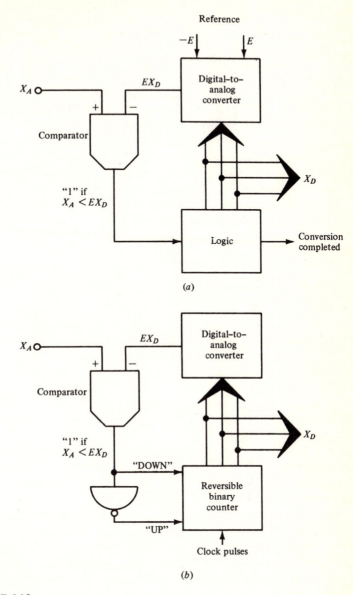

FIGURE 6.19
Principle of (*a*) a feedback-type ADC and (*b*) a "continuous" (reversible-counter-feedback) converter.

converters save an appreciable amount of time if X_A is not a random input (like a multiplexer output) but a time-variable analog voltage *known to increase and decrease no faster than* $R < 2E/2^n$ *V per clock period*. In this case, the first conversion can take up to $2^n R/2E$ clock periods; the converter will then *track the signal* in successive clock periods. We can track faster signals at the expense of circuit complexity if we add comparators

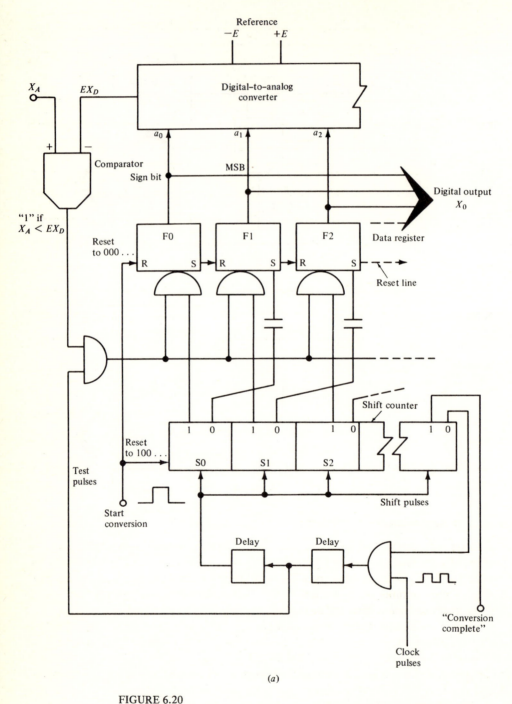

FIGURE 6.20

(a) Successive-approximation ADC; (b) timing diagram showing conversion of successive bits.

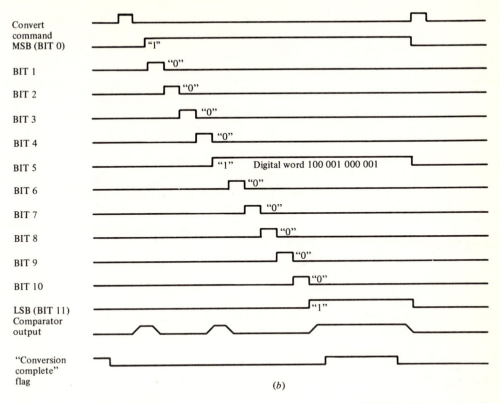

FIGURE 6.20 (*continued*)

which sense large analog errors and transfer the counter input to a higher digit position (variable-increment counter).

Successive-Approximation Converters

For faster conversion, we program the D/A converter register with digital logic to "weigh" X_A by successive binary-digit approximations. We start with $X_D = 0$ to find the sign of X_A. If $X_A > 0$, we next try $EX_D = E/2$ to see if the most significant binary digit is 0 or 1, and continue on to the last binary digit (*successive-approximation converter*). *The required conversion time will be* n *clock periods, one for each bit in* X_D.

Figure 6.20 illustrates the operation of a typical successive-approximation converter. The digital output X_D, and thus the DAC output voltage EX_D, is set by the *ADC (output) register*, comprising n flip-flops F0, F1, F2, ... Successive X_D values will be programmed by a digital 1 passing through an $(n + 1)$-bit *shift counter* (flip-flops S0, S1, S2, ...).

The START CONVERSION pulse resets the shift counter to 1000 ... and the ADC register to 0000 ...; this also releases the test pulses and shift pulses produced by the

ADC clock. Initially, $X_D = 0000 \ldots$. If now $X < EX_D = 0$, the comparator output will be 1, and the first test pulse (suitably delayed to let flip-flops, DAC, and comparator settle) will set the sign-bit flip-flop F0 to 1. The first shift pulse follows and shifts the counter to $0100 \ldots$. We then have

$$X_D = \begin{cases} 1100 \ldots & \text{if } X_A < 0 \\ \\ 0100 \ldots & \text{if } X_A \geqslant 0 \end{cases}$$

and the corresponding DAC output ($-E/2$ or $E/2$) is compared with the analog input X_A. If (and only if) $X_A < -E/2$ or $0 \leqslant X_A < E/2$, the second test pulse will correctly reset the MSB flip-flop F1 to 0. Otherwise, the most significant bit remains, as it should, at 1. Successively following shift pulses and test pulses similarly determine the other bits of X_D in order of decreasing significance. The nth shift pulse shifts the 1 into the $(n + 1)$st shift-counter flip-flop, which serves as a *DONE flag* signaling completion of the conversion.

6.11 SPECIFICATIONS AND COMPARISON OF ANALOG–TO–DIGITAL CONVERTERS

ADC *resolution* refers to the number of bits resolved so that increasing input voltage always increases the digital output ("monotonicity"). ADC *accuracy* ought to match the resolution, so that errors in the digital output are never larger than $\frac{1}{2}$ LSB when the ADC is *calibrated* for zero and full-scale output. Specifications should include *error effects of temperature and power-supply-voltage changes,* or allowed ranges of such changes for a specified accuracy.

The other important ADC specification is the *conversion time* or its reciprocal, the *conversion rate.* Many successive-approximation ADCs permit one to stop the conversion process after a specified number of bit conversions, so that it is possible to trade increased conversion rates for accuracy.

"Instantaneous" (parallel-comparator) ADCs permit conversion rates of at least 10^8 words per second if one is satisfied with 4- to 6-bit resolution. Such fast but coarse conversion is useful for pulse-code-modulation communications. *Cascade encoders* are next in speed. Their resolution is usually limited to 8 or 9 bits, with 10^7 bits or about 10^6 words per second.

The conversion rate of a *ramp-comparator* and *integrating ADC* is limited by analog-switching and comparator-response times, whose unpredictable fluctuations must stay within the specified error percentage of the total conversion time; 7- to 8-bit ramp-comparator circuits used in special instrumentation can convert at up to 10^6 words per second. Digital-voltmeter designers, who rarely require more than 1000 words per second, prize integrating converters for their noise rejection. The accuracy of any ADC using a capacitor for ramp generation or signal integration is limited by dielectric absorption to within 0.01 to 0.02 percent of half-scale $\pm \frac{1}{2}$ LSB (about half the dielectric-absorption effects can be calibrated out). Eleven-bit integrating converters can convert at up to 5000 words per second.

The best accuracies, as well as high conversion rates at extra cost, are obtained with *successive-approximation converters,* which are predominant in hybrid computation. Inexpensive ADCs of this type yield 8 to 12 bits at 1 to 2 μs/bit (i.e., of the order of 10^5 words per second) for less than $300. Up to 16 bits, together with conversion times between 0.1 and 2 μs/bit, can be obtained at higher prices. The DAC used in any feedback-type ADC is subject to the tradeoff between accuracy, settling time, and cost already discussed in Sec. 6.5.

REFERENCES AND BIBLIOGRAPHY

1 KORN, G. A., and T. M. KORN: "Electronic Analog and Hybrid Computers," 2d ed., McGraw-Hill, New York, 1972.

2 HOESCHELE, D. F.: "Analog-to-Digital/Digital-to-Analog Conversion Techniques," Wiley, New York, 1968.

3 SCHMID, H.: "Electronic Analog/Digital Conversions," Van Nostrand Reinhold, New York, 1970.

4 GRAEME, J. G., ET AL.: "Operational Amplifiers," McGraw-Hill, New York, 1971.

5 KORN, G. A.: "Minicomputers for Engineers and Scientists," McGraw-Hill, New York, 1973.

6 PEARMAN, C. R., and A. E. POPODI: How to Design High-speed D/A Converters, *Electronics,* Feb. 21, 1964.

7 PRACHT, C. P.: A New Digital-Attenuator System, *Simulation,* April 1967.

8 AARON, M. R., and S. K. MITRA: Synthesis of Resistive D/A Conversion Ladders for Arbitrary Codes, *IEEETEC,* June 1967.

9 ———: A Note on the Design of D/A Converters, *IEEETEC,* October 1967.

10 FREEMAN, J.: Ladder Networks Are Easy to Design, *Electron. Design,* July 5, 1967.

11 WALKER, M.: Exploit Ladder-Network Design Potential, *EDN,* Feb. 1, 1969.

12 ———: Expand Basic Ladders into Complex Networks, *EDN,* Mar. 1, 1969.

13 HENRY, T.: Binary D/A Converters Can Provide BCD Conversion, *EDN,* August 1973.

14 SCHADE, O. H.: CMOS for Low-power D/A Converters, *Computer Design,* April 1973.

15 GRAEME, J.: Monolithic D/A Improves Conversion Times, *EDN,* Mar. 15, 1971.

16 MARSHALL, W., and C. BROWN: Sixteen-bit Conversion Gets a Lift from IC Technology, *Electronics,* Sept. 25, 1972.

17 DAMROW, R. I.: Settling Time of Operational Amplifiers, *Analog Dialogue,* Analog Devices, Inc., June 1970.

18 SAVITT, A.: A High-speed Analog-to-Digital Converter, *IRETEC,* March 1959.

19 WALDHAUER, F. D.: Analog-to-Digital Converter, U. S. Patent 3,187,325 (assigned to Bell Telephone Laboratories, Inc.), 1965.

20 1-Mc Cyclic A/D Converter, *Electronic Products,* March 1968.

21 KIME, R. C.: The Charge-Balancing ADC, *Electronics,* May 24, 1973.

22 ANDERSON, T. O.: Optimum Control Logic for Successive-Approximation ADCs, *Computer Design,* July 1972.

PROBLEMS (Sec. 6.2)

6.1 Draw a voltage-ladder circuit with SPDT bit switches switching between $-X_1$ and $+X_1$ in the manner of Fig. 6.3c.

6.2 Prove, with the aid of suitable equivalent circuits, that the "inverted" ladder DAC circuit of Fig. 6.7 will, indeed, multiply correctly by

$$Y_D = -a_0 + \frac{1}{2}a_1 + \frac{1}{4}a_2 + \cdots + \frac{1}{2^{n-1}}a_{n-1}$$

6.3 Derive a similar proof for the two current-ladder DACs of Fig. 6.11.

6.4 The accuracy of DAC/operational amplifier circuits will depend on their feedback ratio β. Find β for each of the circuits of Figs. 6.1, 6.3a, b, and c, 6.5c, and 6.7.

6.5 Find the percentage error due to a 0.01 percent error in the input resistors R, $2R$, $4R$, etc., in Fig. 6.3a. This may be due to manufacturing tolerance, temperature effects, or switch ON resistance.

6.6 For the (1, 2, 4, 8) BCD converter of Fig. 6.8, use the equivalent-circuit technique of Fig. 6.4 to prove

$$X_0 = -\alpha X_1 \left[a_1 + \frac{1}{2}a_2 + \frac{1}{4}a_3 + \frac{1}{8}a_4 + \frac{1}{10}\left(b_1 + \frac{1}{2}b_2 + \frac{1}{4}b_3 + \frac{1}{8}b_4 \right) \right.$$
$$\left. + \frac{1}{100}\left(c_1 + \frac{1}{2}c_2 + \frac{1}{4}c_3 + \frac{1}{8}c_4 \right) \right]$$

What is the scaling constant?

INDEX